Flow Cytometry: Applications in Cellular and Molecular Toxicology

AB Pant • Puneet Khare •
Alok Kumar Pandey
Editors

Flow Cytometry: Applications in Cellular and Molecular Toxicology

Springer

Editors
AB Pant
Senior Principal Scientist & Professor (AcSIR) Group Head, Systems Toxicology Group
CSIR-Indian Institute of Toxicology Research
Lucknow, Uttar Pradesh, India

Puneet Khare
Senior Technical Officer Central Facility for Flow Cytometry
CSIR-Indian Institute of Toxicology Research
Lucknow, Uttar Pradesh, India

Alok Kumar Pandey
Principal Scientist Nanomaterial Toxicology Laboratory
CSIR-Indian Institute of Toxicology Research
Lucknow, Uttar Pradesh, India

ISBN 978-981-97-9757-8 ISBN 978-981-97-9758-5 (eBook)
https://doi.org/10.1007/978-981-97-9758-5

This Springer imprint is published by the registered company Springer Nature Singapore Pte Ltd.
The registered company address is: 152 Beach Road, #21-01/04 Gateway East, Singapore 189721, Singapore

Preface

Flow cytometry is an indispensable tool in cellular and molecular biology, and it applies to all facets of fundamental and applied life sciences, including applications in rapid and precise clinical diagnosis. It is a technology used to analyze the characteristics of single cells or particles fluorescently labeled when passed through single or multiple laser beams. The best advantage of flow cytometry is that more than 30 parameters can be done in a single sample, and information on millions of cells can be collected in a few seconds. A thick literature is available on the enormous applications of flow cytometry in identifying biomarkers of exposure, effect, and adverse immune responses in peripheral blood, bone marrow, and other body fluids. Hence, its applicability is well worth and elaborated in multiple disciplines such as immunology, virology, molecular biology, cancer biology, infectious disease monitoring, etc. Besides this, flow cytometry is extensively being used in toxicological perspectives such as in determining reactive oxygen species (ROS), mitochondrial membrane potential (MMP), cell cycle analysis, DNA content, micronucleus (MN) assay, apoptosis, autophagy, immunophenotyping, sorting of heterogeneous cell populations, etc. However, literature on applying flow cytometry in toxicology is scattered and available in bits and pieces. Therefore, this timely published book would be considered an attempt to provide a comprehensive and unparalleled resource of all the relevant information on the applicability of flow cytometry in various facets of toxicology and associated disciplines.

To date, no such book/review/monograph/work has touched upon the application of flow cytometry in toxicology/safety evaluations. This book may be the first of its kind to unravel the journey, current status, various developments, achievements, and lessons learned for applying flow cytometry in toxicology and associated disciplines. With case studies, technical and applied approaches, pictorial representations, informative tables, and simple language, this book would be an invaluable resource for researchers, academia, biopharma industries, graduate and postgraduate students, PhD, and post-doctoral fellows working in the fields of toxicology/biosafety, and biomedical research. At the same time, the book is also an invaluable asset to newcomers and seasoned flow cytometry users who need to strengthen and update themselves with the applications of flow cytometry methods, the underlying mechanical principles, and the interpretation of data about toxicology. The book also instills in the readers the practical application of the discipline, leaving them in awe of the fantastic efforts made by the scientific community to contribute to comprehending the intricacies the topic presents. Furthermore, the content has been organized to reach out to the non-scientific fraternity undergraduates to develop their rudimentary grasp of the subject with better applicability.

This book is a collection of 13 chapters, which range from the historical journey of flow cytometry in toxicology to its scientific acceptance and subsequent implementation as the tool for mechanical, efficacy, and safety studies of all range products. The emphasis has been given to the application and interpretation of flow cytometric data in the areas of cytotoxicity, genotoxicity, immunotoxicity, neurotoxicity, organ-specific toxicity, development of biomarkers, clinical diagnosis for making treatment plans, basic and applied research in toxicology/safety studies, updates in regulatory acceptance of data generated through flow cytometry, etc.

Chapter "Cell Sorting: Underpinnings and Contemporary Developments" presents an in-depth exploration of cell sorting procedures, covering the prerequisites, various cell separation techniques, their mechanisms, and the pros and cons. It concludes with a compelling discussion on the application of cell sorting technology. Chapter "Xenobiotics Mediated Alterations in DNA Content and Cell Cycle: Flow Cytometry-Based Analysis" deals with the application of flow cytometry to evaluate DNA damage using techniques such as H2AX staining, cell cycle analysis, programmed cell death, etc., to establish a detailed analysis of the genotoxic potential of xenobiotics. Chapter "Flow Cytometric Evaluation of Mitochondrial Health" is a detailed elaboration on the bioenergetic system of mitochondria that regulates the entire energy system and cellular functionality under natural and pathophysiological conditions. Further, a collection of flow cytometry-based methodologies has been described to study the mitochondrial health and subsequent events in the cells and tissues. Chapter "Environmental Stressors-Mediated Apoptosis & Autophagy Detected Through Flow Cytometry" aims to provide curated information on applying flow cytometric methods to detect different stages of apoptosis and autophagy, including using pivotal assays, which can detect both early and late apoptosis. Further, the chapter also deals with a description of different dyes or stains, such as monodansylcadaverine, acridine orange, LysoTracker, and autophagy-specific markers, like anti-sequestosome and anti-LC3-II, which are pretty helpful in staining or targeting the autophagic contents or autophagic vacuoles for detection and quantification of autophagic activity in response to different xenobiotic compounds. Chapter "Role of Flow Cytometry in the Rapid Assessment of Organ-Specific Toxicity" narrates the applicability of flow cytometric approaches in the detection of organ-specific toxicity using specific kinds of fluorochromes, dyes, and techniques. Chapter "Flowcytometry-Based Genotoxicity Assays" primarily details flow cytogenetics, which is the analysis and sorting of chromosomes using flow cytometric techniques, cell cycle analysis markers, chromosome genomics, etc. The chapter further discusses the uses, limitations, and potential future developments of flow cytometry in genotoxicity assessment. Chapter "Application of Flow Cytometry in Elucidating the Intricate Mechanism of Neurotoxicity" highlights the multifaceted applications of flow cytometry in advancing our understanding of neurotoxicity. Chapter

"Imagination of Hematotoxicity/Immunotoxicity Without Flow Cytometry" navigates the evolving landscape of hematotoxicity and immunotoxicity assessment within the realm of drug discovery. The chapter narrates the historical overview, highlighting the transformative role of flow cytometry, and unfolds the narrative of its indispensability through the diverse methodologies, pushing the boundaries of our understanding of hepatotoxic and immunotoxic phenomena in the pursuit of more effective drug development. Chapter "Role of Flowcytometry in the Safety/Toxicity Analysis of Cosmetics and Personal Care Products" delves into application of flow cytometry-based immuophenotyping in the detection of surface markers for scrutinizing the ingredients and formulations of cosmetics and personal care products to ensure consumer safety, regulatory compliance, and innovation in product development. Chapter "Flow Cytometric Approaches in Clinical Diagnostics" provides in-depth updates on the application of flow cytometry in the diagnosis of various diseases in clinical samples, which has been a part of day-to-day practice in hospitals and clinics. Chapter "Cellular Uptake of Nanoparticles: A Flow Cytometry Insights" aims to explore the principles and applications of flow cytometry in studying the cellular uptake of nanoparticles/nanoformulations, shedding light on its significance in cellular and molecular toxicology. Chapter "Flowcytometric Capturing of Nanoparticles Internalization in the Cell System" elaborated on the various types of mechanisms involved in cellular internalizing of nanoparticles and their interactions with cell and extracellular matrix. The chapter further reviews the various quantification techniques, highlighting the role of flow cytometry in analyzing nanoparticle uptake and localization. Chapter "Application of Flowcytometry in the Development of New Alternative Model Systems in Biomedical Research" explores how flow cytometry aids could be utilized in characterization of newly developing alternative experimental models by identifying and evaluating cellular heterogeneity, subpopulations, and surface markers. The chapter also addresses the application of flow cytometry in well-established alternative models, viz., *C. elegans*, Zebrafish, and *Drosophila melanogaster*. It describes how flow cytometry helps identify phenotypes, functional indicators, and cellular responses to toxicants. The chapter looks ahead to how new types of flow cytometry, such as mass cytometry (CyTOF), image flow cytometry (IFC), and droplet microfluidics, can be combined to improve analysis.

Editing a book on fundamental and applied aspects of flow cytometry was more challenging than we anticipated. But it could have culminated with the help of many professional and personal friends and colleagues in and out of the workplace. To begin, we would like to express our sincere thanks to all the expert authors for their kind consent and for sparing valuable time to participate in this beautiful collection of curated scientific information on applying flow cytometry in toxicology and associated disciplines. We want to extend our gratitude to everyone who helped us in some way while editing this work, even if their names aren't listed here.

Lucknow, Uttar Pradesh, India

AB Pant
Puneet Khare
Alok Kumar Pandey

Contents

About the Editor

AB Pant is a seasoned toxicologist who started his research career over 32 years back at CSIR-Central Drug Research Institute, Lucknow, India. He is an alumnus of IIT-Roorkee, Uttarakhand, India, and is currently serving as Chief Scientist at CSIR-Indian Institute of Toxicology Research, Lucknow. He also serves as a Professor of the Academy of Scientific & Innovative Research (AcSIR), an Institution of National Importance set up by the Act of Parliament. Professor Pant's elegant research offers a much-sought framework for understanding neurodegenerative disorders, potential therapeutic interventions, and a strong base for interpreting human brain-specific neurotoxicity and developmental neurotoxicity using stem cells as models. His strategic international collaboration with the University of São Paulo, Brazil, has developed a strong roadmap for effective drug development and therapeutic interventions in Amyotrophic Lateral Sclerosis, a fatal neurodegenerative disorder. Being a nodal Scientist, Professor Pant coordinated COVID-19-related activities, including Real-Time PCR-based testing, SARS-Co-V2 genome sequencing, sero-surveillance, and capacity building program at CSIR-IITR. Professor Pant was instrumental in establishing the "Africa Centre of Excellence in Materials, Product Development and Nanotechnology" at Makerere University, Uganda. He also dedicated himself to fostering contemporary science to post-graduates, PhD students, researchers, and university teachers. He is a lead GLP inspector for the Government of India. Also, he serves as an expert for regulatory bodies, such as BIS, CDSCO, NABL, FSSAI, NGCMA, ICMR, DST, Pharmacovigilance, etc. Professor Pant has won several awards and honors, including the Vigyan Ratna Award-2010 (CST UP); Shakuntala Amir Chand Prize-2007 (ICMR); National Bioscience Award-2012 (DBT); Prof. KT Shetty Memorial Oration Award-2017 (Indian Academy of Neurosciences), Toxicology Promotion Award-2018 (National Academy of Sciences, Allahabad), DSc-*h.c.* (HNB Uttarakhand Medical Education University, Dehradun, Uttarakhand-2021), STOX/ASAW Gold Medal (STOX, India), etc. He has been elected Fellow of Academy of Toxicological Sciences, USA, Society of Toxicology, India, Indian Academy of Neurosciences, Academy of Sciences for Animal Welfare, India, Academy of Environmental Biology, India, etc. In view of his outstanding credibility in toxicology, Professor Pant has been included as a global toxicologist in the "European and United Kingdom Register of Toxicologists." He serves as chairman and member of several national and international committees/boards of government institutes and bodies. He also serves as a member of the editorial board of several reputed journals.

Puneet Khare is a Senior Technical Officer in the Central Flow Cytometry Facility at the CSIR-Indian Institute of Toxicology Research, Lucknow, India. He did his PhD in Biochemistry from Babu Banarasi Das University, Lucknow, India. His area of specialization is flow cytometry and in vitro cell culture. In a decade-long research career, Dr. Khare has established himself as one of the pioneers in the area of flow cytometry/in vitro toxicology with special reference to marker-based studies using flow cytometry. Over a period, he enriched himself as an expert for flow cytometric-based detection of toxic responses in primary culture and immortalized cells of human and animal origin and plant and bacterial cells as well. Besides the laboratory work, he has also been involved in fostering contemporary science to post-graduates, PhD students, researchers, and university teachers. Dr. Khare's research efforts have been rewarding in terms of the quantifiable number of research publications in journals of high repute. He was instrumental in organizing various national and international flow cytometry-based workshops.

Alok Kumar Pandey is a Principal Scientist in the Nanomaterial Toxicology Laboratory at CSIR-Indian Institute of Toxicology Research, Lucknow, India. He is the Group Leader of the Biological Instrumentation facility (including Flow cytometry, TEM, SEM, and Confocal Microscopes). He earned his PhD degree in Environmental Sciences from the University of Lucknow. His area of work includes Nanomaterial Toxicology and Genetic toxicology. His current research involves deciphering the cytogenetic, cell cycle, and apoptotic mechanistic insights of nanomaterial toxicity using in vitro and in vivo models. The work is also being carried out to understand the impact of nanomaterials on non-target species in various ecological strata. Dr. Pandey also serves as an Associate Professor in the Academy of Scientific and Innovative Research (AcSIR), an institution of National Importance set up by the Act of Parliament. He is coordinating one of the highly specialized courses for PhD aspirants under AcSIR. He has over 60 research publications in international journals of high repute and many book chapters. He also serves as an editor/reviewer in more than 25 journals. So far, Dr. Pandey has guided 5 PhD students and trained more than 50 MSc/MTech scholars as project trainees. Currently, he is supervising five students for PhD and two for MSc/MTech.

Cell Sorting: Underpinnings and Contemporary Developments

Puneet Khare and AB Pant

Abstract

Cell sorting involves isolating a uniform population from a group of diverse cells. This method employs advanced techniques like fluorescence-activated cell sorting (FACS) to precisely separate cells based on their unique functional and immune-phenotypic properties. Researchers can then conduct additional experiments and analysis on the separated cells, expanding possibilities for important discoveries in areas such as immunology, infectious diseases, regenerative medicine, and oncology. Cell sorting is now widely used in clinical diagnosis and biomedical research. This comprehensive chapter provides an in-depth exploration of cell sorting procedures, covering the prerequisites, various cell separation techniques, their mechanisms, and the pros and cons. It concludes with a compelling discussion on the application of cell sorting technology.

Keywords

Flow cytometry · Cell sorting · Immunology · Cell heterogeneity

P. Khare
Central Flow Cytometry Facility, ASSIST Division, CSIR-Indian Institute of Toxicology Research (CSIR-IITR), Lucknow, Uttar Pradesh, India

AB Pant (✉)
Systems Toxicology & Health Risk Assessment, FEST Division, CSIR-Indian Institute of Toxicology Research (CSIR-IITR), Lucknow, Uttar Pradesh, India
e-mail: abpant@iitr.res.in

AB Pant et al. (eds.), *Flow Cytometry: Applications in Cellular and Molecular Toxicology*, https://doi.org/10.1007/978-981-97-9758-5_1

1 Introduction

In biological research, cell sorting is a crucial procedure that physically separates a single cell or a homogeneous population of cells from a complex mixture. The primary goal of the cell sorting technique is to isolate populations of interest for additional research from a heterogeneous population. The physicochemical and biological characteristics of cells are the basis for sorting them. Physicochemical attributes include size, pH, density, volume, electrical impedance, and light scatter characteristics. The biological attributes comprise an extensive array of elements, such as dimensions, vitality, anatomical parameters, and the manifestation of extracellular and intracellular protein expression (Ohnuma et al., 2006).

In many studies, obtaining a uniform cell population or even single cells is essential to getting meaningful results from the experiments. Sorting cells is therefore, one of the most important stages in sample preparation (Box et al., 2020).

The physical separation may be achieved through various methods, mainly FACS-fluorescence-activated cell sorting (Basu et al., 2010; Liao et al., 2016), MACS-magnetic-activated cell sorting (Mantravadi & Rao, 2024; Pacheco et al., 2020), BACS-buoyancy activated cell sorting (Liou et al., 2015; Wang et al., 2018), and micro-fluidic based cell sorting (Shields 4th et al., 2015; Warkiani et al., 2015; Iqbal et al., 2023).

The FACS method involves labelling target cells with a fluorescent dye using antibodies. On the other hand, the MACs method focuses on using antibody-coated magnetic beads to bind specifically to the target cells, allowing them to be separated from non-labelled cells using a strong magnet.

Various cell isolation techniques are available today, and their essential basis is the density gradient, particle size, adherence, absorbance, dielectric properties, chemoresistance, and antibody binding, among other things.

2 Prerequisites of Cell Isolation Technique

I. **Plan:** Prior to executing any task, meticulous planning plays a crucial role, particularly in the context of sorting. It is imperative to engage in comprehensive discussions with the cell sorter operator well in advance regarding the experiment or samples to be sorted. This may entail deliberating on the stain/fluorescent markers to be utilized, the specific cells for sorting, the desired percentage of cells in the original suspension, the quantity of samples for sorting, the number of cells required per sample for facility processing, and the method of sorting, whether it be into tubes or plates, and the necessity of sterility.

II. **Single-cell suspension should be maintained:** The clumping of suspension cells can be problematic and can significantly impact research results. By addressing the root causes, such as unbound DNA and cell debris, we can prevent clumping and ensure the efficiency of cell cultures. Utilizing DNase I in tissue disaggregation and routine cell culture can effectively prevent clumping in various cell types, especially in the presence of bivalent ions like Mg^{2+} and Ca^{2+}. Single-cell suspension media provides the necessary resources for cell growth, and eliminating clumps can enhance research throughput and accuracy. Furthermore, addressing clumping during the sorting procedure using chelators such as citrate and EDTA can significantly improve the efficiency of the process. Careful handling and the use of appropriate centrifuge settings are crucial when dealing with cell clumps, and minor clumps can be easily removed by gentle pipetting or triturating. By implementing these strategies, we can ensure the reliability and accuracy of our cell culture experiments.

III. **Time:** The duration for sorting cells depends mainly on the number of cells that must go through the equipment. It also depends on the quantity of cells we want to recover and the proportion of the target cell in the original sample. Sorting a rare population of cells will usually take much longer than sorting a population representing 30–50% of your initial sample. The speed at which cells can be sorted is determined by the amount of time available and the number of cells to be sorted. The quality of cell preparation can also affect the speed and may lead to a higher abort rate.

IV. **Aseptic condition:** It is crucial that the aseptic technique establishes a barrier between the microorganisms outside the culture and those within its flask or dish. Therefore, it is imperative that all materials in contact with the culture are sterile. Cells received after sorting are only useful for re-culture if they are sorted under sterile conditions. To achieve it, sterility should be maintained by treating the flow lines, tubes, and filters with at least 70% alcohol, and the hood in which the Sorter is kept should be wiped with 70% alcohol and further sterilized with UV light. Other things involve heap filter and air flow should be on while sorting. The operator should wear a lab coat, gloves, headgear, protective goggles, and a mask. They should also adhere to all aseptic procedures for cell culture. Aseptic technique refers to a set of methods intended to lower the risk of infection. Cell sorters can be housed in biosafety cabinets to lessen the danger to sorter operators and to contain aerosol.

V. **Sorting efficiency:** The efficiency of sorting measures how effectively a sorting machine can separate the cells we want from those we don't want. Sorting efficiency is determined by the ratio of sorted target cells to the total detected target cells. Non-target cells are referred to as conflicts, coincidences, or aborts, while sorted target cells are known as sorts. The sorting efficiency is mainly influenced by the configuration of the sorter and the chosen sorting method. Various methods of sorting include recovery or yield, purity, single-cell mode, and enrichment. The selection of the sorting method depends on the nature of the study and the quantity of cells needed after sorting. To achieve high purity in downstream applications, the sorter must be set to exclude any target events close to non-target events from the sort, typically within one-half of a droplet. Failure to do so may result in non-target events being sorted along with the target events. This type of sorting method is often referred to as "Purity" or "Purify" mode. "Single" or "Single Cell" mode is commonly used when an accurate count of the output cells is crucial for downstream applications, such as single-cell sorting. Finally, if recovering every cell from the sort is more critical than purity, "Yield" or "Recovery" modes are used.

VI. **Recovery/viability:** Cells undergo slight stress during the sorting procedure. The sorted cells should be primarily collected in the vials with media in a sterile environment. Further, the cells should be immediately transferred to the usual culture media under standard culture conditions under a CO2 incubator at 37 °C and 5% CO2 with saturating humidity. The initial sample for cell sorting should be healthy and viable. So, these cells can withstand the pressure generated during the sorting process, and finally, cells recover better.

3 Cell Separation and Cell Isolation Methods

3.1 FACS-Fluorescence-Based Cell Sorting

The FACS approach separates cells based on their fluorescent properties using flow cytometry principles. This process involves tagging specific cell populations with fluorescent dyes or proteins for sorting (McKinnon, 2018). The procedure involves tagging or labelling cells with fluorescent markers, allocation of cells to be sorted and then passing these heterogeneous cells through a cell sorter. Finally, homogenous cells will be the final output (Schmit et al., 2021). One of the critical advantages of FACS is the ability to sort cells based on multiple markers simultaneously, making it widely used in various cell-sorting applications. Now, it has become as the standard method for purifying cell subpopulations. Any cell type that can be converted into a single-cell suspension and has available antibodies for identifying the desired cell population can be isolated using this method. Thus, we can say that when absolute purity of cell populations is vital, FACS emerges as the technique of choice (Fig. 1).

3.2 Mechanism of Cell Sorting

In a nutshell, a flow cytometer with sorting capability and the necessary software is needed for the FACS-based cell separation procedure. In FACS, a stream of suspended cells is passed before a laser in droplets containing a single cell. The system for detecting fluorescence uses the predefined fluorescent characteristics of the cells to identify cells of interest. The instrument charges the droplet holding the cell of interest, and the charged droplets are collected into the proper collection tubes with the help of an electrostatic deflection system (Basu et al., 2010).

Further, a thorough research study comprises that the sheath-a small, quickly flowing liquid stream entangles the sample, which is a cell suspension. This has been accomplished by maintaining sample pressure that is marginally higher than sheath pressure. Before the stream splits into droplets, each cell of interest has its fluorescence measured at a fluorescence measuring station. Cells that satisfy the sorting requirements are briefly charged in the liquid stream right before a droplet

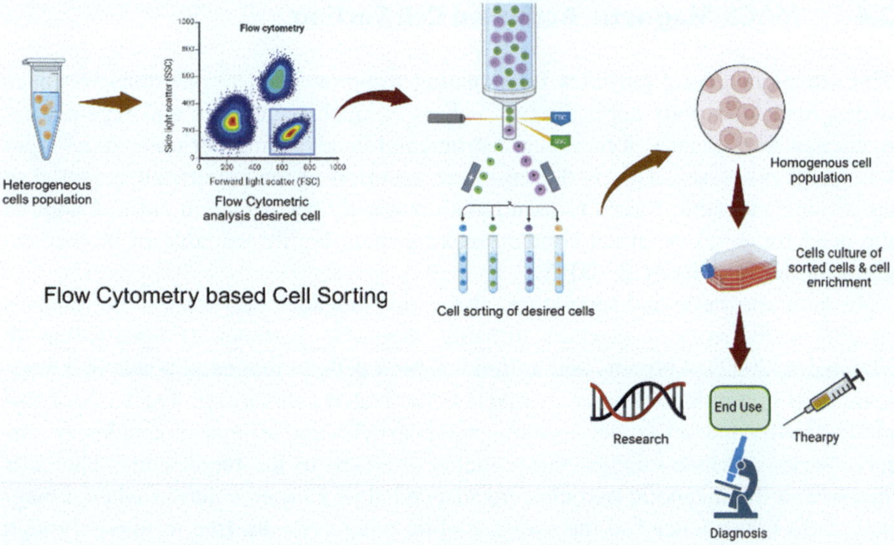

Fig. 1 The schematic representation of the cell sorting technique for the isolation of the homogeneous population. (Parts of the figure were drawn by using pictures from Biorender (https://www.biorender.com/))

carrying the cell is ready to separate from the stream (Ryan et al., 2021). The stream of cells fragments into individual droplets due to a vibrating mechanism. Acoustic waves are introduced into the fluid stream as part of the vibrating mechanism. After passing through two highly charged plates, the charged droplet is diverted from the droplet cascade and collected. The number of droplets produced in a second determines the maximum sorting rate. The nozzle diameter, usually 70 μM, determines the size, and the velocity is linked to the hydrostatic pressure of the liquid entering the nozzle. When pressure rises, and nozzle size decreases, the drop rate also rises (Cossarizza et al., 2019).

3.3 Challenges of Flow Cytometry-Based Cell Sorting

There are several factors to consider when getting ready for a sorting procedure based on flow cytometry, such as the size of the nozzle or cuvette, the pressure inside the sheath and sample, the frequency and timing of the droplet generation, and the time it takes the cell to move from the laser intercept to the droplet generation point. Also liable to harm sensitive cells are the high pressures used in cell sorting. Furthermore, a less pure sample may be obtained by sorting techniques that gather more cells than necessary, mainly when harvesting rare stem cells. Since the sorting technique depends on fluorescence signals, the target population must have a recognized flag distinguishing them from the surrounding cell population.

3.4 MACS-Magnetic-Activated Cell Sorting

The use of magnetic particles for isolating important cell populations has been widespread in various applications, such as clinical diagnostics and regenerative medicine. Additionally, it has been instrumental in advancing our understanding of biological processes. This is due to the recognition of the significant potential of employing magnetic forces to manipulate magnetic particles and entities without the need for wires or direct contact, making them highly valuable in biomedical applications (Plouffe et al., 2015).

Immunomagnetic cell separation, also called magnetic cell separation, removes specific cells from a mixture utilizing magnetic particles (Frenea-Robin & Marchalot, 2022). Antigens and antibodies bind cells to magnetic beads in a magnetic field in a process known as magnetic-activated cell sorting. The basis of this method is the interaction between magnetic particles and affinity molecules, or surface markers, which enables the particles to attach to the target cells. The cells attached to the magnetic particles are suspended in a liquid solution when a magnetic field is introduced to the sample, while other cells are free to move through (Shen et al., 2021) (Fig. 2).

MACS can be used as either a positive or negative selection method, depending on the target cells. Target cells are bound by magnetic particles when specific cell surface proteins come into touch with streptavidin, lectins, enzymes, or antibodies.

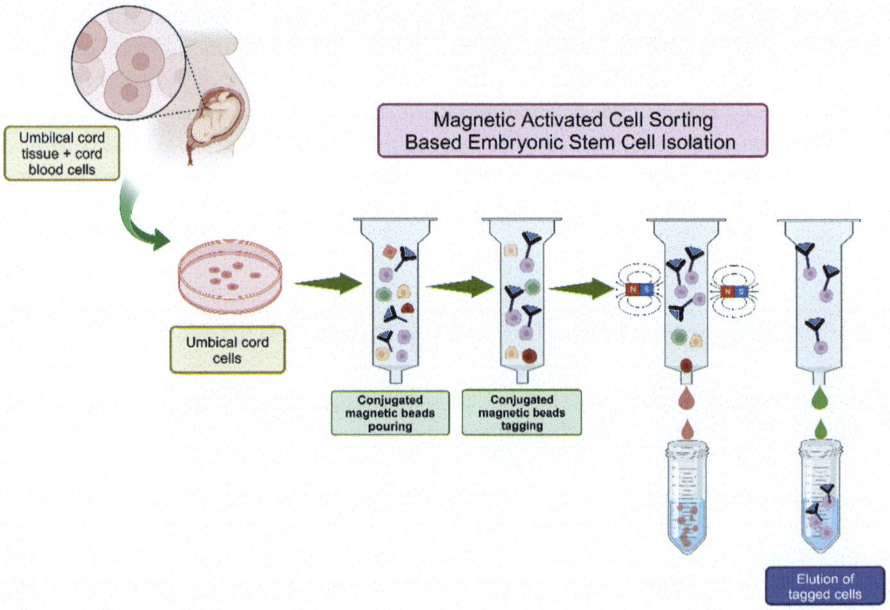

Fig. 2 The schematic representation of the magnetically activated cell sorting-based embryonic stem cell isolation. (Parts of the figure were drawn by using pictures from Biorender (https://www.biorender.com/))

After being labelled with magnetic particles, the cells physically separate from the other cells in the sample when an electromagnetic field pulls them into the sample. This procedure uses magnetic separation to guarantee that the targeted cells are separated from the remaining biological material. Due to its ease of use and speed, magnetic cell separation is frequently used by scientists to obtain highly pure populations of specific cell subsets (Laghmouchi et al., 2020; Weiss et al., 2019).

The two types of magnetic cell separation are column-based and column-free.

3.4.1 Column-Based

The column-based magnetic cell separation process involves passing a sample labelled with magnetic particles through a column matrix in a magnetic field. When a magnetic field is applied to the ferromagnetic spheres in the column, it generates a localized magnetic field that immobilizes the magnetic particles within the sample. This enables the retention of target cells labelled with magnetic particles inside the column through positive selection while allowing non-target cells without magnetic labels to pass through. The target cells can then be collected by flushing a buffer through the column and eliminating the external magnetic field.

3.4.2 Column-Free

The process of magnetic cell separation using columns involves applying a magnetic field to a tube containing a sample that has been tagged with magnetic particles. The magnetically labelled target cells are attracted towards the magnet and collect near the edges of the tube, while the unlabelled cells remain in suspension. The unlabelled cells can then be removed by pipetting or pouring, separating them from the tagged cells. The labelled target cells can be released from the tube by removing them from the magnetic field and can be re-suspended in a buffer for immediate use in other applications.

The MACS method allows cells to be separated by using magnetic nanoparticles coated with antibodies against a particular surface antigen.

4 Buoyancy Activated Cell Sorting (BACS)

BACS, which stands for buoyancy-activated cell sorting, is a technique used to sort cells by utilizing buoyant microbubbles. The specific cells are attracted to the surface of the microbubbles after being coated with antibodies, effectively separating them from the non-target cells. Subsequently, the microbubbles, comprised of a gaseous core often surrounded by polymers, lipids, or proteins, carry the target cells to the top of the sample. Furthermore, the shear stress from a rising bubble and the tension from the buoyancy force are both far below the threshold for cell damage.

Target cells are attracted to the affinity molecules coated on the microbubbles, which cause the cells to rise to the solution's surface. Using vacuum aspiration, the cells can be extracted from the sample as they reach the top, leaving the enriched sample at the bottom (Fig. 3).

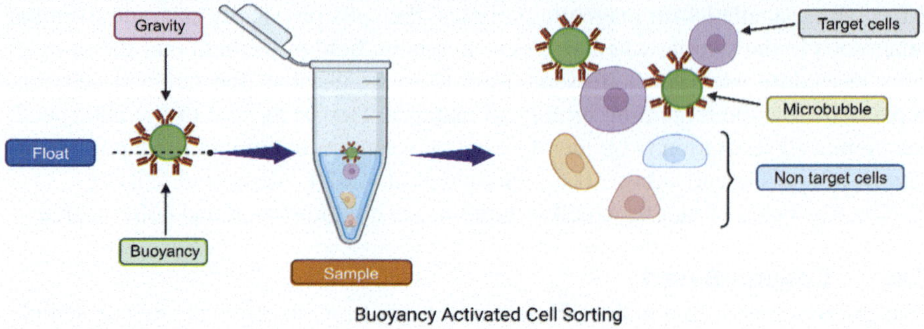

Fig. 3 The schematic representation of the buoyancy-activated cell sorting technique for the isolation of the target cell population. (Parts of the figure were drawn by using pictures from Biorender (https://www.biorender.com/))

Researchers can expand their diagnosis to rare cell populations and make their investigations more scalable with the help of microbubbles. It is a novel approach to target the specific population of viable cells with a high yield, purity and high recovery.

It maintains cell health and physiology for use in the future and is quick, simple, and less expensive than other approaches. It can be employed as a stand-alone isolation approach or in combination with other methods to further purify a material.

The microbubbles' affinity molecules attract target cells, causing them to surface in the solution. BACS offers rapid, simple, cost-effective isolation while preserving cell physiology and vitality for future applications. This versatile procedure can be used alone or with other methods for further material purification. Microbubble production in the sample container streamlines the workflow, making it a more affordable and less technologically demanding approach compared to other methods.

5 Microfluidic-Based Cell Sorters

Microfluidics involves using tiny channels ranging from ten to hundreds of micrometers to handle small volumes of fluids (10^{-9} to 10^{-18} L). It is an interdisciplinary field that combines microelectronics, molecular biology, and analysis. In recent years, the aim of developing microfluidic flow cytometers has been to create smaller, more affordable, easier-to-use, and more self-sufficient flow cytometers compared to traditional ones. These microfluidic-based flow cytometers and cell sorting devices use a chip with microfluidic channels, mechanical gates, or air pressure to divert cells from a central stream. When comparing electrostatic systems to microfluidic-based ones, several advantages become apparent. Electrostatic systems that use high air pressure to produce sheath-contained sample streams might harm delicate cells. In contrast, microfluidic-based sorting methods offer more precision but at slower rates—systems based on microfluidics sort more gently and at lower pressures. Additionally, cell placement is more accurate in microfluidic devices, enabling more precise single-cell collection (Shrirao et al., 2018).

6 Applications of Cell Sorting

Some prominent applications of cell sorting are as follows:

6.1 Drug Discovery & Development

The process of finding new drugs involves identifying screening hits, optimizing medicinal chemistry, and improving factors such as affinity, selectivity, efficacy, potency, metabolic stability, and oral bioavailability while reducing side effects. Drug development can move forward once a chemical that fulfills all of these criteria has been discovered. Clinical studies are then developed if the initial stages are successful. Developing new drugs involves identifying initial screening hits, refining medicinal chemistry, and enhancing characteristics such as affinity, selectivity, efficacy, potency, metabolic stability, and oral bioavailability while minimizing side effects. Once a compound that meets all these criteria is identified, drug development can progress. Successful compounds are then taken into clinical studies. Cell sorting is a crucial aspect of drug development because it simplifies cell-based assays for drug target identification and screening, playing a vital role in every stage of the drug discovery process, from target identification to lead optimization.

6.2 Immunophenotyping

Immunophenotyping uses fluorescent dyes and specific antibodies to assess the expression of specific proteins in a population of cells. The labelled cells are identified and sorted based on their protein expression. It is commonly used to assess CD4-T cell levels when studying specific immunodeficiency diseases and immune-related conditions. In immunophenotyping using flow cytometry, the expression and presence of the target cell protein are detected by scattered light signals that recognize the specific fluorescent dyes.

Immunophenotyping uses flow cytometry to determine if specific cell-surface markers are present. This method is an effective diagnostic tool for various types of leukaemia's and immunodeficiency diseases. Additionally, immunophenotyping can provide further insight into cancer cell lines by revealing the absence of specific cancer cell markers associated with different disease severities.

6.3 Enzyme Engineering

Enzymes are referred to as biocatalysts and are widely used in many industrial processes, especially those that produce bulk chemicals and pharmaceuticals. Enzyme engineering involves altering an enzyme's amino acid sequence to produce a more active enzyme or increase the efficiency of an existing enzyme. Several techniques, such as flow cytometry, de novo enzyme design, enzyme immobilization, and planned divergent evolution, can be employed to generate an enzyme specifically engineered for a given task. Using fluorogenic substrates, FACS can be utilized in enzyme engineering technology to enable the high-specificity screening of enzyme variations (Truppo, 2017).

6.4 Role of Cell Sorting in Cancer Treatment

The technique of cell sorting has revolutionized the ability to isolate and study specific T cell and leukocyte populations from both blood and tumor tissues. Globally, cancer has become a major causative of death. As a result, a major focus of current scientific research is developing innovative substitutes for established cancer treatment approaches. Immunotherapy is one such tactic that focuses on instructing the immune system to identify, target, and eliminate malignant cancer cells. Adoptive cell therapy (ACT) is one of the immunotherapies now used to treat cancer. A type of ATC known as chimeric antigen receptor treatment (CAR therapy) involves genetically modifying autologous T cells to produce CARs (CAR-T cells), which are designed to precisely destroy tumor cells. Patients who have not responded to previous first-line cancer treatments may be able to get treatment with CAR-T cell therapy. The powerful technique of cell sorting has enabled researchers to develop more targeted CAR-T cell therapies, which are now being used to treat a range of cancers (Tian et al., 2024; Robinson et al., 2023). Tumor-infiltrating leukocytes (TILs) are immune cells that surround tumor cells. Studies have suggested that TILs can serve as indicators of survival in different types of cancer. Tumor-infiltrating lymphocytes (TILs) are a naturally occurring group of mononuclear cells that move throughout the solid tumor microenvironment (TME) and may consist of all types of immune cells at the tumor site (Kuroda et al., 2021; Jamiyan et al., 2020).

On the other hand, to support the successful growth and spread of the tumor, tumor cells create the tumor microenvironment (TME), which is a unique ecosystem of host components. The development of 3D culture and advanced bioinformatics tools has enabled the examination of the various components of TME and their interactions at a higher level of detail. It is now feasible to investigate the temporal and spatial modulation of immune therapeutic treatments due to a greater comprehension of immune cell variety, stromal components, repertoire profiling, and TME neoantigen prediction. When identifying patients who will respond to cancer immunotherapy and those who won't, the differences in TME composition between them are crucial. As a result, reprogramming TME components may be able to address the prevalent problem of immunotherapeutic resistance. An indispensable instrument for researching the tumor microenvironment is flow cytometry. In a heterogeneous sample, it enables us to swiftly measure and identify various cell types. Moreover, cell sorting has also facilitated the detailed characterization of tumor-infiltrating leukocytes (TILs), allowing for a better understanding of how these cells interact with cancer cells and the tumor microenvironment.

6.5 Stem Cell Research

Flowcytometry's primary advantage is in the speedy execution of highly multiplexed quantitative measurements on individual cells within a diverse cell population. However, when the target cell type is incredibly uncommon, as most stem and progenitor cells are, multiple artifact sources need to be considered.

Imaging and flow cytometry combined with cell sorting have revolutionized the study of stem cell biology and the use of these cells and their daughter cells in developmental biology, tissue engineering, and cellular treatment. Technological developments in hardware, reagents, and analytical software have led to a rise in the number of parameters and the speed at which they may be measured simultaneously in single cells (Donnenberg et al., 2013).

6.6 Sperm Cell Sorting

The process of sperm sorting helps determine the kind of sperm cell that will fertilize the egg cell-many standard swim-up or centrifugation procedures. The options for sorting sperm are increased by recently discovered techniques and newly applicable technologies like flow cytometry. Sperm can be sorted for later insemination to enable child sex selection. This is done by dividing the sperm based on the number of sex chromosomes using flow cytometric sorting. The process of sorting sperm using flow cytometric sorting is often referred to as particle sorting. Flow sorters are typically used to sort mammalian cells, but they have also been used to sort bacteria, yeast, and phytoplankton. Therefore, calling this process particle sorting is more accurate. At assisted reproductive facilities, a standard procedure known as magnetic-activated cell sorting (MACS) is used to separate sperm from fragmented DNA. A magnetic field is used to pick out sperm cells that are undergoing apoptosis. Unlike other methods, MACS doesn't require the use of fluorescent DNA binding molecules.

7 Conclusion

The discussed fluorescence-based flow cytometry methods involves multipara metric study of even a single cell, unparalleled statistical precision, repeatability, and sensitivity. It enables the simultaneous assessment of multiple parameters on a cell-by-cell basis, which is crucial in understanding various biological processes. The established magnetic separation technique also effectively separates magnetic particles from chemical or biological solutions. Furthermore, buoyant force-based microbubbles have evolved as the preferred technique for cell separation. Cutting-edge fluorescent-based markers, magnetic conjugated antibodies, and conjugated microbubbles are revolutionizing cell isolation and propelling forward biological and clinical research. Despite existing modern cell sorting technologies, even more accessible, cost-effective, rapid, and efficient technologies are on the horizon.

References

Basu, S., Campbell, H. M., Dittel, B. N., & Ray, A. (2010). Purification of specific cell population by fluorescence activated cell sorting (FACS). *Journal of Visualized Experiments: JoVE*, (41), 1546. https://doi.org/10.3791/1546

Box, A., DeLay, M., Tighe, S., Chittur, S. V., Bergeron, A., Cochran, M., Lopez, P., Meyer, E. M., Saluk, A., Thornton, S., & Brundage, K. (2020). Evaluating the effects of cell sorting on gene expression. *Journal of Biomolecular Techniques: JBT, 31*(3), 100–111. https://doi.org/10.7171/jbt.20-3103-004

Cossarizza, A., Chang, H. D., Radbruch, A., Acs, A., Adam, D., Adam-Klages, S., Agace, W. W., Aghaeepour, N., Akdis, M., Allez, M., Almeida, L. N., Alvisi, G., Anderson, G., Andrä, I., Annunziato, F., Anselmo, A., Bacher, P., Baldari, C. T., Bari, S., Barnaba, V., et al. (2019). Guidelines for the use of flow cytometry and cell sorting in immunological studies (second edition). *European Journal of Immunology, 49*(10), 1457–1973. https://doi.org/10.1002/eji.201970107

Donnenberg, V. S., Ulrich, H., & Tárnok, A. (2013). Cytometry in stem cell research and therapy. *Cytometry. Part A, 83*(1), 1–4. https://doi.org/10.1002/cyto.a.22243. PMID: 23281001; PMCID: PMC4293101.

Frenea-Robin, M., & Marchalot, J. (2022). Basic principles and recent advances in magnetic cell separation. *Magnetochemistry, 8*(1), 11. https://doi.org/10.3390/magnetochemistry8010011

Iqbal, M., Mukhamedshin, A., Lezzar, D. L., Abhishek, K., McLennan, A. L., Lam, F. W., & Shevkoplyas, S. S. (2023). Recent advances in microfluidic cell separation to enable centrifugation-free, low extracorporeal volume leukapheresis in pediatric patients. *Blood Transfusion = Trasfusione del Sangue, 21*(6), 494–513. https://doi.org/10.2450/BloodTransfus.506

Jamiyan, T., Kuroda, H., Yamaguchi, R., Nakazato, Y., Noda, S., Onozaki, M., Abe, A., & Hayashi, M. (2020). Prognostic impact of a tumor-infiltrating lymphocyte subtype in triple negative cancer of the breast. *Breast Cancer (Tokyo, Japan), 27*(5), 880–892. https://doi.org/10.1007/s12282-020-01084-1

Kuroda, H., Jamiyan, T., Yamaguchi, R., Kakumoto, A., Abe, A., Harada, O., & Masunaga, A. (2021). Tumor-infiltrating B cells and T cells correlate with postoperative prognosis in triple-negative carcinoma of the breast. *BMC Cancer, 21*(1), 286. https://doi.org/10.1186/s12885-021-08009-x

Laghmouchi, A., Hoogstraten, C., Falkenburg, J. H. F., & Jedema, I. (2020). Long-term in vitro persistence of magnetic properties after magnetic bead-based cell separation of T cells. *Scandinavian Journal of Immunology, 92*(3), e12924. https://doi.org/10.1111/sji.12924

Liao, X., Makris, M., & Luo, X. M. (2016). Fluorescence-activated cell sorting for purification of plasmacytoid dendritic cells from the mouse bone marrow. *Journal of Visualized Experiments: JoVE*, (117), 54641. https://doi.org/10.3791/54641

Liou, Y. R., Wang, Y. H., Lee, C. Y., & Li, P. C. (2015). Buoyancy-activated cell sorting using targeted biotinylated albumin microbubbles. *PLoS One, 10*(5), e0125036. https://doi.org/10.1371/journal.pone.0125036

Mantravadi, K. C., & Rao, D. (2024). In cases with raised sperm DNA fragmentation, can sperm selection by magnetic-activated cell sorting or testicular sperm aspiration help improve reproductive outcomes? *Journal of Assisted Reproduction and Genetics, 41*(6), 1507–1515. https://doi.org/10.1007/s10815-024-03128-3. Epub 2024 May 3. PMID: 38700621; PMCID: PMC11224189.

McKinnon, K. M. (2018). Flow cytometry: An overview. *Current Protocols in Immunology, 120*, 5.1.1–5.1.11. https://doi.org/10.1002/cpim.40

Ohnuma, K., Yomo, T., Asashima, M., & Kaneko, K. (2006). Sorting of cells of the same size, shape, and cell cycle stage for a single cell level assay without staining. *BMC Cell Biology, 7*, 25. https://doi.org/10.1186/1471-2121-7-25

Pacheco, A., Blanco, A., Bronet, F., Cruz, M., García-Fernández, J., & García-Velasco, J. A. (2020). Magnetic-activated cell sorting (MACS): A useful sperm-selection technique in cases of high levels of sperm DNA fragmentation. *Journal of Clinical Medicine, 9*(12), 3976. https://doi.org/10.3390/jcm9123976

Plouffe, B. D., Murthy, S. K., & Lewis, L. H. (2015). Fundamentals and application of magnetic particles in cell isolation and enrichment: A review. Reports on progress in physics. *Physical Society (Great Britain), 78*(1), 016601. https://doi.org/10.1088/0034-4885/78/1/016601

Robinson, J. P., Ostafe, R., Iyengar, S. N., Rajwa, B., & Fischer, R. (2023). Flow cytometry: The next revolution. *Cells, 12*(14), 1875. https://doi.org/10.3390/cells12141875

Ryan, K., Rose, R. E., Jones, D. R., & Lopez, P. A. (2021). Sheath fluid impacts the depletion of cellular metabolites in cells afflicted by sorting induced cellular stress (SICS). *Cytometry. Part A: The Journal of the International Society for Analytical Cytology, 99*(9), 921–929. https://doi.org/10.1002/cyto.a.24361

Schmit, T., Klomp, M., & Khan, M. N. (2021). An overview of flow cytometry: Its principles and applications in allergic disease research. *Methods in Molecular Biology, 2223*, 169–182. https://doi.org/10.1007/978-1-0716-1001-5_13. PMID: 33226595.

Shen, M. J., Olsthoorn, R. C. L., Zeng, Y., Bakkum, T., Kros, A., & Boyle, A. L. (2021). Magnetic-activated cell sorting using coiled-coil peptides: An alternative strategy for isolating cells with high efficiency and specificity. *ACS Applied Materials & Interfaces, 13*(10), 11621–11630. https://doi.org/10.1021/acsami.0c22185

Shields, C. W., 4th, Reyes, C. D., & López, G. P. (2015). Microfluidic cell sorting: A review of the advances in the separation of cells from debulking to rare cell isolation. *Lab on a Chip, 15*(5), 1230–1249. https://doi.org/10.1039/c4lc01246a

Shrirao, A. B., Fritz, Z., Novik, E. M., Yarmush, G. M., Schloss, R. S., Zahn, J. D., & Yarmush, M. L. (2018). Microfluidic flow cytometry: The role of microfabrication methodologies, performance and functional specification. *Technology, 6*(1), 1–23. https://doi.org/10.1142/S2339547818300019

Tian, T., Lin, S., & Yang, C. (2024). Beyond single cells: Microfluidics empowering multiomics analysis. *Analytical and Bioanalytical Chemistry, 416*(9), 2203–2220. https://doi.org/10.1007/s00216-023-05028-4

Truppo, M. D. (2017). Biocatalysis in the pharmaceutical industry: The need for speed. *ACS Medicinal Chemistry Letters, 8*(5), 476–480. https://doi.org/10.1021/acsmedchemlett.7b00114

Wang, G., Benasutti, H., Jones, J. F., Shi, G., Benchimol, M., Pingle, S., Kesari, S., Yeh, Y., Hsieh, L. E., Liu, Y. T., Elias, A., & Simberg, D. (2018). Isolation of Breast cancer CTCs with multi-targeted buoyant immunomicrobubbles. *Colloids and Surfaces B: Biointerfaces, 161*, 200–209. https://doi.org/10.1016/j.colsurfb.2017.10.060. Epub 2017 Oct 23. PMID: 29080504; PMCID: PMC5726926.

Warkiani, M. E., Wu, L., Tay, A. K., & Han, J. (2015). Large-volume microfluidic cell sorting for biomedical applications. *Annual Review of Biomedical Engineering, 17*, 1–34. https://doi.org/10.1146/annurev-bioeng-071114-040818. Epub 2015 Jul 16. PMID: 26194427.

Weiss, R., Gerdes, W., Leonhardt, F., Berthold, R., Sack, U., & Grahnert, A. (2019). A comparative study of two separation methods to isolate monocytes. *Cytometry. Part A: The Journal of the International Society for Analytical Cytology, 95*(2), 234–241. https://doi.org/10.1002/cyto.a.23633

Xenobiotics Mediated Alterations in DNA Content and Cell Cycle: Flow Cytometry-Based Analysis

Kumud Nigam, Juhi Verma, Manish Raj Kulshrestha, Vandana Tiwari, and Anumesh K. Pathak

Abstract

This chapter explores the impact of xenobiotics, or foreign chemicals, on cellular processes, particularly DNA content and cell cycle progression. We first discuss xenobiotics and their diverse effects on the immune system. Further, it ascribes the process of xenobiotic metabolism, which involves the transformation of these compounds within the body. We explore the mechanisms of action by which xenobiotics exert their effects, with a specific focus on their genotoxic potential and the ability to damage genetic material. The genotoxic potential of xenobiotics using flow cytometry, which allows detailed analysis of individual cells, is discussed. The use of flow cytometry to evaluate DNA damage using techniques such as H2AX staining and cell cycle analysis is also compiled. These techniques help to understand how cells respond to xenobiotic exposure, measure changes in DNA content, and identify cells undergoing programmed death or cell cycle arrest. By using flow cytometry and understanding how xenobiotics work, researchers can better grasp the potential health risks of exposure to these chemicals. Overall, the flow cytometry-based approaches presented here serve as valuable tools for evaluating the safety and potential hazards associated with xenobiotic exposure.

Keywords

Xenobiotics · DNA · Cell cycle · Genotoxic · DNA damage

K. Nigam · J. Verma · M. R. Kulshrestha · V. Tiwari · A. K. Pathak (✉)
Department of Biochemistry, Dr. Ram Manohar Lohia Institute of Medical Sciences, Lucknow, Uttar Pradesh, India

AB Pant et al. (eds.), *Flow Cytometry: Applications in Cellular and Molecular Toxicology*, https://doi.org/10.1007/978-981-97-9758-5_2

1 Introduction

Xenobiotics are defined as chemicals that are not produced nor used in the maintenance of normal physiological and biochemical functions but are extrinsic chemicals to which the reference organism is exposed (Stefanac et al., 2021). The word "xeno" is a Greek word that means foreigner, "bio" refers to life, and the "ic" suffix means like. Xenobiotic as a whole means studying the impact of foreign material or chemicals on the living organism (Croom, 2012). Xenobiotics broadly may be divided into two categories: xenobiotics (True xenobiotics) and endogenous xenobiotics (Pseudo xenobiotics). Exogenous xenobiotics are true xenobiotics; organisms having chemicals are sourced from outside and not produced by themselves. A few authors also use the term xenobiotics for the chemical produced by the organism itself that exceeds its normal value and can also be regarded as pseudo xenobiotics (Soucek, 2011).

Bioaccumulation of exogenous or endogenous chemicals/substances may lead to different diseases by mainly affecting the immune system. Urbanization, industrialization, and globalization are the uses of the hour for our countries to uplift lifestyles and improve the living standards and economic stability of their growing population (Gu, 2019). On the other hand, with positivity, there are certain negative impacts. Chemicals emitted from industries as waste, from farmland, and also from household garbage are the true sources of xenobiotics. These pollutants and chemicals are becoming an unavoidable part of our ecosystem (Miglani et al., 2022). The source of xenobiotics is uncleaned water, air pollutants, and healthcare products. Xenobiotics route to living organisms mainly by aeronasal, oral, epidermal, and anorectal, which is depicted in Table 1.

2 Xenobiotics and Their Impact on Immune Functions

Xenobiotic pollutant hazards can be understood by the level of contamination not only to complex organisms but to simple prokaryotes and unicellular organisms. These xenobiotic agents act as antigens to the living and induce immune response in order to maintain normal physiological and biochemical functions (Stefanac et al., 2021).

Xenobiotics agents. → Introduction/accumulation to living → immune response

Immune response → Suppressive effect

Immune response → Stimulative effect

Xenobiotic agents primarily affect the immune system adversely, which is termed "immunotoxicity." Immunotoxicity either can suppress or exaggerate immune responses. Immunotoxicity is often associated with high morbidity and

Table 1 Xenobiotics routes to living

S.N.	Routes	Chemicals	Source	Reference
1	Aeronasal route	2-Butoxyethanol 1-*tert*-Butoxypropan-2-ol	Painting workshop	IARC (2010)
		$PM_{2.5}$ & PM_{10}	Automobile emissions & cooking	Alves et al. (2023)
		NO_2	Traffic emissions and burning of gas oil routs	Nassikas et al. (2024)
		NH_3	Agriculture and livestock maintenance	Wyer et al. (2022)
		SO_2	Fossil fuel	Perera (2018)
		CO	Vehicular emissions	Gireesh Kumar et al. (2021)
2	Oral route	5-Aminosalicylic acid (5-ASA)	Drugs	Collins and Patterson (2020)
		2,4-Dichlorophenoxyacetic acid (2,4-D)	Pesticides	Peterson et al. (2016)
		2-Amino-1-methyl-6-phenylimidazo[4,5-b]pyridine (PhIP)	Food addictive	Felton et al. (2007)
3	Epidermal route	Triclosan (5-chloro-2-(2,4-dichlorophenoxy) phenol, TCS)	Cosmetics	Poole et al. (2016)

mortality (Zerdan et al., 2021). Studies on rodents finds that immune toxicants have overlapping functions at the extent of gene expression, resulting in reduced resistance towards infection, generation of tumors, escaping immune surveillance, or increased incidence of autoimmune disorders. Immunotoxicity affects micromorphologically on immune tissues and organs, altered maturation of immune competent cells, changes in B & T cell populations, altered cytokines, and also functional alteration on immune competent cells (Morsink et al., 2020).

Assays should be able to investigate the stressor's impact at the immune cell level of both innate and adaptive immune systems in response to different stimuli as the responder to bacteria and viruses. The influence of specific chemicals on the activity of immune cells modulation for upkeeping the normal physical and chemical behavior changes are also available (Table 2) (Maddalon et al., 2023).

These assays may be used to investigate the effect of the individual chemicals specifically on keratinocytes (Ramirez et al., 2016), lymphocyte cell lines (Kimura et al., 2018), and dendritic cell lines (Iulini et al., 2022). Assay for peripheral blood mononuclear cells (PBMCs) and whole blood are also available for immune effects (Stølevik et al., 2011; Langezaal et al., 2002). Reporter assay could be engaged against established cell lines for gene activation, though in generic cells, effects implied by particular pathways remain clueless or very little information in living.

Table 2 Immunotoxicity effects by different assays (OECD, 2023)

S.N.	Immunotoxicity assessment method	Immunotoxic effects
1	TG 444A (IL-2 Luc assay)	To evaluating the immunotoxic effects of chemicals on T-cells
2	TG 442D (ARE-Nrf2 luciferase assay)	To assessing keratinocyte activation
3	TG 442E (a series of assay)	To determine dendritic cell activation in the context of skin sensitization

Today's known assays for the chemicals are non-efficient to testing the complexity of cell composition in biological samples advert to cell lines. From the diversity of the peripheral blood cells, the flow cytometric technique may be employed to analyze even a single cell. It allows (flowcytometry) to analyze immune cell types by lineage, and selecting activation numbers and screening of chemicals could be achieved by identifying affected cells and altered or affected signaling pathways (Pierzchalski et al., 2023; Wang et al., 2022; Covey et al., 2010).

Pathological states are the result of xenobiotic interaction with the biological system through a particular molecular mechanism depending on individual xenobiotics. To study the cellular responses to xenobiotics, we must consider changes on a genetic level, protein expression, and morphological alteration on a cellular level. For these studies, we require different tools such as microarray, immunohistochemistry, RT-PCR, histological evaluation, and flow cytometry. The severity of immunotoxicity induced by xenobiotics is multifactorial; some of the factors are dose and duration, nutritional states, pre-existing pathological conditions, previous exposures, mental states, age, and gender.

3 Xenobiotic Metabolism (Biotransformation)

Biotransformation, sometimes referred to as metabolism, is the structural modification of a chemical by enzymes in the body (Fig. 1). Exposure to chemicals (pharmaceutical use) either has some beneficial properties or is toxic in the metabolite form. Metabolism of xenobiotics is generally done by phase I and phase II enzymes working together or separately. Phase I metabolism, composed of oxidases, hydrolases, and reductases, is responsible for the majority of xenobiotic metabolism. Phase I reactions are carried out by oxidative enzymes such as cytochrome P450 and the flavin monooxygenases, as well as hydrolytic enzymes like esterases and amidases (Liska et al., 2006).

Phase I metabolite binds oxygen, hydrogen, water, and amino acid to the lipophilic molecules (drugs) to introduce a hydroxyl (−OH), amino (NH2), sulfhydryl (−SH), or carboxyl (−COOH) and results in xenobiotics (drugs) water solubility. Polar water-soluble metabolites such as amino acid, nucleic acid, sugar, and small

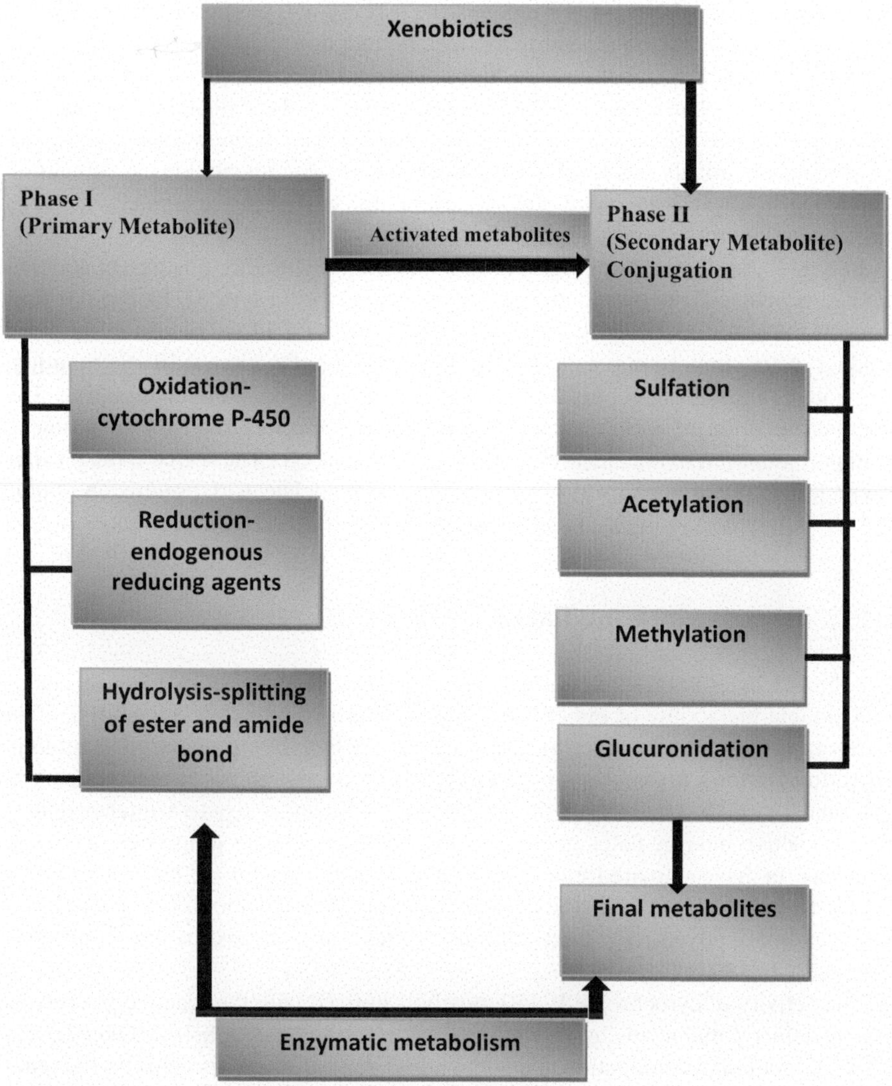

Fig. 1 Metabolism of xenobiotics

organic acids are a must for initial metabolism by phase I enzymes, such as by reactions involved in oxidation, reduction, and hydrolysis (Schroer et al., 2010).

In hydrolysis reaction with water, the xenobiotic compounds are broken into two compounds with parallelly H2O molecules split in two; one is H2 reacting with one compound and hydroxide with another compound. The enzymes of hydrolysis reactions include esterases, peptidases, and amidases. In reduction reactions addition of H2 &/or the removal of O2 will be done (de-hydroxylation, hydrogenation, decarboxylation, amination, methylation). In oxidation reactions, the addition of O2 &/or

removal of H2 will be accomplished, and the process will be (hydroxylation, dehydrogenation, carboxylation, deamination, demethylation) (Guengerich, 2006).

Phase II metabolic reactions are biotransformation reactions, which serve as a detoxifying step of xenobiotics. The phase II reaction is also called the "conjugation reaction," which is a part of detoxifying that also plays an essential role in the toxicity of many chemicals as a result of the formation of toxic metabolites. Conjugation reactions usually activate metabolites by using high energy intermediate, which can be classified into two types—Type I & Type II.

In type I, the activated conjugating agent combines with the substrate to get conjugated products (Glucuronidation & sulfonation), and in type II, firstly, the substrate is activated and then combined with an amino acid to yield a conjugated product. A few important reactions are glucuronide conjugation, sulfo conjugation, acetylation, amino acid conjugation, glutathione conjugation, and methylation. The biotransformation of xenobiotics causes an increase in free radicals, resulting in lipid peroxidation, oxidative stress, and oxidative damage. These free radicals alter intercellular signaling pathways by interacting with nucleic acid, protein, and lipids and ultimately destabilise cellular homeostasis (Djurovic, 2012).

3.1 Reactive Intermediates

Xenobiotics by biotransformation are trends to detoxification but by forming chemically stable metabolite or either way may transformed into toxic metabolites that are more chemically reactive (Srivastava et al., 2010). This process of xenobiotic activation forming toxic forms is referred to as bioactivation or metabolic activation, and the toxic produce of the processes is known as reactive intermediates. These reactive intermediates generally have electron-seeking properties or donate electrons, or have unpaired electrons (Attia, 2010; Lopachin & Decaprio, 2005; Nioi & Hayes, 2004). These ready-to-react and unstable metabolites (reactive intermediate) may adversely affect and could lead to various health conditions (Guengerich, 1992).

The activity of cytochrome P-450 mostly is noticed with the reactive intermediates, and thus, a major role in developing toxicity could be established (Guengerich, 1992). Glutathione conjugation and some other phase II reactions detoxify the reactive intermediates. Toxification and detoxification processes are simultaneous occurring processes in which when balance favours the reactive intermediates causes cell injury and may lead to death (Masubuchi et al., 2007).

3.2 Reactivation

Phase II reactions are associated with detoxification, but in some instances, after conjugation with glutathione may activate again by hydrolytic reaction of glutathione and P-450-dependent biotransformation. This inducing toxic effects is termed reactivation. Reactive intermediates are mostly produced in phase I reactions, but in rare cases, phase II may also generate reactive intermediates (Attia, 2010). Few xenobiotics can be named that, in conjugation with glucuronic acid or sulfate, convert to toxic metabolites (Glatt, 2000).

4 Mechanism for Xenobiotics

Xenobiotic compounds could be organic and inorganic in nature, in which organic are supposed to be more responsible for inducing toxicity in living. The basic constituents for medicines and poisons used are organic xenobiotics (Borzelleca, 2000). Toxic mechanisms include cascades of events from exposure, distribution, and metabolism, and the process is entitled the same as toxicokinetics and toxicodynamics (Evans, 2013). Biotransformation of organic xenobiotics mainly occurs in the liver with the involvement of several other organs. Organic xenobiotics could further be classified as direct-acting or indirect-acting (Kumar et al., 2012). Direct-acting xenobiotics do not require biotransformation in contrast with indirect-acting type, which requires an activation step. Direct-acting xenobiotics usually show electrophilic properties resulting from natural decomposition by hydrolysis, which makes the chemicals directly interact with protein, lipid, and nucleic components and end with a toxic environment for organelles of living. Indirect-acting chemicals are mostly used as medicines and some environmental pollutants. Responsibility for a major chunk of toxic effects can be credited to indirect-acting chemicals (Betharia et al., 2014).

5 Genotoxic Potential of Xenobiotic Compounds by Flow Cytometry

Specific chemicals are able to interact with nucleic acid and cause genetic alteration, leading to the loss of functional activities and capabilities of the gene products (Preston & Hoffmann, 2010). It is well established that mutational events are responsible for diseases like cancer and others (Phillips & Arlt, 2009; Bagyinszky et al., 2014). Keeping in mind xenobiotic toxicity-related issues, we should investigate properly by running genetic and toxicology tests to identify its mutagenic effects before its launching in the market.

A few important techniques have more practical approaches to detecting genotoxic damage, such as image analysis (Francois et al., 2014), cytochalasin B technique (Fenech, 2000), and flow cytometry (Kirsch-Volder et al., 2011). Cytotoxicity may maybe defined as killing of normal or tumour cells by cytotoxic granules. The cells affected by toxic xenobiotics can be examined by flow cytometry. During the process, the target cells, according to their type, have to be prepared detective probes by which the cells emit fluorescence or color (McKinnon, 2018). These prepared cells are suspended in the fluid and injected in a flow cytometer machine to detect the specific cell types or changes within the assessed parameters, such as cell replication, cell viability, apoptosis, genotoxicity, and mutation (Lenzi et al., 2018).

5.1 Real-Time Flow Cytometry (FCM)

The concept of real-time flow cytometry (RT-FCM) was introduced as a result of the inclusion of time as an operative parameter in FCM. This experimental strategy involves the kinetic monitoring of cellular parameter changes within heterogeneous cell populations (Kaur et al., 2015; Felipo-Benavent et al., 2023). RT-FCM involves the sequential analysis of single cells and the progression of the biological process of interest as the sample is processed in the flow cytometer. RT-FCM enables the monitoring of functional changes in a large number of single cells with a theoretical time resolution of less than 1 second and for time windows extending from a few seconds to several minutes. This form of kinetic analysis is particularly pertinent for the investigation of dynamic processes that occur at a rapid pace or are transient in nature, such as those that are characteristic of signal transduction (Felipo-Benavent et al., 2023).

5.2 Individual Cell Sorting

The FCM can analyze a variety of cell populations using fluorescent probes and separate them using the same cytometric detection principles. Cell sorting enables the integration of the intrinsic capabilities of FCM results with information obtained through molecular techniques and image analysis, and it serves as a preparative instrument for the rapid isolation of rare cells of biochemical significance (Sun et al., 2023). Charged electric plates compose the most common and intricate cell sorters, deflecting microdroplets toward specific collection tubes or supports and forming and deflecting these droplets through high-frequency vibration. In general, this form of separator enables the high-speed separation of multiple sub-populations at once (thousands of cells per second). Microfluidics is the foundation of a second family of cell separators. These systems, in general, necessitate a reduced quantity of starting cells; however, they are characterized by inferior performance and speed in comparison to electrostatic separators. Conversely, they pose fewer biological hazards to the operator and are less aggressive toward the cells in the process (Cossarizza et al., 2019).

5.3 Fast Analysis of a Large Number of Live Cells

FCM can investigate a wide range of cell types and samples in different phases of vitality, such as fresh, fixed, and permeabilized cells (Spidlen et al., 2010). Live-cell analysis allows for the identification of biological processes in marginally perturbed intracellular environments and near-physiological conditions. The fast rate of analysis lets forensics look at a lot of individual cells in a reasonable amount of time. This makes it possible to accurately identify and study rare events (as few as one event per 10_8 cells) (Spidlen et al., 2010). Bulk biomolecular determinations, on the other hand, look at parts that have been taken from a group of cells and give an average value of gene expression, enzyme activity, or metabolite concentration.

5.4 Multivariate Data Analysis

The hardware structure and software architecture of modern flow cytometers integrate multiparametric acquisition with multivariate data analysis. This integration enhances the ability to discriminate between cell populations by using the multidimensional correlation of individual properties measured on each cell (Cossarizza et al., 2019; Robinson, 2022). Furthermore, the unique format in which FCM raw data and experimental details are stored as listmode files (uncorrelated data matrices for each cell, in accordance with the flow cytometry standard, FCS) allows for the creation of new parameter correlations and population-gating criteria during the replay of those FCS files (Cossarizza et al., 2019; Robinson, 2022; Spidlen et al., 2010).

6 Flow Cytometry-Based Techniques for DNA Damage

Flow cytometry can be utilized to explore DNA damage in a population of cells. Cell cycle and analysis and γH2AX staining are most commonly used in DNA damage analysis (Johansson et al., 2017).

6.1 Cell Cycle Analysis

During the process of cell division, cells undergo a sequence of distinct phases. The cell cycle can be disrupted by DNA damage, resulting in an accumulation of cells in a specific phase. Flow cytometry can be employed to quantify the DNA content of individual cells by staining cells with a fluorescent dye that binds to DNA, such as propidium iodide or DAPI. This information can be utilized to produce a cell cycle profile, which displays the proportion of cells in each phase of the cell cycle.

6.2 γH2AX Staining

H2AX is a histone protein that is a constituent of chromatin, the complex of DNA and proteins that comprise chromosomes (Johansson et al., 2017). H2AX is phosphorylated to produce γH2AX in response to DNA damage. Using an antibody that is specific to γH2AX, flow cytometry can be employed to detect γH2AX. The quantity of DNA damage in a cell is directly proportional to the intensity of the γH2AX signal. Other methods involve the measurement of DNA fragmentation and the use of fluorescent probes that specifically bind to damaged DNA.

References

Alves, C., Evtyugina, M., Vicente, E., Vicente, A., Rienda, I. C., et al. (2023). $PM_{2.5}$ chemical composition and health risks by inhalation near a chemical complex. *Journal of Environmental Sciences, 124*, 860–874.

Attia, S. M. (2010). Deleterious effects of reactive metabolites. *Oxidative Medicine and Cellular Longevity, 3*, 238–253.

Bagyinszky, E., Youn, Y. C., An, S. S., & Kim, S. (2014). The genetics of Alzheimer's disease. *Clinical Interventions in Aging, 9*, 535–551.

Betharia, S., Corcoran, G. B., & Ray, S. D. (2014). Mechanisms of toxicity. In *Encyclopedia of toxicology* (3rd ed., pp. 165–175). Elsevier Publishing Company.

Borzelleca, J. F. (2000). Profiles in toxicology. Paracelsus: Herald of modern toxicology. *Toxicological Sciences, 53*, 2–4.

Collins, S. L., & Patterson, A. D. (2020). The gut microbiome: An orchestrator of xenobiotic metabolism. *Acta Pharmaceutica Sinica B, 10*(1), 19–32.

Cossarizza, A., Chang, H.-D., Radbruch, A., Acs, A., Adam, D., Adam-Klages, S., Agace, W. W., Aghaeepour, N., Akdis, M., Allez, M., et al. (2019). Guidelines for the use of flow cytometry and cell sorting in immunological studies (second edition). *European Journal of Immunology, 49*, 1457–1973.

Covey, T. M., Cesano, A., & Parkinson, D. R. (2010). Single-cell network profiling (SCNP) by flow cytometry in autoimmune disease. *Autoimmunity, 43*, 550–559.

Croom, E. (2012). Chapter 3: Metabolism of Xenobiotics of human environments. *Progress in Molecular Biology and Translational Science, 112*, 31–88.

Djurovic, J. (2012). Biotransformation of the toxic chemical substances. *Mathematical Models and Methods in Applied Sciences*, 91–95.

Evans, T. J. (2013). Chapter 2: Toxicokinetics and toxicodynamics. In *Small animal toxicology* (3rd ed., pp. 13–19). Elsevier Publishing Company.

Felipo-Benavent, M., Martínez-Romero, A., Rubio-Guerri, C., Álvaro-Álvarez, T., Gil, D., García-Párraga, D., & O'Connor, J. (2023). Flow cytometric kinetic assay of calcium mobilization in whole blood platelets of bottlenose dolphins (*Tursiops truncatus*). *Cytometry. Part A, 103*, 347–352.

Felton, J. S., Knize, M. G., Wu, R. W., Colvin, M. E., Hatch, F. T., & Malfatti, M. A. (2007). Mutagenic potency of food-derived heterocyclic amines. *Mutation Research, 616*, 90–94.

Fenech, M. (2000). The in vitro micronucleus technique. *Mutation Research, 455*, 81–95.

Francois, M., Hochstenbach, K., Leifert, W., & Fenech, M. F. (2014). Automation of the cytokinesis-block micronucleus cytome assay by laser scanning cytometry and its potential application in radiation biodosimetry. *BioTechniques, 57*, 309–312.

Gireesh Kumar, P., Lekhana, P., Tejasvi, M., & Chandrakala, S. (2021). Effects of vehicular emissions on the urban environment – A state of the art. *Materials Today: Proceedings, 45*(7), 6314–6320.

Glatt, H. (2000). Sulfotransferases in the bioactivation of xenobiotics. *Chemico-Biological Interactions, 129*, 141–170.

Gu, C. (2019). Urbanization: Processes and driving forces. *Science China Earth Sciences, 62*, 1351–1360.

Guengerich, F. P. (1992). Metabolic activation of carcinogens. *Pharmacology & Therapeutics, 54*, 17–61.

Guengerich, F. P. (2006). Cytochrome P450s and other enzymes in drug metabolism and toxicity. *The AAPS Journal, 8*, E101–E111.

IARC. (2010). Painting, firefighting, and shiftwork. *IARC Monographs on the Evaluation of Carcinogenic Risks to Humans, 98*, 1–804.

Iulini, M., Maddalon, A., Galbiati, V., & Corsini, E. (2022). The modified THP-1 activation assay for the in vitro identification of drug-inducing systemic hypersensitivity. *Frontiers in Toxicology, 4*, 814050.

Johansson, P., Fasth, A., Ek, T., & Hammarsten, O. (2017). Validation of a flow cytometry-based detection of γ-H2AX, to measure DNA damage for clinical applications. *Cytometry. Part B, Clinical Cytometry, 92*(6), 534–540.

Kaur, A., Haghighatbin, M. A., Hogan, C. F., & New, E. J. (2015). A FRET-based ratiometric redox probe for detecting oxidative stress by confocal microscopy, FLIM and flow cytometry. *Chemical Communications, 51*, 10510–10513.

Kimura, Y., Fujimura, C., Ito, Y., Takahashi, T., Terui, H., & Aiba, S. (2018). Profiling the immunotoxicity of chemicals based on in vitro evaluation by a combination of the Multi-ImmunoTox assay and the IL-8 Luc assay. *Archives of Toxicology, 92*, 2043–2054.

Kirsch-Volder, M., Decordier, I., Elhajouji, A., Plas, G., Aardema, M., & Fenech, M. (2011). In vitro genotoxicity testing using the micronucleus assay in cell lines, human lymphocytes and 3D skin models. *Mutagenesis, 26*, 177–184.

Kumar, V., Abba, A. K., & Aster, J. C. (2012). *Robbins basic pathology* (7th ed., pp. 199–200). Elsevier.

Langezaal, I., Hoffmann, S., Hartung, T., & Coecke, S. (2002). Evaluation and pre validation of an immunotoxicity test based on human whole-blood cytokine release. *Alternatives to Laboratory Animals, 30*, 581–595.

Lenzi, M., Cocchi, V., & Hrelia, P. (2018). Flow cytometry vs optical microscopy in the evaluation of the xenotoxic potential of xenobiotic compounds. *Cytometry Part B (Clinical Cytometry), 94B*, 852–862.

Liska, D. J., Lyon, M., & Jones, D. S. (2006). Detoxification and biotransformation imbalance. *Explore, 2*, 112–140.

Lopachin, R. M., & Decaprio, A. P. (2005). Protein adduct formation as a molecular mechanism in neurotoxicity. *Toxicological Sciences, 86*, 214–225.

Maddalon, A., Iulini, M., Melzi, G., Corsini, E., & Galbiati, V. (2023). New approach methodologies in immunotoxicology: Challenges and opportunities. *Endocrine, Metabolic & Immune Disorders Drug Targets, 23*(14), 1681–1698.

Masubuchi, N., Makino, C., & Murayama, N. (2007). Prediction of in vivo potential for metabolic activation of drugs into chemically reactive intermediate: Correlation of in vitro and in vivo generation of reactive intermediates and in vitro glutathione conjugate formation in rats and humans. *Chemical Research in Toxicology, 20*, 455–464.

McKinnon, K. M. (2018). Flow cytometry: An overview. *Current Protocols in Immunology, 120*, 5.1.1–5.1.11.

Miglani, R., Parveen, N., Kumar, A., Ansari, M. A., Khanna, S., Rawat, G., Panda, A. K., Bisht, S. S., Upadhyay, J., & Ansari, M. N. (2022). Degradation of xenobiotic pollutants: An environmentally sustainable approach. *Metabolites, 12*(9), 818.

Morsink, M. A. J., Willemen, N. G. A., Leijten, J., Ruchi Bansal, S., & Shin, R. (2020). Immune organs and immune cells on a chip: An overview of biomedical applications. *Micromachines, 11*(9), 849.

Nassikas, N. J., McCormack, M. C., Ewart, G., et al. (2024). Indoor air sources of outdoor air pollution: Health consequences, policy, and recommendations: An official American Thoracic Society workshop report. *Annals of the American Thoracic Society, 21*(3), 365–376. https://doi.org/10.1513/AnnalsATS.202312-1067ST

Nioi, P., & Hayes, J. D. (2004). Contribution of NAD(P)H: Quinone oxidoreductase 1 to protection against carcinogenesis, and regulation of its gene by the Nrf2 basic-region leucine zipper and the arylhydrocarbon receptor basic helix-loop-helix transcription factors. *Mutation Research, 555*, 149–171.

OECD. (2023). *OECD guidelines for the testing of chemicals, section 4: Health effects*. Available at: https://www.oecd-ilibrary.org/environment/oecd-guidelines-for-the-testing-of-chemicals-section-4-health-effects_20745788. Accessed November 7, 2023.

Perera, F. (2018). Pollution from fossil-fuel combustion is the leading environmental threat to global pediatric health and equity: Solutions exist. *International Journal of Environmental Research and Public Health, 15*(1), 16.

Peterson, M. A., McMaster, S. A., Riechers, D. E., Skelton, J., & Stahlman, P. W. (2016). 2,4-D past, present, and future: A review. *Weed Technology, 30*(2), 303–345.
Phillips, D. H., & Arlt, V. M. (2009). Genotoxicity: Damage to DNA and its consequences. *EXS, 99*, 87–110.
Pierzchalski, A., Zenclussen, A. C., & Herberth, G. (2023). OMIP-94: Twenty-four-color (thirty-marker) panel for deep immunophenotyping of immune cells in human peripheral blood. *Cytometry. Part A, 103*, 695–702.
Poole, A. C., Pischel, L., Ley, C., et al. (2016). Crossover control study of the effect of personal care products containing triclosan on the microbiome. *mSphere, 1*(3), e00056–e00015.
Preston, R. J., & Hoffmann, G. R. (2010). Genetic toxicology. In C. D. Klaassen & J. B. Watkins III (Eds.), *Casarett & Doull's essentials of toxicology* (2nd ed., pp. 123–134). McGraw-Hill.
Ramirez, T., Stein, N., Aumann, A., Remus, T., Edwards, A., Norman, K. G., et al. (2016). Intra- and inter-laboratory reproducibility and accuracy of the LuSens assay: A reporter gene-cell line to detect keratinocyte activation by skin sensitizers. *Toxicology In Vitro, 32*, 278–286.
Robinson, J. P. (2022). Flow cytometry: Past and future. *BioTechniques, 72*, 159–169.
Schroer, K., Kittelmann, M., & Lütz, S. (2010). Recombinant human cytochrome P450 mono-oxygenases for drug metabolite synthesis. *Biotechnology and Bioengineering, 106*, 699–706.
Soucek, P. (2011). Xenobiotics. In M. Schwab (Ed.), *Encyclopedia of cancer* (pp. 3964–3967). Springer Berlin Heidelberg.
Spidlen, J., Moore, W. E., Parks, D. R., Goldberg, M. L., Bray, C. L., Bierre, P., Gorombey, P., Hyun, B., Hubbard, M., Lange, S., et al. (2010). Data file standard for flow cytometry, version FCS 3.1. *Cytometry. Part A, 77*, 97–100.
Srivastava, A., Maggs, J. L., Antoine, D. J., Williams, D. P., Smith, D. A., & Park, B. K. (2010). Role of reactive metabolites in drug-induced hepatotoxicity. *Handbook of Experimental Pharmacology, 196*, 165–194.
Stefanac, T., Grgas, D., & Dragicevic, T. L. (2021). Xenobiotics—Division and methods of detection: A review. *Journal of Xenobiotics, 11*(4), 130–141.
Stølevik, S. B., Nygaard, U. C., Namork, E., Granum, B., Pellerud, A., van Leeuwen, D. M., et al. (2011). In vitro cytokine release from human peripheral blood mononuclear cells in the assessment of the immunotoxic potential of chemicals. *Toxicology In Vitro, 25*, 555–562.
Sun, G., Qu, L., Azi, F., Liu, Y., Li, J., Lv, X., Du, G., Chen, J., Chen, C.-H., & Liu, L. (2023). Recent progress in high-throughput droplet screening and sorting for bioanalysis. *Biosensors & Bioelectronics, 225*, 115107.
Wang, X., Li, N., Ma, M., Han, Y., & Rao, K. (2022). Immunotoxicity in vitro assays for environmental pollutants under paradigm shift in toxicity tests. *International Journal of Environmental Research and Public Health, 20*, 273.
Wyer, K. E., Kellaghan, D. B., Blane-Vidal, V., Schauberger, G., & Curran, T. P. (2022). Ammonia emissions from agriculture and their contribution to fine particulate matter: A review of implications for human health. *Journal of Environmental Management, 323*, 116285.
Zerdan, M. B., Moussa, S., Atoui, A., & Assi, H. I. (2021). Mechanisms of immunotoxicity: Stressors and evaluators. *International Journal of Molecular Sciences, 22*(15), 8242.

Flow Cytometric Evaluation of Mitochondrial Health

Puneet Khare, Lakshmi Bala, and Alok Kumar Pandey

Abstract

Mitochondria are the leading producers of free radicals, ion homeostasis or redox state regulators, and ATP generators. They are the most critical cellular component for the endurance of aerobic cells. Numerous scientific studies have shown how vital mitochondria are to synthesizing primordial ATP, which is required for eukaryotic cells to survive and proliferate. Thus, these cell organelles are called the "powerhouse" of cell. They possess double membrane and a tiny genome. A system of bioenergetics connections that regulates a large portion of cellular activity revolves around the mitochondrion. In this regard, it is essential to comprehend the intricacy of the mitochondrial compartment and its operation and to create experimental approaches that evaluate its nature, function, and metabolism. The function, structural organization, mitochondrial respiratory chain, transmembrane potential preservation, and production of reactive oxygen species of the mitochondria are all covered in this chapter. Flow cytometry-based probes are highlighted as the primary probes to monitor mitochondrial health. A collection of methodologies for studying mitochondria is included in this chapter to facilitate the researchers in experiment designing, methodology opting tactics, analysis, and result interpretation.

P. Khare
Central Flow Cytometry Facility, ASSIST Division, CSIR-Indian Institute of Toxicology Research, Lucknow, Uttar Pradesh, India

L. Bala
Department of Biochemistry, Babu Banarasi Das College of Dental Sciences, Babu Banarasi Das University, Lucknow, Uttar Pradesh, India

A. K. Pandey (✉)
Nanomaterial Toxicology Laboratory, Drug & Chemical Toxicology Group, FEST Division, CSIR-Indian Institute of Toxicology Research, Lucknow, Uttar Pradesh, India
e-mail: alokpandey@iitr.res.in

AB Pant et al. (eds.), *Flow Cytometry: Applications in Cellular and Molecular Toxicology*, https://doi.org/10.1007/978-981-97-9758-5_3

Keywords

Mitochondria · Flow cytometry · Apoptosis · Probes · Fluorescence

1 Introduction

The cell is the fundamental anatomical and functional component of cellular life. It is a watery space surrounded by a cell membrane that is able to sustain itself on its own and carry out the necessary tasks for life. Both unicellular and multicellular organisms are classified as cellular. All cells fall into one of two primary classes: prokaryotic cells or eukaryotic cells, based on their internal design. Eukaryotic cells contain other membrane-bound organelles in addition to the nucleus, such as the endoplasmic reticulum, the Golgi complex, lysosomes, the mitochondria, microbodies, and vacuoles. Similar unit membrane-bound organelles are absent from prokaryotic cells.

1.1 Mitochondria: An Integral Cellular Organelle of the Cell

Mitochondria are double-membrane organelles that are crucial for eukaryotic cells to produce metabolic energy. The outer membrane is permeable and encloses the inner membrane that has cristae, which are distinctive folds that provide a large surface area for chemical reactions. Mitochondria generate most of the usable power released during the oxidation of fatty acids and carbohydrates. Oxidative phosphorylation converts this energy into ATP. Numerous scientific investigations have demonstrated that mitochondria play a vital role in the energy metabolism necessary for eukaryotic cells to survive and proliferate (Clemente-Suárez et al., 2023).

The process of apoptosis aids in the preservation of equilibrium inside the bodies of multicellular creatures. During evolution, it also plays a role in the formation of organs (Elmore, 2007). Nearly every cell has mitochondria, which are essential to this process. Therefore, a wide range of disorders affecting multiple tissues, including energy-dependent systems like the neurological system, skeletal and cardiac muscles, renal and endocrine systems, can be brought on by mitochondrial malfunction (Wallace, 2018; Chen et al., 2023). For this reason, evaluating mitochondria and their functioning is crucial. Therefore, monitoring mitochondrial morphology can provide valuable insights into the development of multiple diseases and help us learn more about biological processes in general.

1.2 Function of Mitochondria

Mammalian cells primarily obtain energy via glucose and fatty acids' oxidative breakdown. In glycolysis, glucose converts to pyruvate in the cytosol. Once pyruvate is transported into the mitochondria and fully oxidized to CO_2, the majority of the usable energy (ATP) generated by glucose metabolism is produced. Pyruvate gets converted into acetyl CoA and carbon dioxide through the citric acid cycle. The citric acid cycle enzymes, which are present in the matrix of mitochondria, play a crucial role in the oxidative degradation of both carbohydrates and fatty acids. The cycle breaks down acetyl CoA, which is also created during the oxidation of fatty acids. Through the reduction of NAD^+ to NADH and FAD to $FADH_2$, acetyl CoA is converted to CO_2. Oxidative phosphorylation, which occurs within the inner mitochondrial membrane, is responsible for generating the majority of energy through oxidative metabolism. This process involves the transfer of high-energy electrons from FADH2 and NADH to molecular oxygen via carriers in the membrane. The energy produced from these electron transfers is converted to potential energy that is then stored in a proton gradient across the membrane, ultimately driving the synthesis of ATP (Alberts et al., 2002).

1.3 Organization of Mitochondria

Two highly specialized membranes encase each mitochondrion, serving distinct purposes. The inner and outer mitochondrial membranes are separated by an intermembrane gap. The inner membrane contains folds called cristae that extend into the organelle's matrix. The two primary functional compartments of mitochondria are the matrix and the inner membrane, each with a distinct purpose. The enzyme complex responsible for producing ATP, known as ATP synthase (F0-F1 particle), is located in the inner membrane. The matrix contains the main enzymes responsible for oxidative metabolism and the mitochondrial genetic system (Nasrin Afzal et al., 2021).

The surface area of mitochondria increases significantly through folding into cristae (Saneto, 2017). The inner mitochondrial membrane contains a high percentage of proteins (over 70%) that play a vital role in the transportation of metabolites, such as fatty acids and pyruvate, between the mitochondria and the cytosol. Additionally, these proteins are involved in aerobic phosphorylation. The maintenance of the proton gradient that drives oxidative phosphorylation relies on the inner membrane's ability to prevent the majority of ions and small molecules from passing through.

The outside mitochondrial membrane is open to tiny molecules, in contrast to the inner membrane. The outer membrane of the organelle serves as protection and has specific transport proteins known as mitochondrial proteins, which let different chemicals freely enter the inter-membrane area of the mitochondria. The voltage-dependent anion-selective channels (VDAC) found in mitochondrial porins assume an open conformation when the membrane potential is low or zero, and a closed conformation when the potential is higher than 30–40 mV. This is due to the presence of porin proteins, which create channels that allow molecules smaller than 6000 daltons to freely diffuse. In terms of ions and tiny molecules, the intermembrane gap shares many characteristics with the cytoplasm. Consequently, the inner mitochondrial membrane functions as a functional barrier to prevent small molecules from passing between the cytosol and the matrix. It preserves the proton gradient necessary for oxidative phosphorylation. The mitochondrial matrix contains several copies of the genetic material, ds DNA, t RNAs, mitochondrial ribosomes (55S–75S), and other proteins. Most of the dsDNA in mitochondria is circular. Different species also differ significantly in the size of their mitochondrial DNA. With the passage of time, a significant portion of the mitochondrial DNA was horizontally transferred from the organelle to the host nucleus, granting the eukaryotic cell autonomy over the majority of mitochondrial functions.

Mitochondria are semi-autonomous organelles because they have their own genetic materials (DNA) (Pfanner et al., 2019). These organelles also possess their own ribosomes and are able to synthesize some of their own proteins. On the other hand, cytosolic 80S ribosomes synthesize, and nuclear genes encode most mitochondrial proteins.

Because they contain their own genetic material (DNA), mitochondria are semi-autonomous organelles. The organelles can also make some of their own proteins and have their own ribosomes. However, the majority of mitochondrial proteins are produced by cytosolic 80S ribosomes and encoded by nuclear genes.

1.4 Mitochondrial Proteins

The architecture of proteins in the mitochondria is remarkably complicated, as demonstrated by recent functional and proteomic investigations. Protein machinery performs a variety of tasks in dynamic networks, including respiration, protein translocation, metabolite transport, protein quality control, and membrane structure management (Gold et al., 2017). Mitochondrial proteins are synthesized by 80S cytosolic and 70S matrix ribosomes, which are encoded by nuclear genes. Cytosolic membrane-free ribosomes produce 99% of the proteins found in mitochondria. The proteins produced by cytosolic ribosomes are Trans located post-transnationally into the mitochondria and unfolded. Once inside the mitochondria, these proteins can be directed to different locations such as the outer membrane (OMM), inner membrane (IMM), intermembrane space (IMS), or mitochondrial matrix (Gold et al., 2017; Grevel et al., 2019).

1.5 Mitochondrial Membrane Potential

The term "membrane potential" refers to the difference in electrical potential across the plasma membrane of every cell. This potential is established due to the selective migration of ions through ion-specific channels in the membrane, along with the variations in ion concentrations throughout the membrane. When energy is produced in mitochondria, protons are also produced, and their concentration is dispersed throughout the inner and outer membranes of the organelles, resulting in the potential of the mitochondrial membrane (Zorova et al., 2018).

The mitochondrial membrane potential, also known as the proton motive force, is the primary bioenergetics parameter that regulates respiratory rate, ATP synthesis, and the production of reactive oxygen species. It is regulated by electron transport and proton leakage. During oxidative phosphorylation, the mitochondrial membrane potential ($\Delta\Psi m$), produced by proton pumps (Complexes I, III, and IV), is a crucial step in energy storage. The mitochondrial respiratory chain or electron transport chain's proton pump controls the mitochondrial membrane potential, also called the proton motive force (Δp). Protons are forced from the matrix space into the intermembrane space, creating a notable pH and electrical gradient in the mitochondrial inner membrane.

Proper cell function relies on consistent levels of intracellular ATP and $\Delta\Psi m$, which most cells are able to maintain. However, any significant changes to these variables can have harmful effects. Prolonged fluctuations in $\Delta\Psi mv$ levels outside of the norm can result in unintended loss of cell viability and contribute to various diseases. $\Delta\Psi m$ plays a crucial role in maintaining mitochondrial homeostasis by selectively targeting malfunctioning mitochondria (Behl et al., 2021). Additionally, it acts as a driving force for transporting proteins and ions besides H^+ to ensure proper mitochondrial function. Numerous studies have observed a link between abnormal MMP levels and human illnesses, including cancer, Parkinson's disease, and Alzheimer's disease (Sisalli et al., 2020). The integrity of MMP is closely tied to mitochondrial functions. Fluorescent probes that detect MMP are a valuable tool for mitochondrial biology research.

2 Cell and Cell Organelles Health Assessment

Mitochondria are an ideal source of information for assessing the health of a cell. In eukaryotes, mitochondria are involved in energy production, phospholipid and heme synthesis, calcium homeostasis, apoptotic activation, and cell death. For these reasons, mitochondria are necessary for appropriate cellular function (Zemirli et al., 2018). Stress, disease, and mitochondrial function are all correlated with mitochondrial dynamics and morphology (Wang & Zhou, 2020). Since they have the innate capacity to sense their own health and their ability to trigger quality-control mechanisms in response to stress, such as stress-induced hyperfusion or fission and the breakdown of damaged mitochondria (mitophagy) (Rickard et al., 2023). Technologies that can analyze and quantify changes in mitochondrial morphology and dynamic behavior are necessary due to their significance in human disease, stress, and homeostasis. Consequently, the state of the mitochondria has an impact on both health and sickness and is crucial for preserving cellular stability (Chu et al., 2021).

2.1 Fluorochrome-Based Cell Health Assessment

A fluorophore, also known as a fluorochrome, is a chemical molecule that is fluorescent and has the ability to reemit light when exposed to light. Fluorophores are generally composed of many coupled aromatic groups or multiple π-bonded planar or cyclic compounds. Chemical reactions, binding events, or changes in their immediate surroundings can all cause small-molecule fluorescent probes to alter fluorescence emission.

The conventional approach for creating fluorescent chemical probes to observe biological processes linked to mitochondria involves combining an organic fluorophore with the mitochondria-targeting motif through the use of an appropriate linker. The lipophilic cation is primarily thought to be a crucial motif for the chemical's accumulation in the mitochondrial matrix as part of anchoring mitochondria (Alshial et al., 2023). In aromatic rings and other conjugated structures, the presence of a delocalized cationic characteristic is crucial for the accumulation of chemicals in the mitochondria. Therefore, the effectiveness of chemical probes in targeting mitochondria can be affected by changes to their lipophilic nature. To monitor mitochondria or membrane potential, various fluorescent probes, such as MitoTracker dyes, have been developed and are commercially available. These probes contain lipophilic cations in different fluorophores, including acedan, rhodamine, cyanine-based Alexa-fluor, and boron-dipyrromethene (BODIPY), and are based on structural insights.

2.2 Fluorescence Techniques

Fluorescence represents a spectrochemical method of analysis wherein the analyte's molecules are excited by a specific wavelength of radiation, and the radiation emitted at a longer wavelength is detected. With remarkable sensitivity and selectivity, researchers may identify specific elements of intricate biomolecular assemblies, like living cells, using fluorescent probes. This approach to analysis is commonly acknowledged and effective method with a wide range of applications in biotechnology, medicine, pharmacology, and the environment. As a component of the ubiquitous luminescence group of phenomena, fluorescence occurs when molecules that are susceptible to light emit from electronically excited states that are produced by mechanical (such as friction), chemical, or physical mechanisms.

Based on the electrical structure of the state of excitement and the emission route, two types of photoluminescence-fluorescence and phosphorescence—are distinguished when a molecule is stimulated by photons of visible or ultraviolet light. A certain wavelength of light can be absorbed by certain atoms and molecules, which then, after a brief time, release a longer wavelength of light. Fluorescence is the name given to this phenomenon. A mechanism called phosphorescence functions identically to fluorescent but has a much longer lifetime when stimulated.

2.3 Basic Concept

The basic process of fluorescence involves an indicator absorbing light energy or a photon, and then a few nanoseconds later, some of that light energy is released as another photon. Because a certain amount of energy escapes throughout this process, the liberated photons have less energy than the absorbing photon. Light with a short wavelength (toward the blue) has more incredible energy than light with a long wavelength (toward the red). Because of this, an indicator's light wavelength is usually longer than the light wavelength it receives (excitation). This change is known as the Stokes shift. Jablonski energy diagrams, which are used to illustrate radiative and non-radiative energy transfers between a molecule's electronic and vibrational states, explain these events.

Fluorometry, a method of measuring fluorescence, is a quantitative and qualitative analytical technique. The two primary mechanisms that create fluorescence are the virtually instantaneous re-emission at a longer wavelength and the fast molecular absorption of light energy at one wavelength. Since very few molecules naturally fluoresce therefore, fluorescent compounds are frequently formed by adding a fluorescent molecule. An excitation spectrum, which measures the wavelength and absorption efficiency of light, and an emission spectrum, which measures the wavelength and quantity of photons released, are the features of fluorescent substances. These spectra match only one fluorescent chemical and are highly selective. As a result, fluorometry is a particular analytical method.

A fluorometer uses an internal light source to determine the wavelength of light needed to stimulate the substance of interest. This allows for the measurement of fluorescence. Additionally, it chooses the wavelength at which the released light is supposed to be measured. The signal, or the quantity of photons released, directly correlates with the analyte's concentration. Optical filters (filter fluorometer) or monochromators can be used to choose light for excitation and emission. The exceptional sensitivity, high specificity, selectivity, ease of use, qualitative as well as quantitative data, and affordable cost are the benefits of fluorescence-based techniques over other existing analytical approaches (Fig. 1).

2.4 Fluorescent Instrumentation

2.4.1 Flowcytometry

In order to identify and quantify the physical and chemical properties of the cells or particles in a heterogeneous fluid mixture, flow cytometry is a common quick laboratory technique based on laser technology. Fluorescence is a promising approach because of its high sensitivity, signal-to-noise ratio, practicality, ease of use, superior selectivity, and real-time detection capabilities. One way to assess changes in superoxide (ROS), membrane potential, and mitochondrial mass is by flow cytometry. When particles are passed by a laser beam in a stream, the fluorescence that results from the particles' light scattering is the foundation for the fundamental concept of flow Cytometry. Additionally, at work, it is the hydrodynamic focusing principle. Flow cytometers comprise three main parts: optical, fluidics, and electronics.

Flow cytometry enables quick multi-parametric analysis of each cell in solution. Both scattered and fluorescent light signals are produced by lasers in flow cytometers, and photodiodes or other photoreceptors pick them up. Several fluorescent compounds are employed in flow cytometry. Examples are ion indicator dyes,

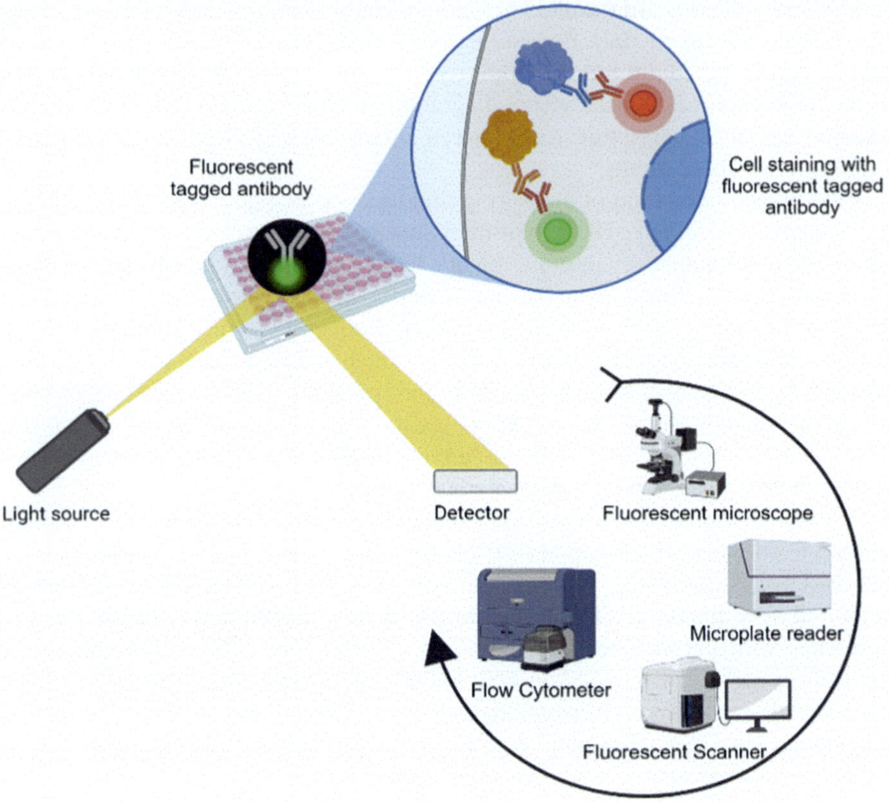

Fig. 1 Image showing the of fluorescent particles and the fluorescent based detectors. (Parts of the figure were drawn by using pictures from Biorender (https://www.biorender.com/))

fluorescent expression proteins, viability dyes, fluorescently coated antibodies, and DNA binding dyes. The electronic signals created from these impulses are analyzed by a computer and saved in a data file with a standard extension (.fcs). One can use fluorescence or light scattering characteristics to assess and purify cell populations.

Flow cytometry is an effective technology used in various domains such as molecular biology, infectious disease surveillance, viral biology, immunological research, and cancer biology. Commonly measured parameters include GFP analysis, CD markers, cell sorting, ROS (Reactive Oxygen Species) analysis, the cell cycle analysis, total DNA, RNA content studies, and cell viability assay.

2.4.2 Fluorescence Microscopy

A fluorescent component in a sample of interest is excited by a significantly more vital light source while using a fluorescence microscope. Instead of the initial light source, this fluorescent species emits a longer wavelength, lower intensity light, resulting in a magnified image. Both organic and inorganic substances can be studied with a fluorescence microscope. Using fluorescence and phosphorescence, fluorescence microscopy looks at materials' spatial distribution and structural organization. It is beneficial for analyzing samples that are too complex to look at with a traditional transmitted-light microscope. Fluorescence microscopy images help study chemicals present in low concentrations and require great sensitivity to identify. A fluorescent microscope's principal components are similar to those of a conventional light microscope. Fluorescent microscopes differ from conventional microscopes in two key ways, though. The first is the kind of light source, and the second is certain filter elements. The samples are illuminated by a brighter light source while using a fluorescence microscope.

2.4.3 Fluorescence Scanners

Fluorescence scanners are devices that display fluorescence for macroscopic objects as a function of space, typically in two to three dimensions. This method is used to measure proteins and nucleic acids using DNA/protein chips, electrophoresis gels, and chromatograms. Microarray readers and other fluorescence scanners are capable of resolving fluorescence for macroscopic objects such as chromatograms, blots, and electrophoresis gels by recording their spatial coordinates in two dimensions.

2.4.4 Spectrofluorometers and Microplate Reader

It is a device that excites a fluorescent sample contained in a microplate when exposed to light at a specific wavelength and allows for detecting and measuring the light photons it releases. Occasionally, it is referred to as a "spectrofluorometer" or a "fluorometer." The average properties of bulk samples (from μL to mL) are measured. Plate readers are devices used to identify biological, chemical, or physical events of samples in microtiter plates. They are also referred to as microplate readers or microplate photometers. In the pharmaceutical and biotechnology industries, as well as in academic institutions, they are extensively utilized in research, drug discovery, quality control, bioassay validation, and manufacturing processes. Microtiter plates with a well format of 1–1536 can be used to analyse sample reactions. A typical response volume per well in a 96-well (8 by 12 matrix) microplate format is between 100 and 200 μL. This format is most frequently employed in university research laboratories and clinical diagnostic laboratories.

3 Quantification of Mitochondrial Membrane Potential

Numerous illnesses, including cancer, heart disease, diabetes, and neurological problems, have been linked to mitochondrial dysfunctions (Bhatti et al., 2017; Elmore, 2007). There are two possible ways that xenobiotic toxicity may affect mitochondrial function: directly or indirectly. Numerous of these substances lower MMP by interfering with a range of macromolecules in the mitochondria and so influencing various processes within the mitochondria. Apoptosis may also be connected to a drop in the MMP (Zorova et al., 2018). These organelles are, therefore, a perfect subject for *in vitro* toxicity research (Sakamuru et al., 2022; Gupta et al.,

2023). There are currently a number of cell membrane permeable fluorescent dyes available to assess changes in MMP, including 3, 3′-dihexyloxacarbocyanine iodide [DiOC6(3)], rhodamine-123 (Rh123), tetraethyl Rhodamine methyl and ethyl esters (TMRM and TMRE), and JC-1.

Estimating the mitochondrial bioenergetic status in living cells can be achieved by measuring the membrane potential of the mitochondria (Δψ) (Esteras et al., 2020; Gerencser et al., 2012; Nikolaev et al., 2023).However, it is not a simple task. Numerous cationic fluorescent dyes can be used to assess each cell's membrane potential since two of its primary properties are limited membrane binding and membrane permeability (Pegoraro et al., 2024). In living cells, cations preferentially accumulate inside the mitochondria due to their high efficacy in overcoming hydrophobic membranes. They are attracted to the inner mitochondrial membrane's negative potential (Zhang et al., 2021).

The distribution throughout the plasma membrane is directly governed by the membrane potential; the greater the potential, the more negatively charged dye accumulates (Sivandzade et al., 2019). Fluorescent lipophilic cations find widespread use in this regard. In a Δψ-dependent manner, these cations enter cells and aggregate mainly in the mitochondrial matrix. A frequently utilized category of probes is "slow cationic dyes," which exhibit a gradual distribution between the internal and external mitochondrial space after being introduced to cells or isolated mitochondria. However, this distribution only sometimes complies with Nernst law. Even while they are unreliable for steady-state measurements or tracking gradual variations in ΔΨm, they cannot react swiftly to abrupt changes in the membrane potential. Here, we outline the semi-quantitative Δψ study of rhodamine 123 (Rhod123) and tetramethylrhodamine methyl ester (TMRM) in living human cells using the cations.

3.1 Mitochondrial Health Indicators Based on MMP as a Target Site

3.1.1 Mitotracker Red

Mitotracker Red is a fluorescent dye that specifically binds to negatively charged and polarized mitochondria. Its binding is thought to depend on the potential of the mitochondrial membrane (ΔΨm). It's important to note that certain stimuli can cause changes in ΔΨm. Mitotracker Red can be used to assess whether the mitochondrial function has increased or decreased (Kholmukhamedov et al., 2013; Lu et al., 2018).

3.1.2 DiOC6(3) [3, 3′-Dihexyloxacarbocyanine Iodide] (Excitation 482 nm; Emission 504 nm)

DiOC6 (3) is a lipophilic dye commonly used to demonstrate the potential of the mitochondrial membrane. It emits a green fluorescent signal specifically at the mitochondria of living cells. It can stain mitochondria at low doses and can also be used to stain interior membranes such as the endoplasmic reticulum at higher concentrations.

3.1.3 Rhodamine 123 [2-(6-Amino-3-imino-3H-xanthen-9-yl) benzoic acid methyl ester] (Excitation 488 nm; Emission 530 nm)

The potential of the mitochondrial membrane can be studied using fluorescent dyes such as Rhodamine or Rh123. Rh123, which is a lipophilic cation, accumulates in the mitochondria in proportion to the potential of the mitochondrial membrane. It is quickly taken up by active mitochondria and can pass through the cell membrane without causing harm to the cells.

3.1.4 JC-1 [5,5′, 6,6′-tetrachloro-1,1′, 3,3′-tetraethylbenzimidazolylcarbocyanine iodide] (Excitement 488 nm; Emission 529 nm Monomer, Green & 590 nm J-aggregate, Red)

The JC-1 dye is a probe that is used to monitor the health of mitochondria and is commonly used in apoptosis studies. It is a potential-sensitive probe that emits green fluorescence at around 529 nm when in its monomeric form. When the concentration increases, it forms red fluorescent J-aggregates, causing the green fluorescence to shift to red at around 590 nm. Therefore, a decrease in the red/green fluorescence intensity ratio indicates mitochondrial depolarization (Dereje et al., 2022; Sivandzade et al., 2019).

3.1.5 JC-9 [3, 3′-dimethyl-β-naphthoxazolium iodide] (Excitation: 514 nm; 585 nm and Emission: 529 nm; 590 nm)

It is a mitochondrial membrane potential indicator. Furthermore, the probe has the potential to emit two different wavelengths. JC-9's red fluorescence significantly increases at hyperpolarized membrane potentials, while its green fluorescence stays stable and does not depend on membrane potential.

3.1.6 Rhodamine 6G [9-[2-(Ethoxycarbonyl) phenyl]-N-ethyl-6-(ethyl amino)-2,7-dimethyl-3H-xanthen-3-iminium chloride] (Excitation: 528 nm; Emission: 551 nm)

The target site for determination of mitochondrial health is mitochondrial membrane potential. Rhodamine 6G is a highly fluorescent rhodamine family dye. It is used to examine the efficiency of P-glycoprotein–mediated exclusion from multidrug-resistant cells.

3.1.7 Merocyanine 540 [3-[(2Z)-2-[(E)-4-(1,3-Dibutyl-4,6-dioxo-2-sulfanylidene-1,3-diazinan-5-ylidene) but-2-enylidene]-1,3-benzoxazol-3-yl] propane-1-sulfonate sodium] (Excitation: 488 nm; Emission: 560 nm)

Polymethine dyes, known as merocyanines, are distinguished by specific structural characteristics. Merocyanines are dyes classified as functional dyes since their beneficial chemical properties decide their applications in addition to their color. These

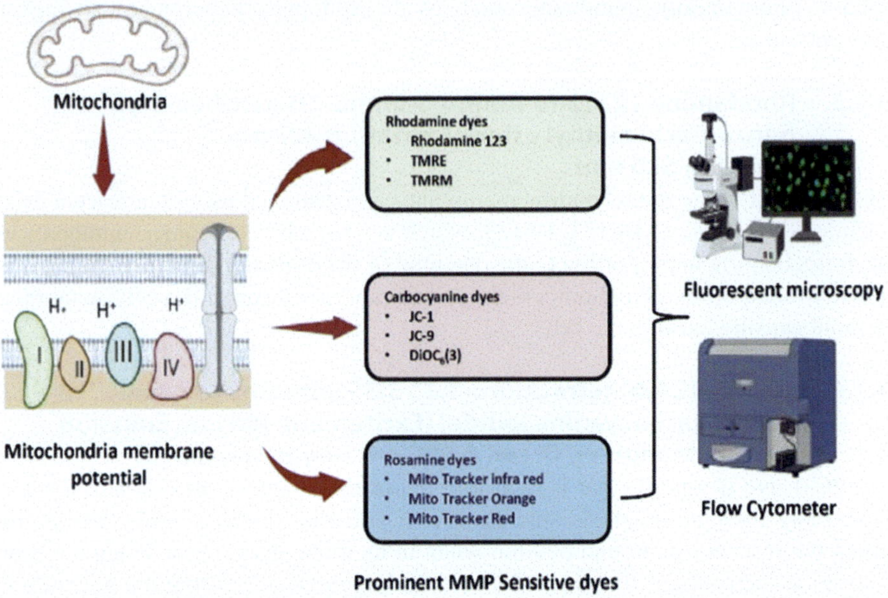

Fig. 2 Image depicting mitochondrial membrane potential (MMP) and MMP sensitive fluorescent detectors. (Parts of the figure were drawn by using pictures from Biorender (https://www.biorender.com/))

dyes typically have high extinction coefficients and vibrant colors. The first fluorescent dye to measure membrane potential was merocyanine 540. It is a fluorescent lipophilic probe called merocyanine 540 that exhibits a preference for binding to membranes that have lipids that are loosely packed. This is because the loss of membrane symmetry leads to an increase in fluorescence. Lipids are less tightly packed on the surface of dying lymphocytes. It has been demonstrated that variations in merocyanine 540 fluorescence are typically correlated with membrane phosphatidyl serine translation from the inner to the outer leaflet of the plasma membrane (Keij et al., 2000) (Fig. 2).

3.2 Mitochondrial Health Indicator Dyes Other Than MMP as a Target Site

3.2.1 Mitotracker Green (Excitation: 490 nm; Emission: 516 nm)

This probe targets mitochondria specifically and reacts with cysteine residues to form covalent bonds with proteins inside the mitochondria. It can be used to visualize the overall mass of mitochondria and is independent of the mitochondrial membrane potential. Specific cysteine residues in mitochondrial proteins have free thiol groups that react with the probe. The probe fluoresces as it accumulates in the lipid environment of the mitochondria, but it does not exhibit fluorescence in an aqueous solution (Monteiro et al., 2020).

3.2.2 Nonyl Acridine Orange [3-N, 3-N, 6-N, 6-N-tetramethyl-10-nonylacridin-10-ium-3, 6-diamine; bromide] (Excitation: 495 nm Emission: 522 nm)

The green fluorescent mitochondrial dye NAO can stain mitochondria independently of the membrane potential. It is independent of mitochondrial membrane potential. Due to its excellent retention in the mitochondria, nonyl acridine orange is a helpful probe for monitoring the mitochondria both during and after cell fusion. It has been used to assess changes in mitochondrial mass during apoptosis and to study resistance to various drugs (Paradies et al., 2019). Research has shown that the dye attaches to the phosphoprotein cardiolipin, which is found in the inner mitochondrial membrane (Fuentes & Morcillo, 2024; Ghazi-Khansari et al., 2006).

3.2.3 Lucigenin (10-methyl-9-(10-methylacridin-10-ium-9-yl) Acridin-10-ium; Dinitrate)

The excitation wavelength for the chemiluminescent probe lucigenin is 455 nm, while the emission wavelength is 505 nm. This probe is known to accumulate in the mitochondria of alveolar macrophages. To perform fluorescent staining, high dye concentrations of about 100 μM are required. However, using moderate quantities of the probe can lead to a chemiluminescent reaction triggered by an increase in superoxide production within the mitochondria. The probe can also be used to measure the chloride content of liposomes, reconstituted membranes, and extracellular medium at a resolution of 5 nm. The target site for assessing mitochondrial health with this probe is the intact mitochondria (Romodin, 2022).

3.2.4 Janus Green B

The dye Janus Green B (JG-B) is a popular choice for staining mitochondria in a process known as vital staining. This technique allows for the staining of living cells without causing harm. Vital staining has been used in a variety of diagnostic and surgical procedures. Janus Green B is an indicator that changes colour based on the presence of oxygen. In the presence of oxygen, the indicator turns blue as it oxidizes. Conversely, in the absence of oxygen, the indicator is reduced and turns pink. Janus Green B is also referred to by alternative names, including Diazin Green S and Union Green B (Checa & Aran, 2020).

4 Examples of Applications

4.1 Mitochondrial ROS (Reactive Oxygen Species)

The assessment of oxidative events and their biological relevance requires measurements of ROS. Reactive species are crucial to the proper functioning of living organisms and are produced constantly during regular cellular processes (de Almeida et al., 2022). The acronym ROS encompasses all chemical entities that fall under the broad category of reactive oxygen species. These species play a critical role in gene activation, cellular proliferation, and various chemical processes that are essential to the cell's defense against bacteria and viruses. They affect many biological functions, including immune system modulation, blood pressure regulation, pain and memory, phagocytosis, cellular signalling, smooth muscle relaxation, peristalsis, platelet aggregation, and energy production. Endogenously generated ROS are vital in producing complex biological molecules, making them indispensable to life (Venditti & Di Meo, 2020; Di Meo et al., 2016).

While mitochondrial respiratory chain complexes are one of the primary sources of ROS in cells, ROS overproduction, whether from the mitochondria or other sources inside the cell, can seriously harm both the mitochondria and the membranes surrounding the mitochondria.

Antioxidant therapy is effective due to the beneficial outcomes. However, if overproduced or the antioxidants are depleted, the reactive species can cause irreversible cellular damage through oxidative stress (Juan et al., 2021).

Exposure to several harmful elements, including pollution, UV light, cigarette smoke, radiation, and some xenobiotics, is well recognized to cause the generation of reactive oxygen species (ROS) (Seo et al., 2023; Schieber & Chandel, 2014). Reactive oxygen species (ROS) are crucial for cellular metabolism and play a significant role in cellular signalling. They are generated by mitochondria and are closely associated with the cellular redox state. There is a direct correlation between increased ROS production, apoptosis, and elevated levels of Ca^{2+} (Fujii et al., 2022). Superoxide radicals are produced as a by-product of the mitochondrial respiratory chain complexes I–IV due to incomplete oxygen reduction when electrons are transferred to oxygen. It is believed that complexes I and III are the primary sources of mitochondrial superoxide radicals. Superoxide dismutase (SOD) can convert these radicals into hydrogen peroxide, which can then be eliminated through catalase processes. While ROS can have physiological effects at low concentrations, excessive ROS production is responsible for various pathologies (Jomova et al., 2023).

In many diseases associated with oxidative stress, mitochondria play a crucial role since they are the main generator of free radicals in most cells. Even though mitochondria are the main generator of ROS, many other radicals in the cell's cytoplasm need to be considered when assessing ROS and RNS in living cells. Reactive species are hard to find because of their short lifetime and the range of antioxidant systems that can trap them in vivo. Because they are so sensitive, fluorescent probes are significant markers of reactive species (Wang & Youle, 2009) (Fig. 3).

4.2 Mitochondrial Function in Apoptosis

There is no proof that mitochondria are involved in a major way in cell apoptosis (McIlwain et al., 2013). Caspases, which are proteases that break numbers of various proteins when activated, are necessary for apoptosis. This results in fast cell death with distinct biochemical and morphological properties.

Normally, caspases are found in healthy cells as zymogens, which are precursor enzymes that have minimal to no protease activity. All apoptotic stimuli, though they may do so in slightly different ways, appear to start processes that end with

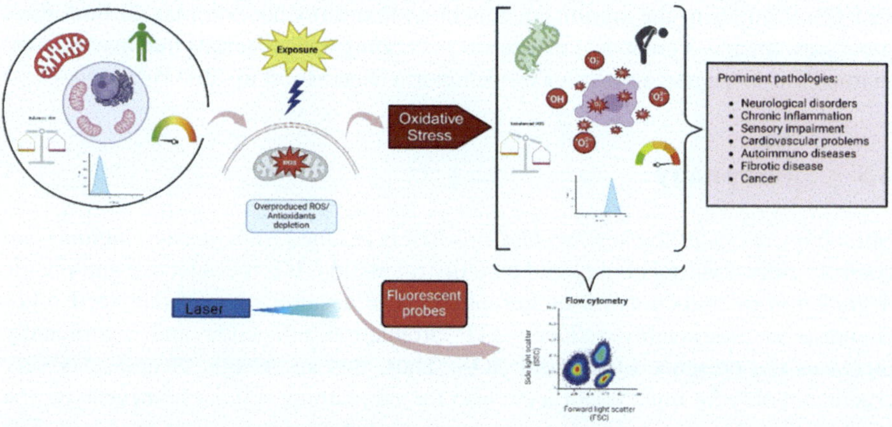

Fig. 3 The cellular changes expressed due to oxidative stress emerged as a result of ROS generation. (Parts of the figure were drawn by using pictures from Biorender (https://www.biorender.com/))

caspase activation (Nicolson, 2014). Typically, there are two ways that caspase activity can be triggered: either via the mitochondrial, also known as the intrinsic pathway of apoptosis, or by cell-surface death receptors, also known as the extrinsic pathway. The initial stages of programmed cell death are characterized by a disturbance in mitochondrial activity. This disturbance can be seen in changes to the oxidation and reduction potential of the mitochondria as well as modifications to the membrane potential, which is a critical aspect of mitochondrial health. Mitochondrial malfunction can lead to depolarization of the membrane, which is a reliable marker of mitochondrial dysfunction (Kowalczyk et al., 2021). The formation of reactive oxygen species, detoxification, and the creation of ATP through oxidative phosphorylation are all reliant on the potential of the mitochondrial inner membrane. Increasingly, pharmacological toxicity is being linked to mitochondrial malfunction (Liemburg-Apers et al., 2015).

The process of apoptosis, which is the programmed cell death, involves the formation of the apoptosome. This process includes the release of cytochrome c, a component of the mitochondrial respiratory chain, and other pro-apoptotic proteins. When cytochrome c is released from the mitochondria, it reduces mitochondrial respiration and ATP production. However, apoptosis itself requires ATP in multiple steps. Therefore, it can be suggested that cytochrome c from a single mitochondrion can promote apoptosis. On the other hand, by preventing the release of cytochrome c, ATP and oxidative phosphorylation will be supported, indicating the potential variability of this process (Endlicher et al., 2023).

The mitochondrial permeability transition is thought to be linked to changes in the membrane potential, ATP to ADP ratio reductions, elevated calcium levels in the mitochondrial matrix, oxidative stress, and the release of cytochrome c into the cytosol. These events are thought to cause disruptions in the homeostasis of ions and small molecules through the mitochondrial permeability transition pore

(MPTP). Similar to the significant functions that mitochondria play in biological processes, there is a great deal of interest in creating new materials that may be used to track the shape and activity of mitochondria (Sanchez et al., 2023).

4.3 Mitophagy

Mitophagy is one of the key mechanisms that is retained from yeast to humans and controls mitochondrial quality and quantity regulation. It is the selective autophagic degradation of mitochondria or mitophagy. The accumulation of damaged mitochondria unquestionably reduces ATP production efficiency and significantly increases the presence of harmful ROS. Thus, this autophagic clearance is very crucial not only for mitochondria but also for maintaining cellular homeostasis. The two main modes of mitophagy are receptor-mediated and ubiquitin-mediated. The collapse of the mitochondrial membrane potential, or mitochondrial depolarization, is the standard trigger for ubiquitin-mediated mitophagy. Hypoxia or food shortage can cause receptor-mediated mitophagy by recruiting mitophagy receptors to dysfunctional mitochondria (Onishi et al., 2021).

Certain mitochondrial outer membrane receptors or ubiquitin molecules bound to proteins on the mitochondrial surface stimulate mitophagy by causing autophagosomes to surround the mitochondria. The removal of mitochondria through mitophagy-mediated mechanisms is crucial for numerous processes, such as cell differentiation, apoptosis, inflammation, and early embryonic development (Song et al., 2022).

The function of mitophagy in intracellular housekeeping is highlighted by the high rates of steady-state mitochondrial turnover observed in various cell types, as revealed by recent advancements in *in vivo* analysis of mitophagy. Since deficiencies in mitophagy are becoming recognized as possible causes of several diseases, therapies aimed at mitophagy may have therapeutic value. A number of clinical diseases, including cancer, heart failure, neurodegeneration, and aging, are correlated with defects in mitophagy (Um et al., 2018).

The small amount of mitochondria that are present in the cell chondriome at any given time makes studies of mitophagy using the method of flow cytometry appear to be extremely difficult. Nonetheless, microscopic methods are very suitable for exposing the physiological alterations of a single organelle diluted in a mitochondrial pool. Investigating the mechanisms that control mitophagy and its function in pathophysiological events is hampered by the lack of an easy approach to measure mitophagy activity (McWilliams et al., 2016).

One effective method for directly seeing autophagosome-engulfed mitochondria is electron microscopy. Nevertheless, there are limits to using electron microscopy to measure mitophagy activity. However, this takes time and an experienced eye for precise quantification, and in heterogeneous tissues, it might be difficult to confidently identify particular cell types (Fang et al., 2017).

Another important method to measure the number of mitochondrial units undergoing mitophagy was to colocalize markers of autophagosomes or lysosomes with mitochondrial fluorescent probes (Williams et al., 2017). With the ability to quantify data for a large number of cells, fluorescence microscopy is a valuable method for co-localizing mitochondria with autophagosomes and lysosomes. But it's crucial to remember that the appearance of aggregates containing light chain 3 (LC3) and green fluorescent protein (GFP)-a protein associated with microtubules-can be deceptive (Montava-Garriga et al., 2020).

If mitophagy is active, it can be measured by Western blot analysis of the breakdown of mitochondrial proteins. However, significant mitophagy might be necessary to identify mitochondrial protein degradation. Furthermore, data interpretation is hampered by the fact that both autophagy and the proteasome degrade proteins found on the outer mitochondrial membrane. On the other hand, Western blot analysis has been suggested as a method for assessing mitophagy through the use of matrix or inner membrane proteins (Elmore et al., 2004). Cells can be co-stained with MitoTracker Green to localize all the mitochondria, LysoTracker Red to label acidic intracellular compartments, and TMRM to monitor the $\Delta\Psi m$ in order to detect autophagic mitochondria.

MitoTraker Green and TMRM co-localize in polarized mitochondria, facilitating a form of fluorescence resonance energy transfer between the two dyes that, when excited by blue light, results in red fluorescence emission. The TMRM is released from individual mitochondria when there is a depolarization of the mitochondria. MitoTracker Green is retained in the depolarized mitochondria due to its covalent attachment to protein thiols during absorption, which causes its fluorescence emission to become unquenched. An autophagic process-induced spontaneous depolarization of the mitochondria is manifested by a chromatic shift from red to green fluorescence. LysoTracker Red adds a level of specificity to the detection of mitophagy (Green et al., 2022).

4.4 Mitochondrial Dynamics and Cell Proliferation

Within cells, mitochondria travel in a dynamic manner, joining together to form networks through fission and fusion. This process is called mitochondrial dynamics, and it allows cells to adapt to changes in their environment. Recent studies have shown that mitochondrial dynamics is essential for understanding many biological processes, and that dysfunction in this process can lead to several human diseases.

Notably, various essential proteins regulate the internal mitochondria's shape, heterogeneity, position/arrangement, and dynamics (fission/fusion, motility). Various diseases and disorders can significantly affect any of these roles, which are essential to the overall physiology of mitochondria. (DiMaio et al., 2018).

The fluorescence intensity increased after being stimulated with the maximum amount of fluorochrome, which occurred at the peak of DNA synthesis, mitotic activity, and an increase in the amount of cellular RNA. This was observed through flow cytometric analysis. The connection between mitochondrial size and cell cycle stages was investigated using flow cytometry and imaging cytometry in rat fibroblast synchronization and staining with Hoechst 33342 and Rh123 (Darzynkiewicz et al., 2010). Two distinct sets of mitochondria were found in viable cells according to the fluorescent pulse width analysis: the smallest mitochondria were found in early G1 phase cells, while the most extensive mitochondria were found in mitotic phase cells. This suggests that mitochondrial size grew as the cell cycle progressed (Kennady et al., 2004).

5 Conclusion

Examining mitochondrial function and malfunction is crucial for understanding the different cellular processes. One of the critical issues in comprehending cellular bioenergetics and energy metabolism in cells and organs is the regulation of mitochondrial function. A simple measurement of the cellular ATP levels is less suggestive of organ function/viability and injury than mitochondrial respiratory activity, which measures the potential for aerobic energy production. Thus, examining mitochondrial activity is critical in fundamental research on mitochondrial physiology, in studies with a clinical focus, and in identifying different metabolic disorders. The understanding of mitochondrial biology has improved with the discovery of different kinds of fluorescent materials targeted at the mitochondria. According to structural insights and systematic research, the control of hydrophobicity and charge is essential for the effectiveness of chemical and peptide-based techniques for mitochondrial targeting. Using chemicals or nanomaterials as mitochondrial targeting probes has opened up new therapeutic possibilities as an agnostic tool for cancer treatment. With the ability to visualize mitochondria and associated biological processes in real-time in both *in vitro* and *in vivo* systems, versatile mitochondrial probes appear valuable for studying mitochondrial biology in both health and sickness. The scientific data gathered from investigations on mitochondrial physiology can be helpful for fundamental and clinical research and for the creation of novel diagnostic techniques and procedures. It will undoubtedly aid in creating novel therapeutic strategies if the molecular pathways involved in mitochondrial physiology and disease are thoroughly characterized.

References

Alberts, B., Johnson, A., Lewis, J., et al. (2002). *Molecular biology of the cell* (4th ed.). Garland Science. The Mitochondrion. Available from: https://www.ncbi.nlm.nih.gov/books/NBK26894/

Alshial, E. E., Abdulghaney, M. I., Wadan, A.-H. S., Abdellatif, M. A., Ramadan, N. E., Suleiman, A. M., Waheed, N., Abdellatif, M., & Mohammed, H. S. (2023). Mitochondrial dysfunction and neurological disorders: A narrative review and treatment overview, 122257. *Life Sciences, 334*. https://doi.org/10.1016/j.lfs.2023.122257. ISSN 0024-3205.

Behl, T., Kaur, G., Sehgal, A., Bhardwaj, S., Singh, S., Buhas, C., Judea-Pusta, C., Uivarosan, D., Munteanu, M. A., & Bungau, S. (2021). Multifaceted role of matrix metalloproteinases in neurodegenerative diseases: Pathophysiological and therapeutic perspectives. *International Journal of Molecular Sciences, 22*(3), 1413. https://doi.org/10.3390/ijms22031413. PMID: 33573368; PMCID: PMC7866808.

Bhatti, J. S., Bhatti, G. K., & Hemachandra Reddy, P. (2017). Mitochondrial dysfunction and oxidative stress in metabolic disorders—A step towards mitochondria based therapeutic strategies. *Biochimica et Biophysica Acta (BBA) – Molecular Basis of Disease, 1863*(5), 1066–1077. https://doi.org/10.1016/j.bbadis.2016.11.010. ISSN 0925-4439.

Checa, J., & Aran, J. M. (2020). Reactive oxygen species: Drivers of physiological and pathological processes. *Journal of Inflammation Research*, (13), 1057–1073. https://doi.org/10.2147/JIR.S275595. PMID: 33293849; PMCID: PMC7719303.

Chen, X., Ji, Y., Liu, R., et al. (2023). Mitochondrial dysfunction: Roles in skeletal muscle atrophy. *Journal of Translational Medicine, 21*, 503. https://doi.org/10.1186/s12967-023-04369-z

Chu, Y., Park, J., Kim, E., & Lee, S. (2021). Fluorescent materials for monitoring mitochondrial biology. *Materials (Basel), 14*(15), 4180. https://doi.org/10.3390/ma14154180. PMID: 34361379; PMCID: PMC8347261.

Clemente-Suárez, V. J., Martín-Rodríguez, A., Redondo-Flórez, L., Ruisoto, P., Navarro-Jiménez, E., Ramos-Campo, D. J., & Tornero-Aguilera, J. F. (2023). Metabolic health, mitochondrial fitness, physical activity, and cancer. *Cancers (Basel), 15*(3), 814. https://doi.org/10.3390/cancers15030814. PMID: 36765772; PMCID: PMC9913323.

Darzynkiewicz, Z., Halicka, H. D., & Zhao, H. (2010). Analysis of cellular DNA content by flow and laser scanning cytometry. *Advances in Experimental Medicine and Biology, 676*, 137–147. https://doi.org/10.1007/978-1-4419-6199-0_9. PMID: 20687474; PMCID: PMC2967208.

de Almeida, A. J. P. O., de Oliveira, J. C. P. L., da Silva Pontes, L. V., de Souza Júnior, J. F., Gonçalves, T. A. F., Dantas, S. H., de Almeida Feitosa, M. S., Silva, A. O., & de Medeiros, I. A. (2022). ROS: Basic concepts, sources, cellular signaling, and its implications in aging pathways. *Oxidative Medicine and Cellular Longevity, 2022*, 1225578. https://doi.org/10.1155/2022/1225578. PMID: 36312897; PMCID: PMC9605829.

Dereje, D. M., Pontremoli, C., Moran Plata, M. J., Visentin, S., & Barbero, N. (2022). Polymethine dyes for PDT: Recent advances and perspectives to drive future applications. *Photochemical & Photobiological Sciences, 21*(3), 397–419. https://doi.org/10.1007/s43630-022-00175-6. Epub 2022 Feb 1. PMID: 35103979.

Di Meo, S., Reed, T. T., Venditti, P., & Victor, V. M. (2016). Role of ROS and RNS sources in physiological and pathological conditions. *Oxidative Medicine and Cellular Longevity, 2016*, 1245049. https://doi.org/10.1155/2016/1245049. Epub 2016 Jul 12. PMID: 27478531; PMCID: PMC4960346.

DiMaio, J., Ruthel, G., Cannon, J. J., Malfara, M. F., & Povelones, M. L. (2018). The single mitochondrion of the kinetoplastid parasite Crithidia fasciculata is a dynamic network. *PLoS One, 13*(12), e0202711. https://doi.org/10.1371/journal.pone.0202711. PMID: 30592713; PMCID: PMC6310254.

Elmore, S. (2007). Apoptosis: A review of programmed cell death. *Toxicologic Pathology, 35*(4), 495–516. https://doi.org/10.1080/01926230701320337. PMID: 17562483; PMCID: PMC2117903.

Elmore, S. P., Nishimura, Y., Qian, T., Herman, B., & Lemasters, J. J. (2004). Discrimination of depolarized from polarized mitochondria by confocal fluorescence resonance energy transfer. *Archives of Biochemistry and Biophysics, 422*(2), 145–152. https://doi.org/10.1016/j.abb.2003.12.031. PMID: 14759601.

Endlicher, R., Drahota, Z., Štefková, K., Červinková, Z., & Kučera, O. (2023). The mitochondrial permeability transition pore-current knowledge of its structure, function, and regulation, and optimized methods for evaluating its functional state. *Cells, 12*(9), 1273. https://doi.org/10.3390/cells12091273. PMID: 37174672; PMCID: PMC10177258.

Esteras, N., Adjobo-Hermans, M. J. W., Abramov, A. Y., & Koopman, W. J. H. (2020). Chapter 9 – Visualization of mitochondrial membrane potential in mammalian cells. In L. A. Pon & E. A. Schon (Eds.), *Methods in cell biology* (Vol. 155, pp. 221–245). Academic. https://doi.org/10.1016/bs.mcb.2019.10.003. ISSN 0091-679X, ISBN 9780128202289.

Fang, E. F., Palikaras, K., Sun, N., Fivenson, E. M., Spangler, R. D., Kerr, J. S., Cordonnier, S. A., Hou, Y., Dombi, E., Kassahun, H., Tavernarakis, N., Poulton, J., Nilsen, H., & Bohr, V. A. (2017). In vitro and in vivo detection of mitophagy in human cells, C. Elegans, and mice. *Journal of Visualized Experiments*, (129), 56301. https://doi.org/10.3791/56301. PMID: 29286376; PMCID: PMC5755444.

Fuentes, J. M., & Morcillo, P. (2024). The role of cardiolipin in mitochondrial function and neurodegenerative diseases. *Cells, 13*(7), 609. https://doi.org/10.3390/cells13070609. PMID: 38607048; PMCID: PMC11012098.

Fujii, J., Homma, T., & Osaki, T. (2022). Superoxide radicals in the execution of cell death. *Antioxidants, 11*(3), 501. https://doi.org/10.3390/antiox11030501

Gerencser, A. A., Chinopoulos, C., Birket, M. J., Jastroch, M., Vitelli, C., Nicholls, D. G., & Brand, M. D. (2012). Quantitative measurement of mitochondrial membrane potential in cultured cells: Calcium-induced de- and hyperpolarization of neuronal mitochondria. *The Journal of Physiology, 590*(12), 2845–2871. https://doi.org/10.1113/jphysiol.2012.228387. Epub 2012 Apr 10. PMID: 22495585; PMCID: PMC3448152.

Ghazi-Khansari, M., Mohammadi-Bardbori, A., & Hosseini, M. J. (2006). Using Janus green B to study paraquat toxicity in rat liver mitochondria: Role of ACE inhibitors (thiol and nonthiol ACEi). *Annals of the New York Academy of Sciences, 1090*, 98–107. https://doi.org/10.1196/annals.1378.010. PMID: 17384251.

Gold, V. A., Chroscicki, P., Bragoszewski, P., & Chacinska, A. (2017). Visualization of cytosolic ribosomes on the surface of mitochondria by electron cryo-tomography. *EMBO Reports, 18*(10), 1786–1800. https://doi.org/10.15252/embr.201744261. Epub 2017 Aug 21. PMID: 28827470; PMCID: PMC5623831.

Green, A., Hossain, T., & Eckmann, D. M. (2022). Mitochondrial dynamics involves molecular and mechanical events in motility, fusion and fission. *Frontiers in Cell and Development Biology, 10*, 1010232. https://doi.org/10.3389/fcell.2022.1010232. PMID: 36340034; PMCID: PMC9626967.

Grevel, A., Pfanner, N., & Becker, T. (2019). Coupling of import and assembly pathways in mitochondrial protein biogenesis. *Biological Chemistry, 401*(1), 117–129. https://doi.org/10.1515/hsz-2019-0310. PMID: 31513529.

Gupta, P., Singh, A., Verma, A. K., Kant, S., Pandey, A. K., Mishra, A., Khare, P., & Prakash, V. (2023). Nanoencapsulation of docetaxel induces concurrent apoptosis and necroptosis in human oral cancer cells (SCC-9) via TNF-α/RIP1/RIP3 pathway. *Indian Journal of Clinical Biochemistry, 38*(3), 351–360. https://doi.org/10.1007/s12291-022-01055-7. Epub 2022 Aug 9. PMID: 37234186; PMCID: PMC10205939.

Jomova, K., Raptova, R., Alomar, S. Y., Alwasel, S. H., Nepovimova, E., Kuca, K., & Valko, M. (2023). Reactive oxygen species, toxicity, oxidative stress, and antioxidants: Chronic diseases and aging. *Archives of Toxicology, 97*(10), 2499–2574. https://doi.org/10.1007/s00204-023-03562-9. Epub 2023 Aug 19. PMID: 37597078; PMCID: PMC10475008.

Juan, C. A., Pérez de la Lastra, J. M., Plou, F. J., & Pérez-Lebeña, E. (2021). The chemistry of reactive oxygen species (ROS) revisited: Outlining their role in biological macromolecules (DNA, lipids and proteins) and induced pathologies. *International Journal of Molecular Sciences, 22*(9), 4642. https://doi.org/10.3390/ijms22094642

Keij, J. F., Bell-Prince, C., & Steinkamp, J. A. (2000). Staining of mitochondrial membranes with 10-nonyl acridine orange, MitoFluor Green, and MitoTracker Green is affected by mitochondrial membrane potential altering drugs. *Cytometry, 39*(3), 203–210. https://doi.org/10.1002/(sici)1097-0320(20000301)39:3<203::aid-cyto5>3.0.co;2-z. PMID: 10685077.

Kennady, P. K., Ormerod, M. G., Singh, S., & Pande, G. (2004). Variation of mitochondrial size during the cell cycle: A multiparameter flow cytometric and microscopic study. *Cytometry. Part A, 62*(2), 97–108. https://doi.org/10.1002/cyto.a.20091. PMID: 15536637.

Kholmukhamedov, A., Schwartz, J. M., & Lemasters, J. J. (2013). Isolated mitochondria infusion mitigates ischemia-reperfusion injury of the liver in rats: Mitotracker probes and mitochondrial membrane potential. *Shock, 39*(6), 543. https://doi.org/10.1097/SHK.0b013e318292300d. PMID: 23680774; PMCID: PMC3759289.

Kowalczyk, P., Sulejczak, D., Kleczkowska, P., Bukowska-Ośko, I., Kucia, M., Popiel, M., Wietrak, E., Kramkowski, K., Wrzosek, K., & Kaczyńska, K. (2021). Mitochondrial oxidative stress-A causative factor and therapeutic target in many diseases. *International Journal of Molecular Sciences, 22*(24), 13384. https://doi.org/10.3390/ijms222413384. PMID: 34948180; PMCID: PMC8707347.

Liemburg-Apers, D. C., Willems, P. H. G. M., Koopman, W. J. H., et al. (2015). Interactions between mitochondrial reactive oxygen species and cellular glucose metabolism. *Archives of Toxicology, 89*, 1209–1226. https://doi.org/10.1007/s00204-015-1520-y

Lu, Y., Kan, H., Wang, Y., Wang, D., Wang, X., Gao, J., & Zhu, L. (2018). Asiatic acid ameliorates hepatic ischemia/reperfusion injury in rats via mitochondria-targeted protective mechanism. *Toxicology and Applied Pharmacology, 338*, 214–223. https://doi.org/10.1016/j.taap.2017.11.023. Epub 2017 Nov 28. PMID: 29196105.

McIlwain, D. R., Berger, T., & Mak, T. W. (2013). Caspase functions in cell death and disease. *Cold Spring Harbor Perspectives in Biology, 5*(4), a008656. https://doi.org/10.1101/cshperspect.a008656. Erratum in: Cold Spring Harb Perspect Biol. 2015 Apr;7(4). pii: a026716. https://doi.org/10.1101/cshperspect.a026716. PMID: 23545416; PMCID: PMC3683896.

McWilliams, T. G., Prescott, A. R., Allen, G. F., Tamjar, J., Munson, M. J., Thomson, C., Muqit, M. M., & Ganley, I. G. (2016). mito-QC illuminates mitophagy and mitochondrial architecture in vivo. *The Journal of Cell Biology, 214*(3), 333–345. https://doi.org/10.1083/jcb.201603039. Epub 2016 Jul 25. PMID: 27458135; PMCID: PMC4970326.

Montava-Garriga, L., Singh, F., Ball, G., & Ganley, I. G. (2020). Semi-automated quantitation of mitophagy in cells and tissues. *Mechanisms of Ageing and Development, 185*, 111196. https://doi.org/10.1016/j.mad.2019.111196. Epub 2019 Dec 13. PMID: 31843465; PMCID: PMC6961211.

Monteiro, L. B., Davanzo, G. G., de Aguiar, C. F., & Moraes-Vieira, P. M. M. (2020). Using flow cytometry for mitochondrial assays. *MethodsX, 7*, 100938. https://doi.org/10.1016/j.mex.2020.100938. PMID: 32551241; PMCID: PMC7289760.

Nasrin Afzal, W., Jonathan Lederer, M., Jafri, S., & Mannella, C. A. (2021). Effect of crista morphology on mitochondrial ATP output: A computational study. *Current Research in Physiology, 4*, 163–176. https://doi.org/10.1016/j.crphys.2021.03.005. ISSN 2665-9441.

Nicolson, G. L. (2014). Mitochondrial dysfunction and chronic disease: Treatment with natural supplements. *Integrative Medicine (Encinitas), 13*(4), 35–43. PMID: 26770107; PMCID: PMC4566449.

Nikolaev, D. M., Mironov, V. N., Shtyrov, A. A., Kvashnin, I. D., Mereshchenko, A. S., Vasin, A. V., Panov, M. S., & Ryazantsev, M. N. (2023). Fluorescence imaging of cell membrane potential: From relative changes to absolute values. *International Journal of Molecular Sciences, 24*(3), 2435. https://doi.org/10.3390/ijms24032435. PMID: 36768759; PMCID: PMC9916766.

Onishi, M., Yamano, K., Sato, M., Matsuda, N., & Okamoto, K. (2021). Molecular mechanisms and physiological functions of mitophagy. *The EMBO Journal, 40*(3), e104705. https://doi.org/10.15252/embj.2020104705. Epub 2021 Jan 13. PMID: 33438778; PMCID: PMC7849173.

Paradies, G., Paradies, V., Ruggiero, F. M., & Petrosillo, G. (2019). Role of cardiolipin in mitochondrial function and dynamics in health and disease: Molecular and pharmacological aspects. *Cells, 8*(7), 728. https://doi.org/10.3390/cells8070728. PMID: 31315173; PMCID: PMC6678812.

Pegoraro, C., Inés, D.-O., Inmaculada, C.-S., & Vicent María, J. (2024). Unlocking the mitochondria for nanomedicine-based treatments: Overcoming biological barriers, improving designs, and selecting verification techniques. *Advanced Drug Delivery Reviews, 207*, 115195. https://doi.org/10.1016/j.addr.2024.115195. ISSN 0169-409X.

Pfanner, N., Warscheid, B., & Wiedemann, N. (2019). Mitochondrial proteins: From biogenesis to functional networks. *Nature Reviews. Molecular Cell Biology, 20*, 267–284. https://doi.org/10.1038/s41580-018-0092-0

Rickard, B. P., Overchuk, M., Chappell, V. A., Kemal Ruhi, M., Sinawang, P. D., Nguyen Hoang, T. T., Akin, D., Demirci, U., Franco, W., Fenton, S. E., Santos, J. H., & Rizvi, I. (2023). Methods to evaluate changes in mitochondrial structure and function in cancer. *Cancers (Basel), 15*(9), 2564. https://doi.org/10.3390/cancers15092564. PMID: 37174030; PMCID: PMC10177605.

Romodin, L. A. (2022). Chemiluminescence detection in the study of free-radical reactions. Part 2. Luminescent additives that increase the chemiluminescence quantum yield. *Acta Naturae, 14*(1), 31–39. https://doi.org/10.32607/actanaturae.11427. PMID: 35441047; PMCID: PMC9013440.

Sakamuru, S., Zhao, J., Attene-Ramos, M. S., & Xia, M. (2022). Mitochondrial membrane potential assay. *Methods in Molecular Biology, 2474*, 11–19. https://doi.org/10.1007/978-1-0716-2213-1_2. PMID: 35294751; PMCID: PMC9850828.

Sanchez, L., Epps, J., Wall, S., et al. (2023). SIM2s directed Parkin-mediated mitophagy promotes mammary epithelial cell differentiation. *Cell Death and Differentiation, 30*, 1472–1487. https://doi.org/10.1038/s41418-023-01146-9

Saneto, R. P. (2017). Genetics of mitochondrial disease. *Advances in Genetics, 98*, 63–116. https://doi.org/10.1016/bs.adgen.2017.06.002. Epub 2017 Sep 1. PMID: 28942795.

Schieber, M., & Chandel, N. S. (2014). ROS function in redox signaling and oxidative stress. *Current Biology, 24*(10), R453–R462. https://doi.org/10.1016/j.cub.2014.03.034. PMID: 24845678; PMCID: PMC4055301.

Seo, Y. S., Park, J. M., Kim, J. H., & Lee, M. Y. (2023). Cigarette smoke-induced reactive oxygen species formation: A concise review. *Antioxidants (Basel), 12*(9), 1732. https://doi.org/10.3390/antiox12091732. PMID: 37760035; PMCID: PMC10525535.

Sisalli, M. J., Feliciello, A., Della Notte, S., Di Martino, R., Borzacchiello, D., Annunziato, L., & Scorziello, A. (2020). Nuclear-encoded NCX3 and AKAP121: Two novel modulators of mitochondrial calcium efflux in normoxic and hypoxic neurons. *Cell Calcium, 87*, 102193. https://doi.org/10.1016/j.ceca.2020.102193. Epub 2020 Mar 6. PMID: 32193001.

Sivandzade, F., Bhalerao, A., & Cucullo, L. (2019). Analysis of the mitochondrial membrane potential using the cationic JC-1 dye as a sensitive fluorescent probe. *Bio-Protocol, 9*(1), e3128. https://doi.org/10.21769/BioProtoc.3128. PMID: 30687773; PMCID: PMC6343665.

Song, C., Pan, S., Zhang, J., Li, N., & Geng, Q. (2022). Mitophagy: A novel perspective for insighting into cancer and cancer treatment. *Cell Proliferation, 55*(12), e13327. https://doi.org/10.1111/cpr.13327. Epub 2022 Oct 5. PMID: 36200262; PMCID: PMC9715364.

Um, J. H., Kim, Y. Y., Finkel, T., & Yun, J. (2018). Sensitive measurement of mitophagy by flow cytometry using the pH-dependent fluorescent reporter mt-Keima. *Journal of Visualized Experiments*, (138), 58099. https://doi.org/10.3791/58099. PMID: 30148491; PMCID: PMC6126785.

Venditti, P., & Di Meo, S. (2020). The role of reactive oxygen species in the life cycle of the mitochondrion. *International Journal of Molecular Sciences, 21*(6), 2173. https://doi.org/10.3390/ijms21062173. PMID: 32245255; PMCID: PMC7139706.

Wallace, D. C. (2018). Mitochondrial genetic medicine. *Nature Genetics, 50*(12), 1642–1649. https://doi.org/10.1038/s41588-018-0264-z. Epub 2018 Oct 29. PMID: 30374071.

Wang, C., & Youle, R. J. (2009). The role of mitochondria in apoptosis*. *Annual Review of Genetics, 43*, 95–118. https://doi.org/10.1146/annurev-genet-102108-134850. PMID: 19659442; PMCID: PMC4762029.

Wang, J., & Zhou, H. (2020). Mitochondrial quality control mechanisms as molecular targets in cardiac ischemia-reperfusion injury. *Acta Pharmaceutica Sinica B, 10*(10), 1866–1879. https://doi.org/10.1016/j.apsb.2020.03.004. Epub 2020 Apr 8. PMID: 33163341; PMCID: PMC7606115.

Williams, J. A., Zhao, K., Jin, S., & Ding, W. X. (2017). New methods for monitoring mitochondrial biogenesis and mitophagy in vitro and in vivo. *Experimental Biology and Medicine (Maywood, N.J.), 242*(8), 781–787. https://doi.org/10.1177/1535370216688802. Epub 2017 Jan 1. PMID: 28093935; PMCID: PMC5407538.

Zemirli, N., Morel, E., & Molino, D. (2018). Mitochondrial dynamics in basal and stressful conditions. *International Journal of Molecular Sciences, 19*(2), 564. https://doi.org/10.3390/ijms19020564. PMID: 29438347; PMCID: PMC5855786.

Zhang, H., He, X., Li, X., Guo, L., Meng, F., Yang, R., Li, C., Zg, L., & Yu, X. (2021). Permeability-controllable potentiometric fluorescent probes enable visually discriminating near-zero and normal situations of cell membrane potential. *Analytical Chemistry, 93*(5), 2728–2732. https://doi.org/10.1021/acs.analchem.0c04928

Zorova, L. D., Popkov, V. A., Plotnikov, E. Y., Silachev, D. N., Pevzner, I. B., Jankauskas, S. S., Babenko, V. A., Zorov, S. D., Balakireva, A. V., Juhaszova, M., Sollott, S. J., & Zorov, D. B. (2018). Mitochondrial membrane potential. *Analytical Biochemistry, 552*, 50–59. https://doi.org/10.1016/j.ab.2017.07.009. Epub 2017 Jul 12. PMID: 28711444; PMCID: PMC5792320.

Environmental Stressors-Mediated Apoptosis & Autophagy Detected Through Flow Cytometry

Meenu Srivastava and AB Pant

Abstract

The chemical substances of synthetic origin and foreign to the biological systems are known as xenobiotics, such as drugs, medicines, pesticides, fertilizers, cosmetics, preservatives, food additives, industrial chemicals, and other environmental pollutants generated either through anthropogenic or geological activities. The xenobiotics may adversely affect human life and result in various disease conditions. Apoptosis and autophagy are two well-regulated evolutionarily conserved self-destructive processes that maintain cellular homeostasis and cell fate. Apoptosis, also known as programmed cell death, is triggered by various stimuli, including xenobiotics. Cellular characteristics of apoptosis include chromatin condensation, cytoplasm shrinkage, and asymmetry in the plasma membrane. Autophagy involves massive accumulation of double-membrane vesicles (autophagosomes) containing damaged or waste cytoplasmic materials, which fuses with lysosomes and results in bulk degradation of the cytosolic components and thus is also considered type II cell death. Since both types of cell death mechanisms are expressed in several cell systems, including the immunological, neurological, hepatic, intestinal, cutaneous, and reproductive, it is appealing as a generic predictor of response. Depending on the stimulus and kind of cell, an apoptotic cell exhibits a wide range of distinctive morphological and biochemical characteristics. Flow and imaging cytometry allows for the quick examina-

M. Srivastava
Systems Toxicology Group, CSIR-Indian Institute of Toxicology Research Vishvigyan Bhavan, Lucknow, Uttar Pradesh, India

AB Pant (✉)
Systems Toxicology Group, CSIR-Indian Institute of Toxicology Research Vishvigyan Bhavan, Lucknow, Uttar Pradesh, India

Academy of Scientific and Innovative Research (AcSIR), Ghaziabad, Uttar Pradesh, India
e-mail: abpant@iitr.res.in

AB Pant et al. (eds.), *Flow Cytometry: Applications in Cellular and Molecular Toxicology*, https://doi.org/10.1007/978-981-97-9758-5_4

tion of the vast majority of traditional apoptotic markers. Thus, cytometry emerged as the preferred tool for various cellular death investigations. The last two decades have seen the development of a wide range of cytometric techniques intended to detect apoptotic cells and investigate the causes behind this type of cell death. Flowcytometric analysis has been used for many years to detect the stages of apoptosis and autophagy. This chapter aims to provide curated information on applying flowcytometric methods to detect different stages of apoptosis and autophagy, including using pivotal assays, which can detect both early and late apoptosis. Further, the chapter also deals with a description of different dyes or stains, such as monodansylcadaverine, acridine orange, LysoTracker, and autophagy-specific markers, like anti-sequestosome and anti-LC3-II, which are pretty helpful in staining or targeting the autophagic contents or autophagic vacuoles for detection and quantification of autophagic activity in response to different xenobiotic compounds.

Keywords

Apoptosis · Autophagy · Flowcytometry · Xenobiotics · Toxic responses

1 Introduction

In all tissues, apoptosis plays a crucial regulatory role. It has been extensively studied in the immune system and tumor cells. DNA fragmentation, cell detachment, mitochondrial fragmentation, externalization of phosphatidylserine (PS) in the plasma membrane, and caspase activation during apoptosis are standard morphological and biochemical features of apoptosis (Guo et al., 2021). Proteins and organelles are transported to the lysosome or vacuole for breakdown through a series of processes collectively called autophagy or "self-eating." Macroautophagy and microautophagy are two autophagy mechanisms highly conserved in eukaryotes (Yamamoto et al., 2023). The most well-known applications of autophagy, a highly evolutionarily conserved system, are metabolism, organelle, protein turnover, and cell quality control. By breaking down cytoplasmic components, lysosome-mediated degradative autophagy eliminates damaged organelles and hazardous protein aggregates while supplying a nutrition and energy source (Raudenska et al., 2021). Assays based on flow cytometry allow for identifying and measuring different cellular markers in either fixed or living cells (Kaminskyy, 2022).

Flow cytometry technique explains the morphology, cellular characteristics, cell cycle stage, and other characteristics of cells. Its main benefit is that, unlike Western blotting, it can examine individual cells within a population without averaging. Furthermore, the flow cytometry variant known as FACS makes it possible to separate the populations of interest for additional analysis. Recent advancements in image processing tools have made it possible to quantify immunofluorescence data akin to flow cytometry data, hence removing the requirement that cells be in suspension for flow cytometry. While flow cytometry works best for cell surface proteins, image analysis is better suited for nucleus proteins, which are easier to identify in individual images (Chantzoura & Kaji, 2017). Cell deaths are identified using morphology-based flow cytometry techniques, which include calculating the percentage of apoptotic cells indicated by nuclear fragmentation, reduced nuclear area, and brilliant detail intensity in the cells. Various staining techniques can be used, such as PI fixation and labeling. Furthermore, using flow cytometry, LC3 puncta labeled and/or the co-localization of fluorescently tagged LC3 provide benefits in an impartial, quantifiable, and statistically reliable manner. Autophagizing cells can also be identified by looking for fluorescently labeled lysosomal markers. Additionally, a morphology-based flow cytometry assay could be used to measure the loss of plasma membrane integrity utilizing cell-impermeant dyes to determine cell viability (Hu et al., 2021).

Worldwide environmental contamination by a complex mixture of xenobiotics has grown to be a significant problem. Many xenobiotic substances have a negative effect on the environment because of their high toxicity, protracted persistence, and restricted capacity for biodegradation (Mishra et al., 2021). Assays that identify apoptotic events are suitable for use as biomarkers of response because they track the series of alterations linked to xenobiotic-induced cell stress and toxicity across several tissues qualitatively and quantitatively (Katalay et al., 2022). This chapter provides an overview of flow cytometry technologies that are frequently used to evaluate xenobiotic-induced caspase activation, cell cycle arrest, and mitochondrial dysfunction, including induction of reactive oxygen species (ROS), alterations in mitochondrial membrane potential (MMP), apoptosis and autophagy.

2 Flowcytometry and Its Application in Research

Flow cytometry was first developed in the 1960s. It is a potent and versatile instrument for single-cell quantitative and qualitative cell analysis (Schmit et al., 2021). Flow cytometry's multiparametric nature makes it possible to integrate many apoptotic traits in a single sample, which makes it an effective tool for deciphering the intricate dynamics of apoptotic death. This very effective technique finds utility across various fields, including cancer biology, immunology, molecular biology, virology, and infectious disease surveillance. For instance, this approach demonstrates high efficacy in investigating the reactions occurring in the immune system of cancer and infectious diseases. This technique enables concurrently characterization of heterogeneous cell populations derived from several sources, including bone marrow, blood, and dissociated solid tissues such as solid malignancies, lymph nodes, spleen, and mucosal tissues. Flow cytometry allows quick analysis of individual cells or particles as they pass in front of one or more lasers while suspended in a buffered salt solution. The visible light scatter and one or more fluorescence characteristics are examined for every particle. To determine the internal complexity or granularity of a cell, light scatter measurement occurred in two different directions: forward (forward scatter, or FSC), which unfolds the cell's relative size, and at 90° (side scatter, or SSC), which unfold the internal complexity or granularity of a cell (McKinnon, 2018).

A flowcytometer is made up of several essential parts: an optical system with one or more lasers for excitation, mirrors, multiple filters, and detectors to catch the scattered and fluorescence signals; an electronics and data acquisition system that is typically integrated within the software for analysis of data; and a fluidic system that regulates the flow of cells or particles through the device. The process of physically separating cells according to different measurable characteristics is called cell sorting. With time, the ability of instruments to sort has increased to the point where very high-speed sorting is now possible, mainly to isolate rare cells but also for fairly common uses like separating sperm. These developments have made it possible to detect and sort cells with greater accuracy and efficiency, opening up a variety of uses in research, diagnosis, and treatment development. The best technique for single-cell recovery is cell sorting, especially when pure populations are required. If necessary, this procedure can be carried out in a sterile environment. Recovery of particular cell populations for use in subsequent processes, including cell culture, genetic analysis, and additional functional research, is made possible by cell sorting. Cell sorters can separate multiple cell populations at once. Even though flow cytometry has advanced significantly since it was first developed, the technology has mainly been employed for fluorescence-based analysis, and its fundamental ideas have evolved little in over 60 years. To examine cellular characteristics and pinpoint particular cell populations, flow cytometry has historically relied on the detection of fluorescence signals released by fluorochrome-labeled probes. This fluorescence-based method has proved essential in characterizing a wide range of biomolecules, such as intracellular markers, nucleic acids, and antibodies, offering critical insights into cells' differentiation, function, and disease states (Robinson et al., 2023). Flow cytometry has become an indispensable tool in assessing primary immune deficiency illnesses. Flow-cytometric techniques have been applied to cell function, with the recent development of methods to analyze intracellular features (Fleisher & Oliveira, 2019).

Apoptotic alterations are characterized by modifications such as the externalization of phosphatidylserine and changes in the permeability of the mitochondrial membrane. The FITC test permits the use of Annexin V in conjunction with fluorescent labeling, such as FITC with the detection of fluorescence microscopy, or an augmentation with alternative methods, such as flow cytometry, after the labeling procedure. False positive results can occur with fluorescence microscopy; hence, flow cytometry can occur. Due to its ability to provide results quickly, flow cytometry is now a potential tool for investigating apoptosis (Obeng, 2020).

3 Xenobiotics

The natural world that envelops us and influences our day-to-day existence on our planet is known as the environment. Life on this planet needs a safe and healthy environment to exist. But in modern times of advanced manufacturing facilities and urbanization, human activities of many kinds are principally to blame for the arrival of dangerous and poisonous chemicals such as environmental xenobiotics

(Rodríguez et al., 2020). Every substance the body comes into contact with has the potential to have positive and negative effects, and xenobiotics are no different. Agents alien to the body or a biological system are called "xenobiotics" (Joseph, 2017). Dr. Howard Mason first used the term "xenobiotic" in the 1960s. The term "xenobiotics" refers to substances that are external to an organism's regular metabolism and that the organism is exposed to (Belizário et al., 2018). Many compounds used to improve everyday life (like dyes, antibiotics, PCPs, pesticides, and additives) were developed in the twentieth century; these substances either do not naturally occur in the environment, or their naturally occurring concentrations are very different from those caused by human activity. Their physicochemical structures, which make them challenging to identify, quantify, and eliminate, are the primary issue. These include ionizability, water solubility, small molecular size, polarity, lipophilicity, and volatility (de Oliveira et al., 2020). Because xenobiotics have a strong attraction for organic materials, following their release into the environment, they can negatively affect people, animals, and natural ecosystems by bioaccumulating throughout the food chain (Maurya & Malik, 2016). Hence, they have the potential to result in serious long-term consequences such as lung bronchitis, immune system damage, respiratory tract infections, endocrine system disturbance, neurological system malfunction, behavioral and developmental abnormalities, and mutagenic and carcinogenic effects (Bertotto et al., 2020; Mishra et al., 2019). Therefore, xenobiotic pollution poses a significant environmental risk and is a continuous human hazard. Such xenobiotic compounds have been treated using various physical and chemical techniques, such as coagulation, adsorption, filtration, chemical precipitation, electrolysis, ozonation, etc. Unfortunately, not all these processes are very effective because of their high cost, problems with waste disposal, and the creation of hazardous byproducts that are sometimes much more deadly than the original substance.

4 Xenobiotic Metabolism

A xenobiotic can go through two alternative routes: (a) it can be removed as unaltered-hydrophilic chemicals via urine, and lipophilic compounds through feces, or (b) it can be kept in the body unaltered for a more extended period. e.g., Organochlorine pesticides, lead, cadmium, etc. There are two possible processes for xenobiotic metabolism: (1) spontaneous transformation or (2) enzymatic metabolism, with the latter being the most typical (Patterson et al., 2010). An intermediary, end product, or parent chemical in a metabolic process can all be considered xenobiotics. The Xenobiotic-Metabolizing Enzymes (XMEs) are responsible for the metabolism of xenobiotics. Phase I (activation), Phase II (conjugation), or Phase III (transporter enzymes) enzymes are the three types of XMEs. Different physiological and pathological states can impact xenobiotic metabolism, specifically biotransformation. A toxic xenobiotic's extra-hepatic metabolism may cause toxicity to specific organs. Medication-induced chemical alteration of xenobiotics facilitates their excretion from the human body (Abdelsalam et al., 2020). About 75% of all

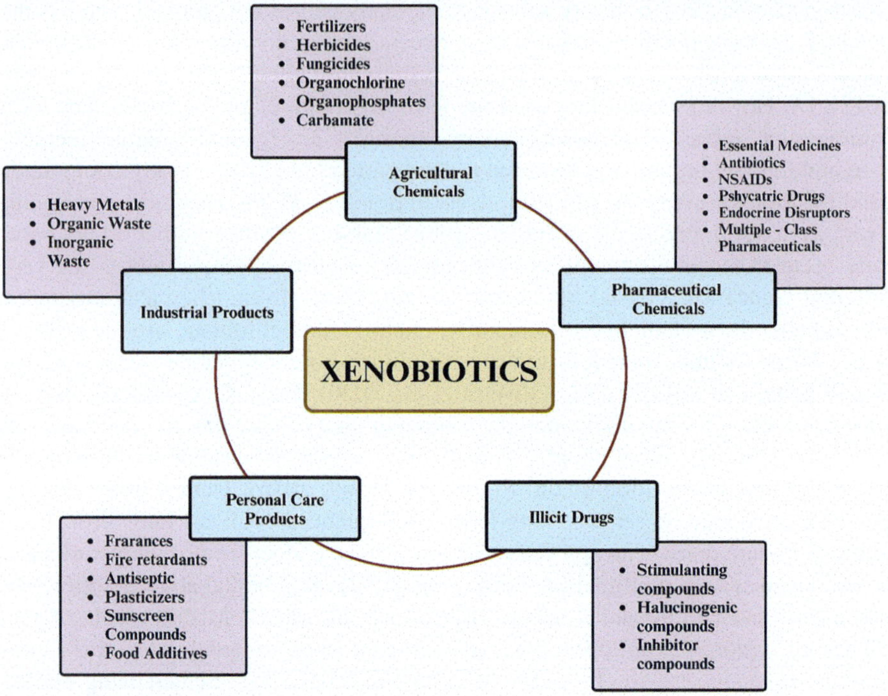

Fig. 1 Various types of xenobiotics and their sources

drug metabolism reactions occur in the human liver, intestine, and kidney, and the primary enzyme system involved in human drug metabolism is the cytochrome P450 (P450) monooxygenase enzyme system. The compound's pharmacokinetics are influenced by the pace at which P450s metabolize drugs and xenobiotics, which may also impact the pharmacodynamic response (Tracy et al., 2016). Any organ in the body can be affected by xenobiotic exposure, with effects ranging from allergic skin responses to potentially fatal repercussions affecting the neurological, reproductive, cardiovascular, and kidney systems; they can also affect individuals from the fetal stage to older people, with differences in metabolism related to gender and age (Fig. 1).

5 Apoptosis

Our bodies lose billions of cells every day. These cells must be eliminated immediately to avoid undesired immunological reactions and preserve a healthy, functional organism. To deal with this ongoing turnover, cells have evolved complex pathways leading to programmed cell death, which brings about the death of the cell (Boada-Romero et al., 2020). Apoptosis is a complicated and tightly controlled cell death that triggers caspase proteases and ultimately destroys cells. The introduction of the theory of programmed cell death, also known as apoptosis, was driven by its biological significance and morphological observations. This development occurred during the period spanning from the 1960s to the 1980s (Lockshin, 1969).

An individual's ability to survive is greatly influenced by apoptosis, which is also considered to be an essential element in several cellular processes, including immune system function, average growth and development, hormone-mediated atrophy, apoptosis, normal embryonic growth and development, and the elimination of unfit cells and cell homeostasis (Singh et al., 2022). Furthermore, during immune responses or in the event of a pathogenic infection, apoptosis may also take place as a defense mechanism. All cells react differently to different stimuli. However, apoptosis is regulated and controlled by various factors (Allawadhi et al., 2021; Khurana et al., 2021). Various events and stimuli can trigger apoptosis, but not all cells respond to the same ones when they die (Navik et al., 2021; D'Arcy, 2019). These stimuli, which cause some cells' DNA to break down, could be radiation or medications used in cancer treatment (Allawadhi et al., 2020).

There are two primary methods by which apoptosis is carried out: the intrinsic pathway, which uses unique properties of the mitochondria to trigger cell death, and the extrinsic system, which is triggered by death receptors on the cell membrane or in the cytoplasm. Some controlled cell death processes, such as pyroptosis and necroptosis, are highly inflammatory and are strong defenses against infection by bacteria, viruses, protozoa, and fungi (Jorgensen et al., 2017). Members of the Bcl-2 family regulate the intrinsic route, sometimes called the mitochondrial pathway, which is involved in developmentally regulated apoptosis caused by genotoxic agents. The proapoptotic effector molecules like Bax and Bak; proapoptotic BH3-only members such as Bim, Bid, Puma, Noxa, Hrk, Bmf, and Bad, and antiapoptotic Bcl-2 family proteins such as Bcl-2, Bcl-B, Mcl1, A1, and Bcl-xL comprise the Bcl-2 family (Czabotar et al., 2014). Not every cell will necessarily undergo apoptosis in response to the same stimulus, even though a wide range of physiological and pathological conditions and stimuli can cause it.

There are various assays for apoptosis, each focusing on a distinct component of the cellular machinery that causes programmed cell death. Assays might be protein-based, examine cell shape alterations, or examine DNA damage and fragmentation. A membrane-bound receptor like the Fas (CD95) receptor can initiate apoptosis (the extrinsic pathway) or be activated by aggregate cell damage (the intrinsic mechanism). The downstream apoptotic events, like executioner caspase activation, remain the same regardless of the triggering source. Mitochondrial transition pores (MPTP) opening in the intrinsic route starts the apoptotic process. Multiple fluorescence probes can analyze this event, such as tetramethyl rhodamine (TMREM), Mitotracker Red, Rhodamine 123, and Mitotracker Deep Red. Regardless of the technique used to induce apoptosis, alterations to the cell membrane can be measured using Annexin-V labeling later in the process. Before phagocytosis, membrane rearrangement can be measured by annexin-V, a protein that labels phosphatidyl serine (Khanal et al., 2011).

6 Flowcytometric-Based Identification of Apoptotic Changes

Annexin V-PI Assay Annexin V was first identified as placental protein 4 (PP4) by Bohn and colleagues (Inaba et al., 1984), who obtained the protein from the human placenta, and termed vascular-anticoagulant-a by Reutelingsperger et al. who retrieved the protein from the umbilical cord due to its anticoagulant effect (Reutelingsperger et al., 1985). The protein was given its name due to its homology with the annexin protein family after the human annexin V cDNA was cloned and sequenced. Phosphatidylserine (PS) on the outer membrane of cells is exposed and labeled using annexin V, a fluorescent probe. A nucleic acid dye called PI is incapable of crossing intact cell membranes. Mainly, PI binds to DNA. The amount of DNA present positively correlates with the amount of PI binding. Dyes can also be used to mark DNA so that the distribution of the cell cycle can be seen. Rhod-2, Indo-1 and other particular fluorescent probes mark the Ca^{2+} in the cells. Flow cytometry is then used to assess the qualitative and quantitative results (Guo et al., 2021).

Recently, fluorescent probes have been created to improve detection accuracy. Anionic phospholipids and phosphatidylserine (PS) exposed on the surface of dead or dying cells are the target of two-related fluorescence probes, which are molecular conjugates made up of one or two zinc dipicolylamine (ZnDPA) coordination complexes with an attached solvatochromic benzothiazolium squaraine dye (Jarvis et al., 2018). Apoptosis may be identified by Annexin V staining with fluorescein isothiocyanate (FITC), which binds to PS in the presence of Ca2 (Hu et al., 2021). This technique has the significant benefit of allowing for the simultaneous characterization of membrane phenotype using additional fluorochrome-matched monoclonal antibodies and differentiating between early apoptotic cells, which are identified as Annexin-V positive only, and late apoptotic cells, which are recognized as double Annexin-V and PI staining and single PI positive necrotic cells. One of the most widely used flow cytometry-based apoptosis detection techniques is the Annexin-V/propidium iodide method (A-V/PI) (Mariotti et al., 2017; Doan et al., 2020).

Using flow cytometry, apoptotic cells provide a single positive result for Annexin V staining. However, because the pyroptotic cells have more membrane permeability, both PI and Annexin V exhibit double-positive staining for PS and DNA due to their ability to enter the cell and stain them; double-positive stained cells can also be recognized as necrotic cells due to membrane rupture. Using commercially available kits, one can witness the phenomenon of binding Annexin V, which is calcium-dependent and has a high affinity for PS residues, by light microscopy or fluorescence flow cytometry (Kari et al., 2022).

More intricate comparisons can then be made between the mechanisms of action of various agents and therapies that cause apoptosis, such as UV radiation, etoposide (ETOP), curcumin, chloroquine (CQ), and staurosporine (STS). A time-dependent study of STS-induced apoptosis from 1 to 30 h was conducted using multiplexing of DAPI for cell viability, Annexin V-FITC, and mitotracker DiIC1(5). This demonstrated alterations in the process of apoptotic culture and the finding that mitochondrial dysfunction starts in living cells (Annexin V-ve/DAPI-ve) within an hour of STS treatment and progressively goes down during the period of investigation (Warnes, 2016). Annexin V binding and measurements of the mitochondrial membrane potential can be employed in further multiplexing experiments to quantify ROS. Diahydroethidium (HE) predominantly detects superoxide generated by the cell during death (Yaffe et al., 2013).

Mitochondrial Toxicity Measurement by Flowcytometry By producing ATP, controlling programmed cell death, and maintaining a balance between the redox states of cells, mitochondria are cellular organelles crucial to maintaining cellular homeostasis (Will & Dykens, 2018). Over the last 10 years, there has been mounting evidence that xenobiotics, which include many medications and environmental pollutants, can cause mitochondrial damage, which may be responsible for the numerous organ toxicities (Duarte-Hospital et al., 2021). Multiparametric flow cytometry was one of these technologies commonly used to evaluate mitochondrial health because it offered a rapid and accurate way to assess mitochondrial function in individual cells (Sheehan et al., 2023). To determine the proton gradient across the inner membrane of the mitochondria, flow cytometry has been utilized to assess MMP; for this purpose, numerous fluorescent probes have been developed. Fluorescent probes have a lyophilic nature and belong to the cation family; hence, they can penetrate the plasma membranes (Will & Dykens, 2018). These fluorescent probes' positive charge allows them to aggregate preferentially in the mitochondrial matrix once inside the cell. The initial generation of fluorescent probes for MMP detection was generated from DiOC6(3) (3,3′-dihexyloxacarbocyanine iodide) and Rhodamine 123 (Rh123), which were used to detect plasma membrane potential. These probes were sensitive to changes in MMP and showed perfect mitochondrial uptake (Yin & Shen, 2022). Numerous cationic and lipophilic fluorescent dyes are membrane-permeable are available to detect MMP and TMP, including Tetramethyl rhodamine methyl and ethyl easter (TMRM and TMRE), rhodamine-123 (Rh123), 5,′6′,6′tetracholor-1,1′,3,3-tetraethylbenzimidazolecarbocy annine (JC-1) and cholormethyl-X-rosamine (CMX-Ros) (Nguyen et al., 2022). These dyes are employed with various conjugates, such as FCCP and PI, to discern between the stages of apoptosis and necrotic versus apoptotic cell death (Kari et al., 2022). TMRM (Tetramethylrhodamine methyl ester) and TMRE (tetramethylrhodamine ethyl ester) are the fluorescent derivatives of Rh123. Flow cytometry uses these probes to find MMP changes in cells (Cano et al., 2021).

Reddish-orange fluorescence may be seen when TMRM is stimulated at 488 nm, and the intensity of this fluorescence is directly correlated with MMP. These two dyes have short loading times (15–20 min), are highly soluble, and have very low cytotoxicity (micromolar suppression of mitochondrial activity is required) compared to rhodamine 123 and JC-1. This is highly suitable for quantitative flow cytometry and quantitative analysis of the potential change of the mitochondrial membrane, with staining concentrations ranging from 0.5 to 30 nM (the concentration of JC-1 must be >0.1 μM). The accumulation in the mitochondria is restricted to changes in membrane potential (Rabinovich-Nikitin et al., 2021; Samuvel et al., 2022).

When the MMP is intact, JC-1 builds up in the mitochondria and generates highly concentrated red fluorescent aggregates. Aggregates split into green fluorescent monomers when JC-1, released from mitochondria during depolarization, is diluted in the cytosol. Because JC-1 can stay in cells for a long time, more than 2 h, it is a good choice for screening bigger batches of compounds (Robinson et al., 2015). As crucial mechanisms linked to xenobiotic-induced mitochondrial toxicity, MMP modification and ROS generation in individual cells may be directly assessed utilizing flow cytometry-based mitochondrial function tests, which gives a reliable platform to examine xenobiotic-induced mitochondrial toxicity. Over the last 10 years, there has been growing data that implies different xenobiotic-induced organ toxicities were partly caused by mitochondrial toxicity. Pharmaceutical companies are becoming more interested in creating assays to find substances that can cause mitochondrial toxicity in vivo (Will & Dykens, 2018). The discovery of more dependable and precise fluorescent probes for mitochondrial ROS (MitoSOX) and MMP (JC-1 or TMRM) has allowed scientists to develop flow cytometry-based techniques for measuring mitochondrial function. In 2020, Agarwal et al. reported that by FCM analysis, in human embryonic kidney cells, Ochratoxin A arrests the cell cycle at phases S and G0, ROS production, MMP loss by mitochondrial impairment and promotes cell death (Agarwal et al., 2020).

Caspase Activation A class of proteases known as caspases (Cysteine-ASPartic proteASES) cleaves their substrates following particular short tetrapeptide motifs that conclude in an aspartic acid residue. Caspases are divided into two classes based on their functions: apoptotic caspases, which are engaged in apoptosis, and inflammatory caspases, which control inflammatory reactions. Further classification of the apoptotic caspases can be made into initiator caspases (human caspase −8, −9, and −10) and executioner caspases (human caspase −3, −6, and −7). Every caspase is produced as an inactive zymogen composed of a small subunit, a big subunit, and a pro-domain. The pro-domain of an executioner caspase is composed of less than thirty amino acids. In contrast, the pro-domain of an initiator caspase is longer (between 150 and 200 amino acids) and includes a caspase activation and recruitment domain (CARD) or two death effector domains (DED). An essential component of apoptosis is the cascade of initiator and executioner caspases (Sun, 2023).

The potential use of caspases as therapeutic targets for various disorders has garnered much attention due to their crucial function as mediators of apoptosis (Van Opdenbosch & Lamkanfi, 2019). Thus, chemical substrates have been devised to track caspase activity in live cells. Death receptors, which belong to the tumor necrosis factor superfamily (TNFSF), initiate the extrinsic apoptotic pathway. Because death receptors include DED domains, caspase-8 activity is enhanced. The release of cytochrome c and Apaf-17 is necessary for activating caspase-9 in the intrinsic apoptotic pathway, which occurs after apoptosome formation. Activating either caspase-8 or caspase-9, the initiator caspase, causes the executioner caspase-3, caspase-6, and caspase-7 to cleave and become active. The presence of active executioner caspases suggests that the cells are going through apoptosis, and this activation is thought to be crucial in determining the mechanism of cell death (Jan, 2019).

Additionally, caspase activation is crucial for controlling inflammation and triggering different types of programmed cell death. On activation of Caspase 1, it directs the maturation of cytokines, which results in the pore formation at the plasma membrane and cleavage of gastrin D. The loss of gastrin D pore membrane integrity can lead to a kind of cell death called pyroptosis. Necroptosis is the caspase-independent cell death process that is inhibited by caspase-8 activity. In necroptosis, receptor-interacting serine/threonine protein kinase 1 (RIPK1) is the critical regulator (Tsuchiya et al., 2019). Jing Tong recently reported that caspase activity measurement was done using a fluorometric assay or flow cytometry. This study shows how to employ a fluorogenic substrate with flow cytometry in either a population-based or single-cell test. Different amino acid sequences can be utilized to determine which caspases are active with the proper controls. These tests have been used to detect the simultaneous loss of apoptosis protein inhibitors in response to tumor necrosis factor (TNF) activation, which mainly causes apoptosis in macrophages as opposed to other types of cell death (Tong et al., 2023).

Autophagy The word "autophagy" literally means "self-eating," it is a highly conserved, natural degradative process by which cells remove unnecessary and damaged or dysfunctional cellular components like organelles, macromolecules, proteins, etc., and recycle the breakdown products for energy homeostasis. It plays a crucial role in cell survival and the maintenance of cells. In this physiological process, autophagosomes transport cell constituents to the lysosome for breakdown (Ichimiya et al., 2020; Liu et al., 2023). Autophagy is a dynamic cellular process, and it includes three discrete stages: sequestration of cytosolic components and formation of autophagosome (Ravanan et al., 2017), formation of autolysosome by the fusion of the autophagosome with lysosome (Levine & Kroemer, 2019), and degradation of autolysosomal contents (Ichimiya et al., 2020; Liu et al., 2023). Moreover, it provides a valuable chance for the cell to eliminate broken and unnecessary organelles, misfolded proteins, or even invading microbes. All cells have low basal levels to carry out essential homeostatic functions. Still, it can quickly become upregulated in response to nutrient shortages, such as starvation, or when cells need to eliminate damaged organelles and cell components, such as during infections or protein aggregate accumulations. Thus, it is not unexpected that autophagy is conserved mainly throughout evolution and can even be found in yeast (Liu et al., 2023), where fundamental autophagy research has been conducted since the 1970s (Harnett et al., 2017).

Microautophagy, macroautophagy, and chaperon-mediated autophagy are the three subtypes of autophagy (Galluzzi et al., 2017). The cargo of all three types is delivered to the lysosome for disintegration and recycling, notwithstanding their structural differences. In a nutshell, lysosomal membrane protrusions or invasions' ability to seize cargo defines microautophagy. Chaperones are used in chaperone-mediated autophagy (CMA) to recognize cargo proteins designated by a specific pentapeptide pattern. Double membrane vesicles are synthesized from scratch in the case of macroautophagy (Liu et al., 2023). Recently, the autophagic processes have been investigated by flow cytometry, with an emphasis on the identification and semi-quantification of the autophagosome-resident autophagic biological marker, microtubule-associated protein LC3-II (Ichimiya et al., 2020; Liu et al., 2023). The autophagosome, a vesicle joined by two membranes, is the precursor of the autolysosome, a lysosome that produces energy when it fuses with other lysosomes (Sorice, 2022; Liu et al., 2023). The so-called "autophagic flux," which allows for the throughput of these autophagic-related structures, appears to be facilitated by this fusion process, which produces additional lysosomes during the autophagic process. LysoTracker dyes are helpful for measuring the amount of lysosomes produced during autophagy since they can measure lysosomal mass, and their fluorescent signal increases during autophagy. By comparing the median fluorescence intensities (MFI) of autophagic cells produced by different autophagy inducers, such as serum starvation, rapamycin, and chloroquine, and comparing this to that observed in untreated cells, LC3-II during the autophagic process may be semi-quantified (Ichimiya et al., 2020; Liu et al., 2023). Involvement of autophagy has also been reported in various human disease conditions, including neurodegeneration, cancer, cardiovascular diseases, and viral and bacterial infections, and both positive and negative effects of autophagic mechanism have been reported in different disease conditions (Liu et al., 2023; Klionsky et al., 2021). There are several FDA-approved chemicals/drugs available in the market that are known to modulate the activity of autophagy by regulating different signaling pathways, such as rapamycin, chloroquine, and hydroxychloroquine, etc., and used for the treatment of various disease conditions (Mohamed et al., 2022). The autophagic activity is usually found to promote tumor survival and chemoresistance. Thus, autophagy inhibitors like 3-methyl-adenine, chloroquine, and hydroxychloroquine are used to enhance the antitumor effect of drugs like trastuzumab by sensitizing chemoresistant cells to chemotherapeutics and promote tumor apoptosis (López-Méndez et al., 2023). Chloroquine and hydroxychloroquine are FDA-approved as they inhibit autophagy by hampering the fusion of autophagosomes with lysosomes (Mauthe et al., 2018; Xu et al., 2018). 3-methyl-adenine selectively blocks autophagy by preventing PI3K production required for autophagosome formation at an early stage of autophagy under starvation (Mauthe et al., 2018). However, there is substantial evidence that suggests that 3-methyl-adenine also enhances autophagy flux under nutrient-rich conditions, and such effects of 3-methyl-adenine are due to its differential temporal effect on class I and III PI3K inhibitor (Xu et al., 2018). Besides 3-methyl-adenine, other chemicals like wortmannin and LY294002 also serve as inhibitors of class III PI3K and, thus, widely used autophagy inhibitors (Chicote et al., 2020).

Different chemicals and drugs may act during any of the events of autophagy, and by modulating the autophagic process, they may serve as potential drugs or poisons (Tavakol et al., 2019). Some anti-tumor/anti-cancer and immunosuppressive drugs like rapamycin, dactylitis, metformin, simvastatin, and carbamazepine, which are known to activate autophagy, and some of these drugs, are under clinical trials (Tavakol et al., 2019). Thus, flow cytometry serves as the best tool for the rapid screening of several chemicals simultaneously. This allows simultaneous monitoring of numerous processes in the autophagic/endolysosomal process and also helps in the robust interpretation of each stage/phase of autophagy by utilizing different fluorescence probes, stains, dyes, etc., such as fluorescent protein (GFP) tagged autophagic probes, including GFP-LC3, SQSTM1/p62 and lysotrackers (Wang et al., 2017). Flow cytometry serves as a screening platform that provides multiparametric and quantitative detection of several samples at high rates and generates data that can be statistically analyzed. It is notably, using flow cytometry to integrate various route sensors allowed for the exact determination of autophagic flux without using lysosomal inhibitors. It also helps to measure the time-dependent variations in autophagic and endolysosomal activity using automated sampling (Zappavigna et al., 2017; Wang et al., 2021a, b).

Autophagy Detection by Labeling with Intracellular Anti-LC3-II The primary biological marker for autophagy detection is LC3-II. LC3 is a normal cytoplasmic microtubule protein; during autophagy, it undergoes cleavage and lipidation with phosphatidyl-ethanolamine (PE) to form LC3B and is found inside the autophagosomes. By incubating with frozen and permeabilized autophagically processed cells, rabbit polyclonal anti-LC3-II, also known as rabbit immunoglobulin (isotype control), is used for indirect intracellular labeling. These cells are then treated with the goat anti-rabbit Alexa Fluor-647 fluorescent tag. The advantage of the indirect labeling approach is that it increases the fluorescence signal from LC3-II, which is present in cells in relatively modest numbers. It is possible, therefore, to detect the LC3-II signal above that of the isotype control tube. After that, the resultant cells are examined using a red laser on a flow cytometer. The relative elevation in the LC3-II autophagic signal above that of resting control cells was then calculated using the cells' median fluorescence intensity (MFI) of the LC3-II-AF-647 signal. Further, fluoresce intensity can be enhanced by GFP (fluorescence is acid sensitive) and RFP (acid resistant) tagging of LC3B for flow cytometric detection (Zappavigna et al., 2017).

There are scientific reports where flow cytometry has been used as an effective tool for studying autophagic mechanisms in response to various chemicals and toxicants. Rotenone is a colorless and odorless natural isoflavone-based broad-spectrum insecticide known to affect the functioning of mitochondrial Complex-I and impair autophagy. Quantification of lapidated LC3B-positive cells after labeling cells with Anti-LC3-II and using flowcytometry effectively show changes in the expression of enhanced/reduced levels of LC3B-positive cells (Quintero-Espinosa et al., 2023; Perez-Abshana et al., 2023). Quintero-Espinosa et al. (2023) have reported impaired autophagy mechanism in response to rotenone-induced toxicity in dopaminergic-like neurons differentiated from menstrual stromal cells by using quantification of lapidated LC3B-positive cells by flow cytometry (Quintero-Espinosa et al., 2023).

Similarly, Perez-Abshana et al. (2023) have also identified alteration in autophagy using flowcytometric quantification of anti-LC3II fluorescent autophagosome marker in human embryonic kidney derived HEK-293 Cell Line in response to rotenone toxicity (Perez-Abshana et al., 2023).

7 Autophagy Detection by Labeling with Intracellular Anti-SQSTM1 & LysoTracker

SQSTM1 Measurement SQSTM1/p62 or sequestosome-1 is a selective autophagy cargo or receptor protein incorporated into the autophagosomes, usually through its interaction with LC3. It is considered as an indicator of autophagic flux. Accumulation of SQSTM1/p62 suggests an impaired autophagic mechanism. Thus, selective antibodies were developed against SQSTM1/p62 to measure its expression using flow cytometry. (Hargarten et al., 2023; Koepke et al., 2020).

LysoTracker A flow of related organelles is autophagously generated throughout the autophagic process. Autophagosomes with double membranes are the first to form, eventually combining with lysosomes to create autolysosomes. The cell produces more lysosomes to support the autophagic flux maintenance mechanism. Thus, LysoTracker dyes may quantify this increase in lysosomal mass within individual cells throughout the autophagic process (Zappavigna et al., 2017; Ichimiya et al., 2020). LysoTracker Green is given to the cells and incubated to assess cell viability. The cells are then rinsed in PBS, and DAPI is added. After that, the resultant cells are examined using violet and blue (488 nm) lasers on a flow cytometer. Next, the percentage rise in lysosome mass in the autophagically-undergoing cell populations is determined by comparing the LysoTracker Green signal's median fluorescence intensity (MFI) in the living cells. When a cell is stressed by internal and external environmental factors, such as a shortage of nutrients, autophagy takes place as a cell survival mechanism (Zappavigna et al., 2017).

Multispectral Imaging Flowcytometry for Measurement of LC3 Puncta and Autolysosome Formation in Cells Multispectral imaging flow cytometry is an advanced and combined version of flow cytometry and fluorescent microscopy. It can acquire images of every cell or particle, demonstrating high throughput, high quantitative power, and multiplexing potential in flow cytometry. In traditional flow cytometry, non-specific staining and fluorescent dye background may complicate the measurement and analysis of autophagy signal intensity. However, multispectral imaging flow cytometry allows detailed examination, counting, and analysis of co-localization of the autophagosomes and lysosomes (puncta formation) within each cell with the help of acquired images of cells with the help of specific markers like anti-LC3 and anti- SQSTM1 co-localization along with lysosome staining dye like Lysotracker (Pugsley, 2017).

8 Measurement of Autophagy Using Dyes or Stains with Flow Cytometry

Monodansylcadaverine is a lysosomotropic compound used as a specific fluorescent marker for labeling autophagic vacuoles. Monodansylcadaverine is an autofluorescent compound that conjugates with cadaverine and accumulates in the acidic autophagic vacuoles due to interaction with membrane lipids and an ion-trapping mechanism. MDC staining is used as a rapid and convenient approach for autophagic assay, specifically within in vivo systems. (Zhang et al., 2015; Xiao et al., 2022).

Acridine Orange is another fluorescent dye commonly used to measure the levels of DNA and RNA within cells. However, it is also efficiently used to measure autophagy. Within autolysosomes, acridine orange interacts with acidic granules and is trapped within the vesicles by forming aggregates after protonation; these aggregates emit bright red fluorescence and thus allow detection and quantification of autophagy by measuring the intensity of red signal (Murugan & Amaravadi, 2016). The degree of autophagy in cells can be measured using biochemical techniques like immunoblotting by quantifying the signal for autophagy proteins LC3I and LC3-II. Typically found in the cytoplasm as LC3I, LC3-II is incorporated into the autophagosome after being cleaved and lipidated by phosphatidylethanolamine (Ichimiya et al., 2020; Liu et al., 2023). By fluorescent tag addition through transfections with GFP-RFP, LC3-II can be imaged and subjected to flow cytometric analysis. This is useful because LC3-II-GFP-RFP detection is selective for autophagosomes, while LC3-II-RFP-only vesicles are specific for autolysosomes since the acidic conditions seen in autolysosomes destroy GFP fluorescence. (Zappavigna et al., 2017). Autophagosomes and autolysosomes can be identified using an antibody against LC3-II for flow cytometric examination of puncta or the median fluorescence intensity of labeled cells, respectively. The technique is significantly less onerous when the median fluorescence values of LC3-II antigen levels grow flow cytometrically as opposed to increases in the number of puncta labeled with anti-LC3-II, which can only be determined via a laborious image analysis method (Kaizuka et al., 2016; Zappavigna et al., 2017).

Autophagy in cells has also been measured using LysoTracker probes by microscopy and, to a limited extent, flow cytometry (Kaizuka et al., 2016; Zhang et al., 2022). Using imaging cytometry, the novel autophagy dye Lyso-ID from Enzo tracks autophagy co-localizing with LC3-II during the autophagic process (Phadwal et al., 2012). The combined effect of chloroquine-inducing autophagy and suppressing it by restricting lysosomal fusion with the autophagosome through vesicular acidification disruption—leads to an accumulation of lysosomes and autophagosomes inside the cell. More lysosomes appear to be produced during the autophagic process, which fuses with the autophagosomes to generate autolysosomes. This enables flow cytometric quantification of the autophagic process indirectly in living cells. This facilitates concurrently assessing several cellular activities, including autophagy, ER, and mitochondrial activity (Zappavigna et al., 2017). The capacity to investigate the mechanisms and processes behind autophagy-mediated cell death in primary cells separated from cell lines and patients will be improved by developing an inexpensive, repeatable, and simple flow cytometric-based approach to assessing the level of autophagy in living cells.

9 Xenobiotic-Induced Apoptosis Assessed by Flow Cytometry: A Comprehensive Overview

Identifying ROS production, mitochondrial membrane potential, and caspase activation through *in vitro* exposure to xenobiotics is a significant component of investigations using the flow cytometry technique (Fallico et al., 2020). To ensure that cell aggregates, dead, or dying cells are not included in the cell function analysis process using flow cytometry, it is necessary to identify individual living cells accurately (Ortolani, 2022). Additionally, a different way to evaluate the hazardous potential of xenobiotics is to recognize living cells using flow cytometry (Fallico et al., 2020; Rooprai et al., 2020). When considering human or mammalian cells, the flow cytometry processes for single live-cell identification are generally simple (Fallico et al., 2020). However, because bacteria are smaller than eukaryotic cells and have a distinct permeability to viability dyes, the procedures become more complex (Jávega et al., 2022).

Through flow cytometry, cells undergoing necrosis and apoptosis can be distinguished using the Annex V/propidium iodide test. This method involves detecting phosphatidylserine externalizations using FITC-conjugated Annex V protein (fluorescein isothiocyanate), an early indicator of apoptosis. Simultaneously, nuclear DNA binding to propidium iodide (PI) allows cell membrane damage detection. Hence, flow cytometry is the preferred method for determining individuals' viability, apoptosis, and necrosis. The scatter analysis of cell populations typically provides a clear separation between non-viable and viable cells. To distinguish between different types of cells, this scattered approach can be advantageously paired with labeling using Annexin V and propidium iodide (PI). This DNA-binding dye does not pierce the intact cell membranes. Necrotic cells are Annexin-negative and PI-positive, whereas Annexin-V-positive and PI-negative cells are believed to be early apoptotic. Double-positive cells for Annexin V and PI are classified as late apoptotic (Salama et al., 2018). In this regard, a wealth of information indicates that xenobiotics mediate or induce apoptosis, which can be detected through flow cytometry techniques.

In a study conducted by Silva et al., the toxicity of three commonly used pesticides—herbicide (glyphosate, i.e., GLY), insecticide (imidacloprid, i.e. IMD), and fungicide (imazalil i.e. IMZ)- was compared to uncover their potential effects on Caco-2 and HepG2 cells to investigate the potential toxicity of accidental or intentional oral exposure to these chemicals as food contaminants. The toxicity of pesticides was similar in both cell lines, with a rank of GLY < IMD < IMZ; however, HepG2 cells exhibited higher toxicity than Caco-2 cells. Increased ROS, decreased mitochondrial membrane potential, and apoptotic cell death are the three major processes measured by flow cytometry analysis (Silva et al., 2022). The study performed by Mehatap et al. analyzed the oxidative stress-related apoptotic processes in H9c2 cardiomyoblast cells and human neuroblastoma (SH-SY5Y) cells. Benomyl, a benzimidazole fungicide, was found to have cardiotoxic effects. Benomyl is classified as a xenobiotic compound and has been observed to exert its hazardous effects through redox reactions. The ability of benomyl to induce apoptosis or necrosis in H9c2 cells and SH-SY5Ycells was assessed using a flow cytometer and Annexin V with PI. Higher concentrations of Benomyl were associated with a considerable increase in both early and late apoptosis (Kara et al., 2020; Mehtap et al., 2021). Furthermore, flow cytometric analysis revealed that both benomyl and carbendazim were found to arrest MCF-7 cells in the G2/M phase and increased the number of sub-G 0/G 1 phase cells, indicating the presence of a considerable number of apoptotic cells (Kawaratani et al., 2015).

Yuan et al. reported a study in which cadmium (Cd) significantly impairs nervous system function by causing neuronal death. A hazardous element called cadmium can permeate neurons and the brain's parenchyma, altering neurological function. The flow cytometer was used to analyze the apoptosis rate. Early apoptosis (annexin V+/PI−, Q4) and late apoptosis (annexin V+/PI+, Q2) were quantified to determine the percentage of apoptosis. These findings imply that the Cd-induced neuronal apoptosis may be related to the activation of Fas/FasL. This study reports that Cd-induced neural apoptosis mostly depends on the Fas/FasL-mediated mitochondrial apoptotic pathway (Yuan et al., 2018). In a survey by Olszowski et al., apoptosis triggered via mitochondrial pathways was observed to assess how cadmium affects apoptotic processes in macrophages generated from THP-1 (acute monocytic leukemia cell line) at low ambient (nanomolar) concentrations. PI double labeling and Annexin V/FITC flow cytometry were utilized in this work to determine the overall amount of apoptosis in the cell population. This work shows that cadmium causes mitochondrial damage in THP-1 macrophages, even at low ambient exposures. Exposure to extremely low doses for 48 h reduces cell viability and causes cell necrosis and apoptosis. Cadmium-induced apoptosis is caused by several mitochondrial processes, including decreased Bcl-2 mRNA expression, increased ROS generation, increased Bax, and decreased mitochondrial membrane potential (Olszowski et al., 2015).

Ahamed et al. reported reduced Graphene Oxide (Rgo) and lead (Pb) combined toxicity in A549 cells. Flow cytometry analysis revealed that rGO at a moderate dosage (<50 μg/mL) did not induce toxicity in A549 cells. In contrast, Pb exposure resulted in decreased cell viability, LDH leakage, MMP loss, cell cycle arrest, and increased activity of the enzymes caspase-3 and -9. The co-exposure of rGO attenuated the toxicity caused by Pb in A549 cells. Furthermore, the co-exposure of rGO with Pb significantly inhibited the production of pro-oxidants produced by Pb and the depletion of antioxidants (Ahamed et al., 2020). Lead (Pb) poisoning poses a severe environmental threat and has disastrous impacts on human health (Kumar et al., 2020). Karri et al. tested for Pb, Cd, As, and MeHg and found that HT-22 cells underwent apoptosis with long-term exposure. Flow cytometry analysis using PI and annexin V labeling revealed a concentration-dependent pattern of heavy metal-induced apoptosis in HT-22 cells (Karri et al., 2018). Pukanha et al. showed the levels of Pb exposure in work environments and the detrimental consequences of such exposure levels on immune system function. The regulatory T (Treg) cells (CD4 + CD25brightCD127dim) were assessed using flow cytometry. Blood Pb levels in every individual exhibited a positive association with the percentages of IL-4 ($r = 0.473$, $p = 0.041$) and Treg cells ($r = 0.843$, $p < 0.001$), but an inverse relationship with the percentages of Tc cells ($r = -0.563$, $p = 0.015$). These results point to a possible link between long-term high Pb exposure and a shift in the humoral immune response and cellular immunity suppression, which may raise the risk of cancer in Pb-exposed workers (Pukanha et al., 2020). Huang et al. revealed that $MnCl_2$ may induce mitophagy in human neuroblastoma SH-SY5Y cells, increasing ROS production and dose-dependent apoptosis. The effects of treatment with 200, 400, and 800 μM MnCl2 over 24 h were found to cause a concentration-dependent drop in the mitochondrial membrane potential. Furthermore, the FITC Annexin V was utilized to ascertain the apoptosis rate (Huang et al., 2021). In an investigation performed by Vidhyapriya et al., the cytotoxic capability of MnCORM was assessed on various cancer cell lines, including colon (HCT-15), cervical (HeLa), lung (A549), breast (MDA MB-231) and normal cells (HEK 293). Propidium iodide (PI) stained cells underwent flow cytometric examination, which revealed a notable proportion of sub-G1 hypodiploid cells following MnCORM treatment. Reactive oxygen species (ROS), mainly superoxide radicals, were produced in response to MnCORM. Thus, the potential of the mitochondrial membrane was lost (Vidhyapriya et al., 2018). Additionally, Althobiti et al. reported the results of PF127-coated Mn2O3 NP (derived from the leaf extract of *Glycyrrhiza uralensis* (GU)) on breast cancer cells. Their research revealed apoptotic cell death and nuclear damage, which resulted in an arrest of the cell cycle in the S phase. This was observed through EtBr/AO dual staining, DAPI staining, and Annexin-V-FITC-based flow cytometry analysis (Althobiti et al., 2023).

Alhadlaq et al. reported the potential toxicity of MnO_2 NPs in human fibrosarcoma epithelial (HT1080) and breast cancer epithelial (MCF-7) cells. Flow cytometry data from this study showed that MnO_2 NPs caused both MCF-7 and HT1080 cells to undergo apoptosis. A statistically significant decrease in the S and G2/M stages was also seen in both cell types (Alhadlaq et al., 2019).

Another study by Guo et al. showed the effects of $NiCl_2$ exposure on broiler chickens. Five broiler chickens in each group were exposed to $NiCl_2$ to measure kidney cell apoptosis using flow cytometry at 14, 28, and 42 days. Ni is a metal that has commercial and industrial uses, but it is hazardous for both human and animal health. According to the current study, feeding $NiCl_2$ at 300+ mg/kg causes broiler hens to undergo tubular apoptosis through caspase-dependent and mitochondria-mediated mechanisms. These findings offer new perspectives into the *in vivo* and *in vitro* toxicity of nickel and nickel compounds in the kidneys of broiler chickens (Guo et al., 2016).

Salama et al. utilized the flow cytometry technique to study the effects of Trichloroethylene (TCE). This typical volatile organic solvent is widely used in consumer and commercial items like varnishes and ink as a cleaning and degreasing agent. It is also used extensively in industrial applications. Their study revealed a notable rise in late apoptosis and necrosis in the neurospheres treated with TCE

(0.25 μM) compared to the control cells. Thus, this study showed that a low dose of TCE (0.25 μM) harms neurospheres by causing various forms of cell death (Salama et al., 2018). In a study reported by Ali et al., TCE-induced cellular toxicity results in the upregulation of molecular pathways in HaCaT (human epidermal keratinocyte) cells. In this study, AnnCy3 cytotag and pre-genomic apoptosis processes were assessed to confirm if exposure to TCE may initiate apoptosis. Through flow cytometry, it is seen that the number of cells undergoing apoptosis was much higher in TCE-exposed cells, as indicated by a considerable rise in annexin V positive and 6-CFDA positive cells, and the possibility of necrosis was eliminated (annexin V positive and 6-CFDA negative). Moreover, a considerably more significant percentage of nuclei (48%) showed a sub-diploid peak, indicative of apoptotic cells, when the cells were sorted for DNA content following TCE exposure. Nevertheless, just 10% of the sub-diploid genome was visible in the unexposed cells. These findings indicated that exposure to TCE leads to the fragmentation of the diploid genome (Ali et al., 2016).

In Wang et al.'s study, urban particulate matter (PM) was applied to human bronchial epithelial cells (HBECs) to measure the inflammatory response. The primary source of air pollution in metropolitan areas globally is PM. PM causes inflammation in the tissues of the lungs and is a significant risk factor for several respiratory illnesses. HBECs were stimulated with PM at various intervals and dosages to measure the amount of ROS production. Flow cytometry analysis showed that after treating HBECs with varying concentrations of PM for 24 h, PM significantly increased the generation of ROS in a dose-dependent way. Early PM exposure resulted in a considerable rise in ROS, which may play a vital role as a mediator in the inflammatory response triggered by PM (Wang et al., 2021a, b). In the study by Morozesk et al., MRC-5 cells were subjected to flow cytometry to assess PM cellular absorption and death induction. The outcomes of the cytotoxic studies showed that exposure to PM2.5 and PM10 raises the rates of necrosis and apoptosis and reduces viable cell populations. The granularity of MRC-5 cells increased in response to the investigated PM fractions, indicating particle internalization (Morozesk et al., 2021).

Guan et al. studied the combined harmful effects of lead (Pb) and carbon black nanoparticles (CBNPs) on autophagy flux and apoptosis in rat alveolar macrophages. The results demonstrated that whereas CBNPs or Pb^{2+} did not cause cell apoptosis at their respective no observed effect concentrations (NOEC), their combined exposure considerably raised the level of apoptosis by causing MMP degradation and triggering apoptotic signal pathways as analyzed by flow cytometry analysis (Guan et al., 2020).

The study performed by Chen et al. reported the potential of lathyrol, an unidentified natural substance extracted from Semen Euphorbiae, a traditional Chinese medicinal, that may have anti-tumor properties. The potential of lathyrol as a novel medication candidate for the treatment of lung cancer was emphasized in their research. The cell cycle analysis was performed using flow cytometry, revealing that the treatment with lathyrol induced G2/M phase cell cycle arrest. Using a DCFH-DA cell-permeable probe, the presence of reactive oxygen species (ROS) was measured to investigate the anti-tumor mechanism of lathyrol. Increased ROS were seen following lathyrol therapy. In addition to being a significant contributor to cell cycle arrest, ROS-induced DNA damage is also a major cause of cell apoptosis (Chen et al., 2023).

Repositioning drugs is a contemporary tactic for developing novel anticancer therapies regarding their mechanisms of action, safety, and adverse effects (Kirtonia et al., 2021). In this regard, Delgado-Waldo et al. demonstrated that the known medications metformin, doxorubicin, and sodium oxamate-also known as triple therapy, or TT target vital pathways for cell growth and survival, killing cancer cells in three distinct cervical cancer cell lines (Hela, SiHa and CaSki cells). Through low cytometry, viable and dead cells can be quantitatively distinguished using propidium iodide (PI) and Annexin V (AV) double labeling. The findings of this study revealed that TT induced apoptosis through the intrinsic route via caspase three activation (Delgado-Waldo et al., 2023). García-Castillo et al. reported that doxorubicin, metformin, and oxamate were combined in a triple-negative xenograft model to target critical molecular pathways in cancer. This combination produced an effective and quick tumor growth inhibitor. The conclusions of this study are confirmed by the flow cytometry data, which demonstrated that DoxMetOx increased the effect of apoptosis starting 48 hours after treatment and maintained the same activity level for 72 h (García-Castillo et al., 2017). In the investigation by Shashiraj et al., it was shown that by using PI and Annexin-V, the apoptotic effects of Ls-AgNPs (silver nanoparticles from *Lagerstroemia speciosa* flower bud extract) on MG-63 (bone osteosarcoma cell line) cells were assessed to ascertain the degree of late and early apoptosis. The tumor cells were stained with FITC Annexin V/PI, and flow cytometry employing fluorescence-activated cell sorting (FACS) confirmed the identification of early and late apoptotic cells. The investigation results showed that apoptosis was induced in the cells when Ls-AgNPs were applied to cancer cells for a full day at an IC50 of 37.57 μg/mL. The number of early (89.88%) and late (39.8%) apoptotic cells increased significantly in the treated cells compared to the untreated cells, which showed no appreciable apoptosis (Shashiraj et al., 2023). Niu and colleagues found comparable apoptotic effects between free and DOX/MLT-loaded nanocarrier treatments on MG-63 cells for 24 h. The apoptosis produced by the nano-carrier was 28.7% and 50.4% in the late apoptosis and 3.05% and 28.78% in the early apoptosis, as assessed by flow cytometry (Niu et al., 2021). According to a study by Iram et al., there was an unavoidable consequence when Tb_2O_3 nanoparticles at a dosage of 0.102 μg/mL were administered. PI/FITC uptake was used to quantify the cells with impaired membrane integrity. The concentration-dependent absorption of PI/FITC and early apoptotic rates of 5.3% and 6.4%, respectively, were shown by the MG-63 cells exposed to Tb_2O_3 NPs. The Annexin V-FITC/PI investigation aimed to distinguish between cells undergoing early and late phases of apoptosis (Iram et al., 2016). Similar results were obtained by Nagaraja et al., who investigated the effects of Tb_2O_3 nanoparticles on the Pa-1 ((Human ovarian teratocarcinoma cancer) cell line. Flow cytometry analysis showed a significant population of cells undergoing early apoptosis (12.91%) and late apoptosis (45.03%) (Nagaraja et al., 2023). Pan et al. investigated the effects of Z-ajoene (found in crushed garlic) on human NSCLC A549 and H1299 cell lines. They employ a FACScalibur flow cytometer to detect apoptosis in cells using a propidium iodide (PI) apoptosis detection kit and an annexin-V fluorescein isothiocyanate (FITC). Z-ajoene therapy significantly increased tumor cell apoptosis in a dose-dependent manner compared to the control group. Z-ajoene may cause A549 and H1299 cells to undergo apoptosis via the caspase signal pathway. This novel medication is anticipated to be positively connected with the mass action concentration and may be used to treat lung cancer (Pan et al., 2023).

10 Xenobiotic-Induced Autophagy Assessed by Flow Cytometry: A Comprehensive Overview

Exposure to different xenobiotic compounds may affect other physiological and biochemical parameters of the cells. It may alter cellular homeostasis, induce cell death by impairing different cell survival pathways, and induce/activate necrosis, ferroptosis, autophagy, or apoptotic pathways (Ornatowski et al., 2020). Flow cytometry is an advanced technique for selective and simultaneous identification/measurement of different markers in different types of cells regulating cell death (necrosis, ferroptosis, autophagy, or apoptotic) on exposure to xenobiotics (Sica et al., 2016). The advanced analytical flow cytometry known as fluorescence-activated cell sorting (FACS) also enables sorting a heterogeneous mixture of biological samples based on their physical and biological characteristics (Shashiraj et al., 2023).

Flow cytometry allows the identification of impairments at the specific stage of autophagy by measuring the expression of particular markers regulating each stage simultaneously, such as Beclin1, ATGs, LC3 I/II, SQSTM1 P62, etc. (Sica et al., 2016). Besides, several fluorescent stains/dyes also facilitate the analysis of autophagy, like Monodansylcadavrine, acridine orange, lysoTracker, and CYTO-ID dye. There are many scientific reports where flow cytometry has been used as an effective tool for the study of autophagic mechanisms in response to various chemicals and toxicants. Rotenone is a colorless and odorless natural isoflavone-based broad-spectrum insecticide known to affect the functioning of mitochondrial Complex-I and impair autophagy. Quantification of lapidated LC3B-positive cells after labeling cells with Anti-LC3-II and using flowcytometry effectively show changes in the expression of enhanced/reduced levels of LC3B-positive cells (Quintero-Espinosa et al., 2023; Perez-Abshana et al., 2023). Quintero-Espinosa et al. (2023) have reported impaired autophagy mechanism in response to rotenone-induced toxicity in dopaminergic-like neurons differentiated from menstrual stromal cells by using quantification of lapidated LC3B-positive cells by flow cytometry (Quintero-Espinosa et al., 2023). Similarly, Perez-Abshana et al. (2023) have also identified alteration in autophagy using flowcytometric quantification of anti-LC3II fluorescent autophagosome marker in human embryonic kidney derived HEK-293 Cell Line in response to rotenone toxicity (Perez-Abshana et al., 2023).

LysoTracker dye is a widely used autophagy marker in flow cytometry that is effectively used for the study of alterations in the autophagy mechanism on exposure to several chemicals such as rotenone, monocrotophos, zinc by automatic quantification of LysoTracker positive cells (Wei et al., 2018; Perez-Abshana et al., 2023). A susceptible and efficient flow cytometry technique has also been used for the quantification of lysotracker-monodansylcadavrine and lysootracker-mitotracker positive cells by Srivastava et al. (2018) while studying the toxicity mechanism of Monocrotophos in human mesenchymal stem cells and SHSY-5Y Cells (Srivastava et al., 2018). Wei et al. (2018) have also done flow cytometric quantification of lysootracker red fluorescence while studying the role of autophagy in zinc-induced impairments in primary hepatocyte culture (Wei et al., 2018). Further, Perez-Abshana et al. (2023) have also used lysootracker dye for the quantification of changes in autophagy mechanism in response to rotenone toxicity in human embryonic kidney-derived HEK-293 Cell Line using flowcytometry (Perez-Abshana et al., 2023).

Quantification of Lysotracker- Monodansylcadavrine and lysotracker-mitotracker positive cells using flow cytometer has been done for the study of toxicity mechanism of monocrotophos in human mesenchymal stem cells and SHSY-5Y Cells and results exhibit enhanced levels of Lysotracker- Monodansylcadavrine positive cells (Srivastava et al., 2018).

The use of acridine orange for quantitative analysis of autophagic mechanism in response to exposure to chemicals like zinc and cadmium has been reported (Li et al., 2020; Wei et al., 2018). In 2020, Li et al. used acridine orange solution for the study of the role of autophagy in cadmium-induced toxicity on HK-2 cells using the flow cytometry technique (Li et al., 2020). Further, Wei et al. (2018) have also used acridine orange stain and flow cytometry techniques to quantify the autophagy mechanism in zinc-induced impairments in primary hepatocyte culture (Wei et al., 2018).

In 2022, Loubaki et al. also used flow cytometry for the study of autophagy in cannabinoid response in human gingival carcinoma cell line—Ca9-22 cells and human polyclonal oral epithelial cell line—GMSM-K cell line using autophagy probe (Red stain/probe) (Loubaki et al., 2022). Besides, CYTO-ID dye is also used to quantify autophagy using flow cytometry. Rapid cell partitioning characterizes this cationic amphiphilic tracer dye. The dye cannot build up inside the lysosomes of its titratable functional moieties. It reduces lysosome staining, thus removing background, unlike other dyes, and enabling vacuole labeling, which is associated with the autophagy pathway (Guo et al., 2015). CYTO-ID stains autophagic vacuoles and measures autophagic flux by selectively labeling the accumulated autophagic vacuoles. The excitable wavelength for this dye is 488 nm, and as soon as it enters into pre-autophagosomes, autolysosomes, and autophagosomes, it releases a brilliant fluorescence. In 2022, Jih-Jung et al. used CYTO-ID dye for the study of autophagy mechanism in response to anticancer drug- neogambogic acid (NGA) and ceria nanoparticles (CNPs) on breast carcinoma cells- MCF-7 using flowcytometry technique (Jih-Jung et al., 2022).

11 Summary and Future Prospects

With advancements in science and technology, several new chemicals have been introduced into the environment to improve human health and livelihood, such as drugs, pesticides, fertilizers, etc. However, besides their beneficial effects, these xenobiotics chemicals or compounds may have detrimental effects on human health and the environment. Therefore, the development of practical tools and techniques for the screening of toxic potential xenobiotic compounds and their specific targets is essential for the development of drugs and the prevention of their possible adverse effects on humans. Flow cytometry is an analytical technique used for rapid quantitative detection, analysis, and sorting of cells or biological particles by following the principles of fluid dynamics, optics, and electronics. A typical flow cytometer measures the scattering and emission of light of multiple fluorescences by cells or microscopic particles suspended in an isotonic liquid medium, aligned in a single file passage, and presented to illumination sources (lasers) one by one at high speed. It allows simultaneous detection and quantification of many structural and functional characteristics of every particle or cell present within a heterogeneous population at rates of thousands of events per second by measuring interaction patterns of light and matter and fluorescence with the help of illumination sources and light detectors (Jávega et al., 2022). Computerized data provides a comprehensive descriptive analysis of the biological features of samples (cells or particles). With recent advancements in algorithms for multispectral fluorescence unmixing and using different fluorescent reagents, quantifying up to 30–50 parameters is possible. Being a multiparametric and rapid technique for identifying and detecting various types of cell populations with high accuracy and precision, flow cytometry is an invaluable tool for analyzing scarce or rare samples (Robinson, 2019; Jávega et al., 2022).

Apoptosis and autophagy are well-explored mechanisms of cell survival mechanisms initiated in response to various physiological or chemical stressors and also play essential roles in disease conditions. Several drugs and xenobiotic chemicals adversely affect these pathways and induce cellular damage. Flow cytometry allows rapid detection, analysis, and even sorting of cells lying under different stages of apoptosis or autophagy by using specific markers, fluorescent probes, or stains. Thus, flow cytometry is an essential tool for biochemical analysis in Cellular and Molecular Biology, Biotechnology, Drug Discovery, Environmental Studies, and Toxicology for studying highly diverse cell types and samples, either fresh, fixed, or permeabilized. It is routinely used in clinical diagnosis and prognosis, especially in hematology, immunology, and oncology (Cossarizza et al., 2019). Besides, flow cytometry also allows live-cell analysis with minimally perturbed intracellular environments near physiological conditions and helps to determine associated biological processes and infrequent events. Flow cytometry also allows the simultaneous isolation or separation of different sub-populations of cells present within a heterogeneous population, called cell sorting (Sun, 2023). Different cell sorters or separators usually do individual cell sorting. In most commonly used cell sorters, high-frequency vibration forms microdrops where electrically charged plates help deflect droplets toward specific collection tubes. Further, microfluidics-based cell separators are also used, which allow starting of small numbers of cells (Cossarizza et al., 2019).

With advanced hardware structure and software design, flow cytometry allows multivariate data analysis and multiparametric data acquisition. Thus, a single-tube assay can identify and select specific cell subsets within morphologically or functionally heterogeneous populations of cells using particular structural and functional parameters of the individual cells or particles (Robinson, 2019). The advanced software enhances accuracy and discriminating power by using multidimensional correlation of the particular properties of each cell within a heterogeneous population of cells (Jávega et al., 2022). Further, real-time flow cytometry is used to monitor changes in cellular and functional parameters with a resolution time <1 s and for time windows ranging from seconds to several minutes within heterogeneous cell populations. This kinetic analysis allows the detection of very fast or transient dynamic processes such as those typically involved in signal transduction. Thus, flow cytometry is a beneficial technique for rapidly detecting, isolating, and analyzing cells based on their different cellular and morphological characteristics.

References

Abdelsalam, N. A., Ramadan, A. T., ElRakaiby, M. T., & Aziz, R. K. (2020). Toxicomicrobiomics: The human microbiome vs. pharmaceutical, dietary, and environmental xenobiotics. *Frontiers in Pharmacology, 390.*

Agarwal, P., Singh, D., Raisuddin, S., & Kumar, R. (2020). Amelioration of ochratoxin-a induced cytotoxicity by prophylactic treatment of N-acetyl-L-tryptophan in human embryonic kidney cells. *Toxicology, 429*, 152324.

Ahamed, M., Akhtar, M. J., Khan, M. M., & Alhadlaq, H. A. (2020). Alleviating effects of reduced graphene oxide against lead-induced cytotoxicity and oxidative stress in human alveolar epithelial (A549) cells. *Journal of Applied Toxicology, 40*(9), 1228–1238.

Alhadlaq, H. A., Akhtar, M. J., & Ahamed, M. (2019). Different cytotoxic and apoptotic responses of MCF-7 and HT1080 cells to MnO2 nanoparticles are based on similar modes of action. *Toxicology, 411*, 71–80.

Ali, F., Khan, A. Q., Khan, R., & Sultana, S. (2016). Trichloroethylene-mediated cytotoxicity in human epidermal keratinocytes is mediated by the rapid accumulation of intracellular calcium: Interception by naringenin. *Human & Experimental Toxicology, 35*(2), 147–161.

Allawadhi, P., Khurana, A., Allwadhi, S., Navik, U. S., Joshi, K., Banothu, A. K., & Bharani, K. K. (2020). Potential of electric stimulation for the management of COVID-19. *Medical Hypotheses, 144*, 110259.

Allawadhi, P., Singh, V., Khurana, I., Rawat, P. S., Renushe, A. P., Khurana, A., Navik, U., Allwadhi, S., Karlapudi, S. K., Banothu, A. K., & Bharani, K. K. (2021). Decorin as a possible strategy for the amelioration of COVID-19. *Medical Hypotheses, 152*, 110612.

Althobiti, M. M., Alzahrani, B., Elderdery, A. Y., Alzerwi, N. A., Rayzah, M., Elkhalifa, A. M., et al. (2023). In vitro anti-cancer and antimicrobial effects of manganese oxide nanoparticles synthesized using the Glycyrrhiza uralensis leaf extract on breast cancer cell lines. *Green Processing and Synthesis, 12*(1), 20230063.

Belizário, J. E., Faintuch, J., & Garay-Malpartida, M. (2018). Gut microbiome dysbiosis and immunometabolism: New frontiers for treatment of metabolic diseases. *Mediators of Inflammation, 2018, 1–12.*

Bertotto, L. B., Catron, T. R., & Tal, T. (2020). Exploring interactions between xenobiotics, microbiota, and neurotoxicity in zebrafish. *Neurotoxicology, 76*, 235–244.

Boada-Romero, E., Martinez, J., Heckmann, B. L., & Green, D. R. (2020). The clearance of dead cells by efferocytosis. *Nature Reviews Molecular Cell Biology, 21*(7), 398–414.

Cano, M., Datta, S., Wang, L., Liu, T., Flores-Bellver, M., Sachdeva, M., Sinha, D., & Handa, J. T. (2021). Nrf2 deficiency decreases NADPH from impaired IDH shuttle and pentose phosphate pathway in retinal pigmented epithelial cells to magnify oxidative stress-induced mitochondrial dysfunction. *Aging Cell, 20*(8), e13444.

Chantzoura, E., & Kaji, K. (2017). Flow cytometry. In *Basic science methods for clinical researchers* (pp. 173–189). Academic.

Chen, P., Li, Y., Zhou, Z., Pan, C., & Zeng, L. (2023). Lathyrol promotes ER stress-induced apoptosis and proliferation inhibition in lung cancer cells by targeting SERCA2. *Biomedicine & Pharmacotherapy, 158*, 114123.

Chicote, J., Yuste, V. J., Boix, J., & Ribas, J. (2020). Cell death triggered by the autophagy inhibitory drug 3-methyladenine in growing conditions proceeds with DNA damage. *Frontiers in Pharmacology, 11*, 580343.

Cossarizza, A., Chang, H. D., Radbruch, A., Acs, A., Adam, D., Adam-Klages, S., et al. (2019). Guidelines for the use of flow cytometry and cell sorting in immunological studies (second edition). *European Journal of Immunology, 49*(10), 1457–1973. https://doi.org/10.1002/eji.201970107

Czabotar, P. E., Lessene, G., Strasser, A., & Adams, J. M. (2014). Control of apoptosis by the BCL-2 protein family: Implications for physiology and therapy. *Nature Reviews Molecular Cell Biology, 15*(1), 49–63.

D'arcy, M. S. (2019). Cell death: A review of the major forms of apoptosis, necrosis and autophagy. *Cell Biology International, 43*(6), 582–592.

de Oliveira, M., Frihling, B. E. F., Velasques, J., Magalhães Filho, F. J. C., Cavalheri, P. S., & Migliolo, L. (2020). Pharmaceuticals residues and xenobiotics contaminants: Occurrence, analytical techniques and sustainable alternatives for wastewater treatment. *Science of the Total Environment, 705*, 135568.

Delgado-Waldo, I., Contreras-Romero, C., Salazar-Aguilar, S., Pessoa, J., Mitre-Aguilar, I., García-Castillo, V., & Jacobo-Herrera, N. J. (2023). A triple-drug combination induces apoptosis in cervical cancer-derived cell lines. *Frontiers in Oncology, 13*, 1106667.

Doan, P., Musa, A., Murugesan, A., Sipilä, V., Candeias, N. R., Emmert-Streib, F., Ruusuvuori, P., Granberg, K., Yli-Harja, O., & Kandhavelu, M. (2020). Glioblastoma multiforme stem cell cycle arrest by alkylaminophenol through the modulation of EGFR and CSC signaling pathways. *Cells, 9*(3), 681.

Duarte-Hospital, C., Tête, A., Brial, F., Benoit, L., Koual, M., Tomkiewicz, C., Kim, M. J., Blanc, E. B., Coumoul, X., & Bortoli, S. (2021). Mitochondrial dysfunction as a hallmark of environmental injury. *Cells, 11*(1), 110.

Fallico, V., Rea, M., Stanton, C., Ilestam, N., & McKinney, J. (2020). Next-generation multiparameter flow cytometry assay improves the assessment of oxidative stress in probiotics. *Food Microbiology, 91*, 103501.

Fleisher, T. A., & Oliveira, J. B. (2019). Flow cytometry. In *Clinical immunology* (pp. 1239–1251). Elsevier.

Galluzzi, L., Baehrecke, E. H., Ballabio, A., Boya, P., Bravo-San Pedro, J. M., Cecconi, F., Choi, A. M., Chu, C. T., Codogno, P., Colombo, M. I., Cuervo, A. M., & Kroemer, G. (2017). Molecular definitions of autophagy and related processes. *The EMBO Journal, 36*(13), 1811–1836.

García-Castillo, V., López-Urrutia, E., Villanueva-Sánchez, O., Ávila-Rodríguez, M. Á., Zentella-Dehesa, A., Cortés-González, C., López-Camarillo, C., Jacobo-Herrera, N. J., & Pérez-Plasencia, C. (2017). Targeting metabolic remodeling in triple negative breast cancer in a murine model. *Journal of Cancer, 8*(2), 178.

Guan, S., Tao, S., Huang, Y., Jin, Y., Hu, Y., & Lu, J. (2020). Combined toxic effects of CBNPs and Pb on rat alveolar macrophage apoptosis and autophagy flux. *Ecotoxicology and Environmental Safety, 205*, 111062.

Guo, S., Liang, Y., Murphy, S. F., Huang, A., Shen, H., Kelly, D. F., Sobrado, P., & Sheng, Z. (2015). A rapid and high content assay that measures cyto-ID-stained autophagic compartments and estimates autophagy flux with potential clinical applications. *Autophagy, 11*(3), 560–572.

Guo, H., Cui, H., Fang, J., Zuo, Z., Deng, J., Wang, X., Zhao, L., Wu, B., Chen, K., & Deng, J. (2016). Nickel chloride-induced apoptosis via mitochondria-and Fas-mediated caspase-dependent pathways in broiler chickens. *Oncotarget, 7*(48), 79747–79760.

Guo, M., Lu, B., Gan, J., Wang, S., Jiang, X., & Li, H. (2021). Apoptosis detection: A purpose-dependent approach selection. *Cell Cycle, 20*(11), 1033–1040.

Hargarten, J. C., Hu, G., Elsegeiny, W., & Williamson, P. R. (2023). Measurement of SQSTM1 by flow cytometry. *Autophagy*, 1–11.

Harnett, M. M., Pineda, M. A., de Laté, P. L., Eason, R. J., Besteiro, S., Harnett, W., & Langsley, G. (2017). From Christian de Duve to Yoshinori Ohsumi: More to autophagy than just dining at home. *Biomedical Journal, 40*(1), 9–22.

Hu, X. M., Li, Z. X., Lin, R. H., Shan, J. Q., Yu, Q. W., Wang, R. X., Liao, L. S., Yan, W. T., Wang, Z., Shang, L., & Xiong, K. (2021). Guidelines for regulated cell death assays: A systematic summary, a categorical comparison, a prospective. *Frontiers in Cell and Developmental Biology, 9*, 634690.

Huang, Y., Wen, Q., Huang, J., Luo, M., Xiao, Y., Mo, R., & Wang, J. (2021). Manganese (II) chloride leads to dopaminergic neurotoxicity by promoting mitophagy through BNIP3-mediated oxidative stress in SH-SY5Y cells. *Cellular & Molecular Biology Letters, 26*(1), 23.

Ichimiya, T., Yamakawa, T., Hirano, T., Yokoyama, Y., Hayashi, Y., Hirayama, D., Wagatsuma, K., Itoi, T., & Nakase, H. (2020). Autophagy and autophagy-related diseases: A review. *International Journal of Molecular Sciences, 21*(23), 8974.

Inaba, N., Sato, N., Ijichi, M., Fukazawa, I., Nito, A., Takamizawa, H., Lüben, G., & Bohn, H. (1984). The immunocytochemical location of two membrane-associated placental tissue proteins in human and cynomolgus monkey placentae. *Tumour Biology: The Journal of the International Society for Oncodevelopmental Biology and Medicine, 5*(2), 75–85.

Iram, S., Khan, S., Ansary, A. A., Arshad, M., Siddiqui, S., Ahmad, E., Khan, R. H., & Khan, M. S. (2016). Biogenic terbium oxide nanoparticles as the vanguard against osteosarcoma. *Spectrochimica Acta Part A: Molecular and Biomolecular Spectroscopy, 168*, 123–131.

Jan, R. (2019). Understanding apoptosis and apoptotic pathways targeted cancer therapeutics. *Advanced Pharmaceutical Bulletin, 9*(2), 205.

Jarvis, T. S., Roland, F. M., Dubiak, K. M., Huber, P. W., & Smith, B. D. (2018). Time-lapse imaging of cell death in cell culture and whole living organisms using turn-on deep-red fluorescent probes. *Journal of Materials Chemistry B, 6*, 4963–4971.

Jávega, B., Herrera, G., & O'Connor, J. E. (2022). Flow cytometric analysis of oxidative stress in Escherichia coli B strains deficient in genes of the antioxidant defence. *International Journal of Molecular Sciences, 23*(12), 6537.

Jih-Jung, C., Ming-Jen, C., Lee, T. H., Kuo, Y. H., & Chao-Tsen, L. (2022). Correction: Chen et al. Secondary metabolites with anti-inflammatory from the roots of Cimicifuga taiwanensis. *Molecules, 27*(18), 6035.

Jorgensen, I., Rayamajhi, M., & Miao, E. A. (2017). Programmed cell death as a defence against infection. *Nature Reviews Immunology, 17*(3), 151–164.

Joseph, P. (2017). Transcriptomics in toxicology. *Food and Chemical Toxicology, 109*, 650–662.

Kaizuka, T., Morishita, H., Hama, Y., Tsukamoto, S., Matsui, T., Toyota, Y., Kodama, A., Ishihara, T., Mizushima, T., & Mizushima, N. (2016). An autophagic flux probe that releases an internal control. *Molecular Cell, 64*(4), 835–849.

Kaminskyy, V. O. (2022). A quantitative flow cytometry–based method for autophagy detection across the cell cycle. *Autophagy and Cancer: Methods and Protocols*, 65–74.

Kara, M., Oztas, E., Ramazanoğulları, R., Kouretas, D., Nepka, C., Tsatsakis, A. M., & Veskoukis, A. S. (2020). Benomyl, a benzimidazole fungicide, induces oxidative stress and apoptosis in neural cells. *Toxicology Reports, 7*, 501–509.

Kari, S., Subramanian, K., Altomonte, I. A., Murugesan, A., Yli-Harja, O., & Kandhavelu, M. (2022). Programmed cell death detection methods: A systematic review and a categorical comparison. *Apoptosis, 27*(7), 482–508.

Karri, V., Kumar, V., Ramos, D., Oliveira, E., & Schuhmacher, M. (2018). Comparative in vitro toxicity evaluation of heavy metals (lead, cadmium, arsenic, and methylmercury) on HT-22 hippocampal cell line. *Biological Trace Element Research, 184*, 226–239.

Katalay, S., Guner, A., Dagdeviren, M., Yigitturk, G., Yavasoglu, A., Gunal, A. C., Karabay Yavasoglu, N. U., & Oltulu, F. (2022). Oxidative stress-induced apoptotic changes after acute exposure to antifouling agent zinc pyrithione (ZnPT) in Mytilus galloprovincialis Lamark (Mediterranean mussels) tissues. *Chemistry and Ecology, 38*(4), 356–373.

Kawaratani, Y., Matsuoka, T., Hirata, Y., Fukata, N., Nagaoka, Y., & Uesato, S. (2015). Influence of the carbamate fungicide benomyl on the gene expression and activity of aromatase in the human breast carcinoma cell line MCF-7. *Environmental Toxicology and Pharmacology, 39*(1), 292–299.

Khanal, G., Chung, K., Solis-Wever, X., Johnson, B., & Pappas, D. (2011). Ischemia/reperfusion injury of primary porcine cardiomyocytes in a low-shear microfluidic culture and analysis device. *Analyst, 136*(17), 3519–3526.

Khurana, I., Allawadhi, P., Khurana, A., Srivastava, A. K., Navik, U., Banothu, A. K., & Bharani, K. K. (2021). Can bilirubin nanomedicine become a hope for the management of COVID-19? *Medical Hypotheses, 149*, 110534.

Kirtonia, A., Gala, K., Fernandes, S. G., Pandya, G., Pandey, A. K., Sethi, G., Khattar, E., & Garg, M. (2021). Repurposing of drugs: An attractive pharmacological strategy for cancer therapeutics. In *Seminars in cancer biology* (Vol. 68, pp. 258–278). Academic.

Klionsky, D. J., Petroni, G., Amaravadi, R. K., Baehrecke, E. H., Ballabio, A., Boya, P., & Pietrocola, F. (2021). Autophagy in major human diseases. *The EMBO Journal, 40*(19), e108863.

Koepke, L., Winter, B., Grenzner, A., Regensburger, K., Engelhart, S., van der Merwe, J. A., Krebs, S., Blum, H., Kirchhoff, F., & Sparrer, K. M. (2020). An improved method for high-throughput quantification of autophagy in mammalian cells. *Scientific Reports, 10*(1), 12241.

Kumar, A., Kumar, A., Cabral-Pinto, M. M. S., Chaturvedi, A. K., Shabnam, A. A., Subrahmanyam, G., Mondal, R., Gupta, D. K., Malyan, S. K., Kumar, S. S., et al. (2020). Lead toxicity: Health hazards, influence on food chain, and sustainable remediation approaches. *International Journal of Environmental Research and Public Health, 17*(7), 2179.

Levine, B., & Kroemer, G. (2019). Biological functions of autophagy genes: A disease perspective. *Cell, 176*(1), 11–42.

Li, J. R., Ou, Y. C., Wu, C. C., Wang, J. D., Lin, S. Y., Wang, Y. Y., Chen, W. Y., Liao, S. L., & Chen, C. J. (2020). Endoplasmic reticulum stress and autophagy contributed to cadmium nephrotoxicity in HK-2 cells and Sprague-Dawley rats. *Food and Chemical Toxicology, 146*, 111828.

Liu, S., Yao, S., Yang, H., Liu, S., & Wang, Y. (2023). Autophagy: Regulator of cell death. *Cell Death & Disease, 14*(10), 648.

Lockshin, R. A. (1969). Programmed cell death. Activation of lysis by a mechanism involving the synthesis of protein. *Journal of Insect Physiology, 15*(9), 1505–1516.

López-Méndez, T. B., Sánchez-Álvarez, M., Trionfetti, F., Pedraz, J. L., Tripodi, M., Cordani, M., & González-Valdivieso, J. (2023). Nanomedicine for autophagy modulation in cancer therapy: A clinical perspective. *Cell & Bioscience, 13*(1), 1–17.

Loubaki, L., Rouabhia, M., Zahrani, M. A., Amri, A. A., & Semlali, A. (2022). Oxidative stress and autophagy mediate anti-cancer properties of cannabis derivatives in human oral cancer cells. *Cancers, 14*(19), 4924.

Mariotti, S., Pardini, M., Teloni, R., Gagliardi, M. C., Fraziano, M., & Nisini, R. (2017). A method permissive to fixation and permeabilization for the multiparametric analysis of apoptotic and necrotic cell phenotype by flow cytometry. *Cytometry Part A, 91*(11), 1115–1124.

Maurya, P. K., & Malik, D. S. (2016). Bioaccumulation of xenobiotics compound of pesticides in riverine system and its control technique: A critical review. *Journal of Industrial Pollution Control, 32*(2), 580–594.

Mauthe, M., Orhon, I., Rocchi, C., Zhou, X., Luhr, M., Hijlkema, K. J., Coppes, R. P., Engedal, N., Mari, M., & Reggiori, F. (2018). Chloroquine inhibits autophagic flux by decreasing autophagosome-lysosome fusion. *Autophagy, 14*(8), 1435–1455.

McKinnon, K. M. (2018). Flow cytometry: An overview. *Current Protocols in Immunology, 120*(1), 5–1.

Mehtap, K., Ezgi, Ö., Tugce, B., Fatma, K. E., & Gul, O. (2021). Benomyl induced oxidative stress related DNA damage and apoptosis in H9c2 cardiomyoblast cells. *Toxicology In Vitro, 75*, 105180.

Mishra, V. K., Singh, G., & Shukla, R. (2019). Impact of xenobiotics under a changing climate scenario. In *Climate change and agricultural ecosystems* (pp. 133–151). Woodhead Publishing.

Mishra, S., Lin, Z., Pang, S., Zhang, W., Bhatt, P., & Chen, S. (2021). Recent advanced technologies for the characterization of xenobiotic-degrading microorganisms and microbial communities. *Frontiers in Bioengineering and Biotechnology, 9*, 632059.

Mohamed, M. A., Elkhateeb, W. A., & Daba, G. M. (2022). Rapamycin golden jubilee and still the miraculous drug: A potent immunosuppressant, antitumor, rejuvenative agent, and potential contributor in COVID-19 treatment. *Bioresources and Bioprocessing, 9*(1), 65.

Morozesk, M., da Costa Souza, I., Fernandes, M. N., & Soares, D. C. F. (2021). Airborne particulate matter in an iron mining city: Characterization, cell uptake and cytotoxicity effects of nanoparticles from PM2. 5, PM10 and PM20 on human lung cells. *Environmental Advances, 6*, 100125.

Murugan, S., & Amaravadi, R. K. (2016). Methods for studying autophagy within the tumor microenvironment. *Tumor Microenvironment: Study Protocols*, 145–166.

Nagaraja, S. K., Kumar, R. S., Chakraborty, B., Hiremath, H., Almansour, A. I., Perumal, K., Gunagambhire, P. V., & Nayaka, S. (2023). Biomimetic synthesis of silver nanoparticles using Cucumis sativus var. hardwickii fruit extract and their characterizations, anticancer potential and apoptosis studies against Pa-1 (Human ovarian teratocarcinoma) cell line via flow cytometry. *Applied Nanoscience, 13*(4), 3073–3084.

Navik, U., Sheth, V. G., Khurana, A., Jawalekar, S. S., Allawadhi, P., Gaddam, R. R., Bhatti, J. S., & Tikoo, K. (2021). Methionine as a double-edged sword in health and disease: Current perspective and future challenges. *Ageing Research Reviews, 72*, 101500.

Nguyen, P., Doan, P., Murugesan, A., Ramesh, T., Rimpilainen, T., Candeias, N. R., Yli-Harja, & Kandhavelu, M. (2022). GPR17 signaling activation by CHBC agonist induced cell death via modulation of MAPK pathway in glioblastoma. *Life Sciences, 291*, 120307.

Niu, G., Yousefi, B., Qujeq, D., Marjani, A., Asadi, J., Wang, Z., & Mir, S. M. (2021). Melatonin and doxorubicin co-delivered via a functionalized graphene-dendrimeric system enhances apoptosis of osteosarcoma cells. *Materials Science and Engineering: C, 119*, 111554.

Obeng, E. (2020). Apoptosis (programmed cell death) and its signals – A review. *Brazilian Journal of Biology, 81*, 1133–1143.

Olszowski, T., Baranowska-Bosiacka, I., Gutowska, I., Piotrowska, K., Mierzejewska, K., Korbecki, J., Kurzawski, M., Tarnowski, M., & Chlubek, D. (2015). The effects of cadmium at low environmental concentrations on THP-1 macrophage apoptosis. *International Journal of Molecular Sciences, 16*(9), 21410–21427.

Ornatowski, W., Lu, Q., Yegambaram, M., Garcia, A. E., Zemskov, E. A., Maltepe, E., Fineman, J. R., Wang, T., & Black, S. M. (2020). Complex interplay between autophagy and oxidative stress in the development of pulmonary disease. *Redox Biology, 36*, 101679.

Ortolani, C. (2022). *Flow cytometry today: Everything you need to know about flow cytometry*. Springer.

Pan, D., Zheng, M., Liu, J., Sun, Z., & Shi, X. (2023). Garlic extract participates in the proliferation and apoptosis of nonsmall cell lung cancer cells via endoplasmic reticulum stress pathway. *Evidence-Based Complementary and Alternative Medicine*.

Patterson, A. D., Gonzalez, F. J., & Idle, J. R. (2010). Xenobiotic metabolism: A view through the metabolometer. *Chemical Research in Toxicology, 23*(5), 851–860.

Perez-Abshana, L. P., Mendivil-Perez, M., Velez-Pardo, C., & Jimenez-Del-Rio, M. (2023). Rotenone blocks the glucocerebrosidase enzyme and induces the accumulation of lysosomes and autophagolysosomes independently of LRRK2 kinase in HEK-293 cells. *International Journal of Molecular Sciences, 24*(13), 10589.

Phadwal, K., Alegre-Abarrategui, J., Watson, A. S., Pike, L., Anbalagan, S., Hammond, E. M., et al. (2012). A novel method for autophagy detection in primary cells: Impaired levels of macroautophagy in immunosenescent T cells. *Autophagy, 8*(4), 677–689.

Pugsley, H. R. (2017). Quantifying autophagy: Measuring LC3 puncta and autolysosome formation in cells using multispectral imaging flow cytometry. *Methods (San Diego, CA), 112*, 147–156.

Pukanha, K., Yimthiang, S., & Kwanhian, W. (2020). The immunotoxicity of chronic exposure to high levels of lead: An ex vivo investigation. *Toxics, 8*(3), 56.

Quintero-Espinosa, D. A., Sanchez-Hernandez, S., Velez-Pardo, C., Martin, F., & Jimenez-Del-Rio, M. (2023). LRRK2 knockout confers resistance in HEK-293 cells to rotenone-induced oxidative stress, mitochondrial damage, and apoptosis. *International Journal of Molecular Sciences, 24*(13), 10474.

Rabinovich-Nikitin, I., Rasouli, M., Reitz, C. J., Posen, I., Margulets, V., Dhingra, R., Khatua, T. N., Thliveris, J. A., Martino, T. A., & Kirshenbaum, L. A. (2021). Mitochondrial autophagy and cell survival is regulated by the circadian Clock gene in cardiac myocytes during ischemic stress. *Autophagy, 17*(11), 3794–3812.

Raudenska, M., Balvan, J., & Masarik, M. (2021). Crosstalk between autophagy inhibitors and endosome-related secretory pathways: A challenge for autophagy-based treatment of solid cancers. *Molecular Cancer, 20*, 1–27.

Reutelingsperger, C. P., Hornstra, G., & Hemker, H. C. (1985). Isolation and partial purification of a novel anticoagulant from arteries of human umbilical cord. *European Journal of Biochemistry, 151*(3), 625–629.

Ravanan, P., Srikumar, I. F., & Talwar, P. (2017). Autophagy: The spotlight for cellular stress responses. *Life Sciences, 188*, 53–67. https://doi.org/10.1016/j.lfs.2017.08.029

Robinson, J. P., Li, N., & Narayanan, P. K. (2015). High throughput-based mitochondrial function assays by multi-parametric flow cytometry. *Current Protocols in Cytometry, 73*, 9.48.1–9.48.9.

Robinson, J. P. (2019). Spectral flow cytometry-Quo vadimus? *Cytometry. Part A, 95*(8), 823–824. https://doi.org/10.1002/cyto.a.23779

Robinson, J. P., Ostafe, R., Iyengar, S. N., Rajwa, B., & Fischer, R. (2023). Flow cytometry: The next revolution. *Cells, 12*(14), 1875.

Rodríguez, A., Castrejón-Godínez, M. L., Salazar-Bustamante, E., Gama-Martínez, Y., Sánchez-Salinas, E., Mussali-Galante, P., Tovar-Sánchez, E., & Ortiz-Hernández, M. L. (2020). Omics approaches to pesticide biodegradation. *Current Microbiology, 77*, 545–563.

Rooprai, H. K., Lawrence, P., Keshavarz, S., Yashod, P., Gullan, R. W., Selway, R. P., & Davies, D. (2020). Draq7 as an alternative to mtt assay for measuring viability of glioma cells treated with polyphenols. *Anticancer Research, 40*(10), 5427–5436.

Salama, M. M., El-Naggar, D. A., Abdel-Rahman, R. H., & Elhak, S. A. (2018). Toxic effects of trichloroethylene on rat neuroprogenitor cells. *Frontiers in Pharmacology, 9*, 741.

Samuvel, D. J., Li, L., Krishnasamy, Y., Gooz, M., Takemoto, K., Woster, P. M., Lemasters, J. J., & Zhong, Z. (2022). Mitochondrial depolarization after acute ethanol treatment drives mitophagy in living mice. *Autophagy, 18*(11), 2671–2685.

Schmit, T., Klomp, M., & Khan, M. N. (2021). An overview of flow cytometry: Its principles and applications in allergic disease research. *Animal Models of Allergic Disease: Methods and Protocols*, 169–182.

Shashiraj, K. N., Hugar, A., Kumar, R. S., Rudrappa, M., Bhat, M. P., Almansour, A. I., et al. (2023). Exploring the antimicrobial, anticancer, and apoptosis inducing ability of biofabricated silver nanoparticles using Lagerstroemia speciosa flower buds against the Human Osteosarcoma (MG-63) cell line via flow cytometry. *Bioengineering, 10*(7), 821.

Sheehan, H. C., Tilly, J. L., & Woods, D. C. (2023). Assaying mitochondrial function by multiparametric flow cytometry. In *Cell viability assays: Methods and protocols* (pp. 65–80). Springer.

Sica, V., Maiuri, M. C., Kroemer, G., & Galluzzi, L. (2016). Detection of apoptotic versus autophagic cell death by flow cytometry. *Programmed Cell Death: Methods and Protocols*, 1–16.

Silva, A. M., Martins-Gomes, C., Ferreira, S. S., Souto, E. B., & Andreani, T. (2022). Molecular physicochemical properties of selected pesticides as predictive factors for oxidative stress and apoptosis-dependent cell death in caco-2 and HepG2 cells. *International Journal of Molecular Sciences, 23*(15), 8107.

Singh, V., Khurana, A., Navik, U., Allawadhi, P., Bharani, K. K., & Weiskirchen, R. (2022). Apoptosis and pharmacological therapies for targeting thereof for cancer therapeutics. *Sci, 4*(2), 15.

Sorice, M. (2022). Crosstalk of autophagy and apoptosis. *Cells, 11*(9), 1479.

Srivastava, A., Singh, S., Rajpurohit, C. S., Srivastava, P., Pandey, A., Kumar, D., & Pant, A. B. (2018). Secretome of differentiated PC12 cells restores the monocrotophos-induced damages in human mesenchymal stem cells and SHSY-5Y cells: Role of autophagy and mitochondrial dynamics. *Neuromolecular Medicine, 20*, 233–251.

Sun, G. (2023). Death and survival from executioner caspase activation. In *Seminars in cell & developmental biology*. Academic.

Tavakol, S., Ashrafizadeh, M., Deng, S., Azarian, M., Abdoli, A., Motavaf, M., Poormoghadam, D., Khanbabaei, H., Ghasemipour Afshar, E., Mandegary, A., & Pardakhty, A. (2019). Autophagy modulators: Mechanistic aspects and drug delivery systems. *Biomolecules, 9*(10), 530.

Tong, J., Rufli, S., & Wong, W. W. L. (2023). Measuring caspase activity using a fluorometric assay or flow cytometry. *JoVE (Journal of Visualized Experiments), 193*, e64745.

Tracy, T. S., Chaudhry, A. S., Prasad, B., Thummel, K. E., Schuetz, E. G., Zhong, X. B., et al. (2016). Interindividual variability in cytochrome P450–mediated drug metabolism. *Drug Metabolism and Disposition, 44*(3), 343–351.

Tsuchiya, K., Nakajima, S., Hosojima, S., Thi Nguyen, D., Hattori, T., Manh Le, T., Hori, O., Mahib, M. R., Yamaguchi, Y., Miura, M., Kinoshita, T., & Suda, T. (2019). Caspase-1 initiates apoptosis in the absence of gasdermin D. *Nature Communications, 10*(1), 2091.

Van Opdenbosch, N., & Lamkanfi, M. (2019). Caspases in cell death, inflammation, and disease. *Immunity, 50*(6), 1352–1364.

Vidhyapriya, P., Divya, D., Bala, M., & Sakthivel, N. (2018). Photoactivated [Mn (CO) 3Br (μ-bpcpd)] 2 induces apoptosis in cancer cells via intrinsic pathway. *Journal of Photochemistry and Photobiology B: Biology, 188*, 28–41.

Wang, J., Huang, J., Wang, L., Chen, C., Yang, D., Jin, M., Bai, C., & Song, Y. (2017). Urban particulate matter triggers lung inflammation via the ROS-MAPK-NF-κB signaling pathway. *Journal of Thoracic Disease, 9*(11), 4398–4412.

Wang, W., Liu, T., Liu, Y., Yu, L., Yan, X., Weng, W., Lu, X., & Zhang, C. (2021a). Astaxanthin attenuates alcoholic cardiomyopathy via inhibition of endoplasmic reticulum stress-mediated cardiac apoptosis. *Toxicology and Applied Pharmacology, 412*, 115378.

Wang, Z., Wu, Q., Li, C., Sun, S., Li, Z., & Wu, J. (2021b). Quantitative determination of autophagy flux by probes. In *Methods in cell biology* (Vol. 164, pp. 157–165). Academic.

Warnes, G. (2016). The multiplexing of assays for the measurement of early stages of apoptosis by polychromatic flow cytometry. In *Flow cytometry select topics. On-line* (pp. 85–99). InTech Open.

Wei, C. C., Luo, Z., Hogstrand, C., Xu, Y. H., Wu, L. X., Chen, G. H., Pan, Y. X., & Song, Y. F. (2018). Zinc reduces hepatic lipid deposition and activates lipophagy via Zn2+/MTF-1/PPARα and Ca2+/CaMKKβ/AMPK pathways. *The FASEB Journal, 32*(12), 6666–6680.

Will, Y., & Dykens, J. A. (Eds.). (2018). *Mitochondrial dysfunction caused by drugs and environmental toxicants*. Wiley.

Xiao, Y. L., Wang, X. R., Ye, K., Chen, W. Z., Zheng, B. R., Huang, Y. H., & Wu, Y. Y. (2022). Autophagy mediates MMP-2 expression in glaucomatous trabecular meshwork cells. *Journal of Ophthalmology, 2022, 1–9.*

Xu, R., Ji, Z., Xu, C., & Zhu, J. (2018). The clinical value of using chloroquine or hydroxychloroquine as autophagy inhibitors in the treatment of cancers: A systematic review and meta-analysis. *Medicine, 97*(46), e12912.

Yaffe, P. B., Doucette, C. D., Walsh, M., & Hoskin, D. W. (2013). Piperine impairs cell cycle progression and causes reactive oxygen species-dependent apoptosis in rectal cancer cells. *Experimental and Molecular Pathology, 94*(1), 109–114.

Yamamoto, H., Zhang, S., & Mizushima, N. (2023). Autophagy genes in biology and disease. *Nature Reviews. Genetics, 24*, 382–400. https://doi.org/10.1038/s41576-022-00562-w

Yin, Y., & Shen, H. (2022). Common methods in mitochondrial research. *International Journal of Molecular Medicine, 50*(4), 1–29.

Yuan, Y., Zhang, Y., Zhao, S., Chen, J., Yang, J., Wang, T., Zou, H., Wang, Y., Gu, J., Liu, X., & Bian, J. (2018). Cadmium-induced apoptosis in neuronal cells is mediated by Fas/FasL-mediated mitochondrial apoptotic signaling pathway. *Scientific Reports, 8*(1), 8837.

Zappavigna, S., Lombardi, A., Misso, G., Grimaldi, A., & Caraglia, M. (2017). Measurement of autophagy by flow cytometry. *Adult Stem Cells: Methods and Protocols*, 209–216.

Zhang, L., Zhao, Y., Ding, W., Jiang, G., Lu, Z., Li, L., Wang, J., Li, J., & Li, J. (2015). Autophagy regulates colistin-induced apoptosis in PC-12 cells. *Antimicrobial Agents and Chemotherapy, 59*(4), 2189–2197.

Zhang, L., Zhao, J., Ding, W. X., & Xia, M. (2022). GFP-LC3 high-content assay for screening autophagy modulators. In *High-throughput screening assays in toxicology* (pp. 83–89). Springer.

Role of Flow Cytometry in the Rapid Assessment of Organ-Specific Toxicity

Jahnavy Madhukar Joshi, Shweta Verma, Raghavendra Upadhya, Raviraja Neelavar Seetharam, and Abhishek Kumar Singh

Abstract

Flow cytometry has been an integral part of toxicity testing in biological models. The simultaneous assessment of cellular parameters using fluorescent dyes aids in the detection of responses to various endobiotic or xenobiotic toxicants at cell, tissue, and organ levels. The effects of toxicants are monitored by the decrease in cell viability, inflammation, expression of receptors on the cell surface, and cell death, and these parameters are analyzed for a comprehensive understanding of the toxicological impacts. Given its wide range of applications, flow cytometry has been extensively utilized in *in vitro*, pre-clinical, and clinical toxicity testing. The high-throughput analysis of samples makes it an essential technique for developing new and improved toxicity testing methodologies, aiding in toxicological and pharmacological advancements.

Keywords

Flow cytometry · Toxicity · Apoptosis · Immunophenotyping

1 Introduction

The necessity to explore microscopic organisms has led to the development of flow cytometry, a technique that aided scientists in accessing cells. This advanced technique provides in-depth information about cell constituents and characteristics, enabling researchers to gain a more thorough understanding of cellular processes. A

J. M. Joshi · S. Verma · R. Upadhya · R. N. Seetharam · A. K. Singh (✉)
Manipal Centre for Biotherapeutics Research, Manipal Academy of Higher Education, Manipal, Karnataka, India
e-mail: abhishek.singh@manipal.edu

AB Pant et al. (eds.), *Flow Cytometry: Applications in Cellular and Molecular Toxicology*, https://doi.org/10.1007/978-981-97-9758-5_5

flow cytometer was first developed by Moldavan in 1934 that focused the cells through a capillary tube and analyzed via a simple microscope with an eyepiece connected to a photoelectric device (Moldavan, 1934). This setup was later modified by Crosland-Taylor using a fluid called sheath fluid and a capillary with a larger diameter (Crosland-Taylor, 1953). This led to the development of hydrodynamic focusing, which is present in all modern flow cytometers. The first patented device for analyzing blood cells was introduced by Coulter in 1953 (Coulter, 1953). Subsequently, in 1959, a group of scientists outlined a protocol employing the dyes to stain the whole blood cells flowing through a glass channel with a single suspension, focusing light on one cell at a time, and measuring the transmitted light using the photomultiplier tube (Parker & Horst, 1959). In recent times, the use of flow cytometers has expanded into various fields. One significant application involves assessing toxicity at both cellular and organ levels (Fig. 1).

In the context of drug toxicity testing, the primary goal of the Centre of Drug Evaluation and Research at the FDA is to review and approve safe and effective drugs and certain biological agents, and some clinical trials evaluate these agents for specific symptoms over short periods and with a limited number and category of participants. Once the drug is approved by a larger population for a certain period, an unidentified adverse reaction may be seen (Wysowski & Swartz, 2005). According to the Adverse Event Recording System of the FDA, between 1969 to 2002, 2.3 million adverse drug effect reports were submitted for 6000 registered compounds (Wysowski & Swartz, 2005). During this time, 11 drugs or products were given

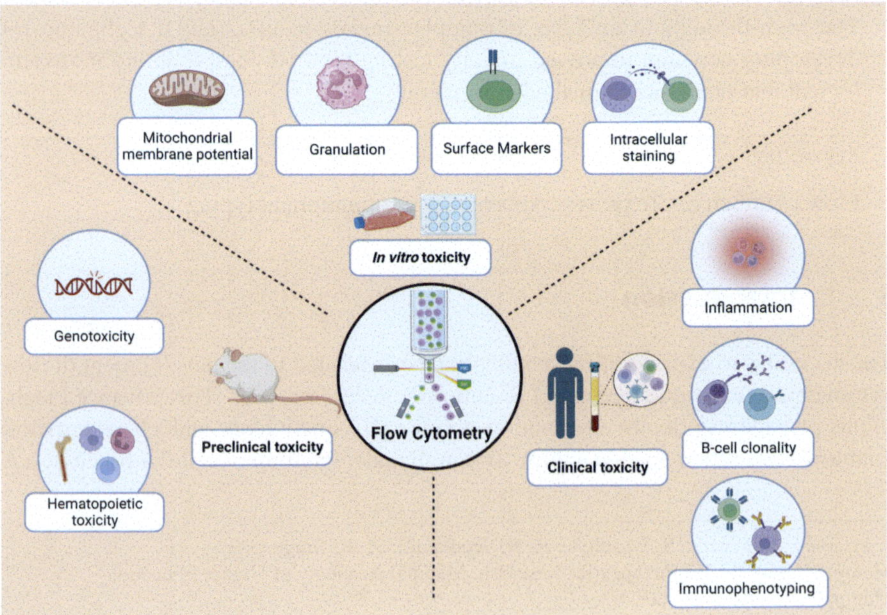

Fig. 1 Graphical abstract depicting the various applications of a flow cytometer in assessing *in vitro*, preclinical, and clinical toxicity

special instructions or restrictions, and 75 drugs/products were removed from the market (Wysowski & Swartz, 2005). Therefore, the flow cytometer emerges as an invaluable instrument for the multi-dimensional analysis of the diverse impacts of various toxicants both *in vitro* and *in vivo* settings.

2 Different Types of Toxicity

The study of the toxicity of a substance with known and unknown toxicity is necessary to assess the risk potential. Before initiating the study, the protocol must be prepared in accordance with national and international guidelines provided by the Organization for Economic Cooperation and Development (OECD), World Health Organization (WHO), and European Union (EU), which defines the unit of investigation and conditions for investigation. The main objective of following the guidelines is to reduce errors and enhance the transparency of procedures and results by following good laboratory practice (GLP). A toxicity study is necessary to establish a dose response curve, to ensure the safety of any new chemical entities to be used before their registration in their respective field. Also, toxicity studies are important to study the mechanism of toxic effects, epidemiological studies on the population, and to validate methods for investigation (Chinedu et al., 2013).

Acute Toxicity Acute toxicity refers to the impact of a substance resulting from single or multiple exposures over a brief period. The aim of assessing acute toxicity is to compile the action profile of substances in high doses within 24 hours of their administration or through multiple exposures within 14 days of initial exposures (Heimann & Doughty, 2014). Various routes of administration, such as oral, dermal, and nasal routes, are followed. LD50, or the dose that causes the death of 50% of the population, is a measure used to describe acute toxicity and serves as a guideline for determining doses or planning other testing studies. Therefore, acute toxicity testing holds significance in identifying target organ toxicity, formulating guidelines for populations at risk of exposure, selecting doses for prolonged studies, and investigating any adverse effects on humans, animals, or the environment.

The evaluation of acute toxicity is done through 4 distinct methods. The first method is the graphical method developed by Miller and Tainter, also known as the classical LD_{50} test, which originated in 1920. It involves administering the same volumes of different test substances to various groups, with one group receiving only a vehicle while other groups are treated with test substances dissolved in the vehicle. The assessment involves plotting log doses against the percentage of prohibitions on a graph. The second method is the arithmetical approach introduced by Reed and Muench in 1938, which calculates the percentage of survival and LD_{50} based on the count of live and deceased individuals. The third method is Karber's technique, introduced in 1931, which involves administering a single vehicle in one group and varied doses in multiple groups. The drug administration is consistent within a group but increases from one group to another. The key parameter in this method is the mean mortality in each group and the variation in dose across the groups. The key parameter of this method is the mean mortality in each and the difference in dose across groups. The lethal dose (LD) is calculated by: $LD50 = LD100 - \sum (a * b/n)$ Where LD50 = mean lethal dose, and LD100 = least dose required for 100% mortality, a = dose difference, b = mean mortality, n = group population.

The last and the fourth is Lorke's method, introduced in 1944, and is carried out in two phases. In Phase 1, animals are treated with 10, 100, and 1000 mg/kg body weight of test substance in triplicates. After administration, groups are observed at regular intervals to monitor the effect of the substance for 24 hours. In the second phase, based on phase 1, the dose concentration is increased or decreased and observed for 24 h and 7–14 days. The lethal dose is calculated by: $LD50 = \sqrt{(D0 * D100)}$ Where D0 = highest dose with no mortality, and D100 = lowest dose with mortality (Chinedu et al., 2013; Lorke, 1983).

Sub-Acute Toxicity Building on acute toxicity data, this study is designed to evaluate organs affected by different doses of experimental substances. Unlike acute toxicity studies, these focus more on the nature of the drug under realistic conditions. This basically encompasses 3 main different doses, which include a high dose aimed at mitigating toxicity signs with minimal harm, a low dose where no toxicity is discerned, and an intermediate dose. Dose concentration is primarily derived from LD50, and the dose-response curve is established in acute toxicity studies (Chinedu et al., 2013; Lorke, 1983). It provides information on the profile for extended studies, specifically for chronic or sub-chronic investigations, thereby preventing unnecessary redundancy or repetition of doses.

Sub-Chronic and Chronic Toxicity These toxicity studies are formulated based on subacute toxicity studies. However, the difference between these two lies in the duration of the treatment period, where sub-chronic toxicity is studied for 13 weeks, and chronic toxicity studies are carried out for a longer period from 26 to 52/105 weeks (Heimann & Doughty, 2014). In essence, this study helps in identifying the effects that require a latency period before showing the impact. After toxicological profiling, chronic toxicology defines NOAEL (no observed adverse effect level). NOALs of subacute, chronic, and sub-chronic toxicity studies deduce the threshold limit of ADI (acceptable daily intake), AOEL (acceptable operator exposure level), etc.

3 Organ-Specific Toxicity

Addressing the challenges that are associated with drugs is a monumental task for the pharmaceutical industry, which impacts the clinical trials at their various stages. Numerous drugs face non-release status due to their accumulation in particular organs, causing toxicity. Thus, identifying those susceptibilities related to organ toxicities should be achieved through screening models. In this regard, flow cytometry-based experimental design would help us to discern such organ-specific toxicity by identifying specific markers at cellular and molecular levels (Table 1 and Fig. 2).

Neurotoxicity The incidence of neurogenerative diseases due to the toxicity of environmental pollutants has increased steadily over the past two decades. These pollutants encompass a variety of chemicals which are then released into various

Table 1 Biomarkers commonly used in different types of organ-specific toxicity

Toxicity type	Markers	References
Dermatotoxicity	Histamine, ROS, RNS, flaggrin, keratin-1, CGRP, cyclooxygenase-2	Heck et al. (2014)
Neurotoxicity	p-glycoprotein, ROS, RNS, S100β, prolactin, isofuran, neurofuran, 4-hydroxy-2-nonenal, nitrotyrosine, AGE	Maciejska et al. (2022)
Pneumotoxicity	Surfactant protein B, HIF-1α, albumin, KL-6, alkaline phosphatase, thrombomodulin, N-acetyl glucosaminidase, lysozyme, protein-C, desmosine	Kodavanti (2014)
Cardiotoxicity	cTnI, cTnT, NT-proBNP, IL-6, soluble CD40, Galectin-3	Muller and Dieterle (2009)
Ocular toxicity	MMP-9, proline-rich protein 4, lipophilin A, calgranulin S100A8, ocular mucin, BDNF, CD44	Kontadakis et al. (2019)
Hematological toxicity	Heinz bodies in RBCs, bilirubin, δ-aminolevulinic acids, corpropoporphyrin, carboxyhemoglobin levels, PIVKA, c-reactive protein/albumin	Gwaltney-Brant (2014)
Hepatotoxicity	ALT, bilirubin, glutathione, ATP, cell viability, mitochondrial functions, alkaline phosphatase, aspartate aminotransferase, albumin, ammonia, triglycerides, urobilinogen	Kaur (2019)
Nephrotoxicity	Creatinine, BUN, glomerular filtration rate, alkaline phosphatase, GST, KIM-1, $β_2$-microglobulin, cystatin C, NGAL, IL-18, OPN, exosomal fetuin-A	Vaidya et al. (2008)
Gut microbiome toxicity	LPS, Bile acids, Indolelatic acid, vitamin E, Phenylalanine, Taurine, Glycine	Tu et al. (2020)
Reproductive toxicity	SPACA4, FAM71F1, SPATA3, TMEM225, GSG1cathepsin D, total sialic acid, DNA degradation index of sperms, mucin 1, complement C3, follistatin, hCG	Wahid et al. (2017)

Abbreviations: *ROS* reactive oxygen species, *RNS* reactive nitrogen species, *CGRP* calcitonin gene-related peptide, *AGE* advanced glycosylated end products, *HIF-1α* hypoxia inducible factor-1α, *KL-6* krebs von den lungen-6, *cTnI* cardiac troponin I, *cTnT* cardiac troponin T, *NT-proBNP* N terminal natriuretic peptide B, *IL-6* interleukin-6, *MMP-9* matrix metalloprotease-9, *BDNF* brain derived neurotrophic factor, *PIVKA* proteins induced by vitamin K antagonism, *ALT* alanine aminotransferase, *BUN* Blood urea nitrogen, *GST-α* glutathione-S-transferase, *KIM1* Kidney injury molecule 1, *NGAL* neutrophil gelatinase associated lipocalin, *IL-18* interleukin-18, *OPN* osteopontin, *SPACA4* sperm acrosome associated 4, *FAM71F1* family with sequence similarity 71 member F1, *SPATA3* spermatogenesis associated protein 3, *TMEM225* transmembrane protein 225, *GSG1* germ cell associated 1, *hCG* human chorionic gonadotropin

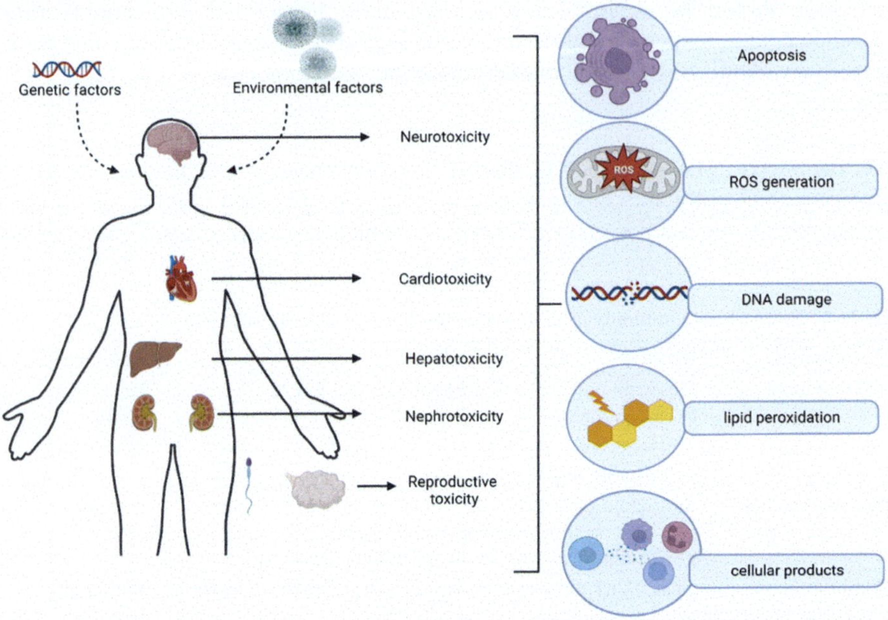

Fig. 2 Organ-specific toxicity causing various alterations at cellular and molecular levels in various organs, which can be detected by the flow cytometer technique

natural sources such as water, air, and soil through anthropological activities like chemical industries, waste disposal, vehicles, improper agriculture, etc. (Maciejska et al., 2022). These environmental hazards comprise numerous chemicals heavy metals, hydrocarbons, polyphenyls, pesticides, etc. Within the human body, they tend to accumulate within the brain, breaching the blood-brain barrier and causing neurotoxicity. In addition to this, the prolonged presence of such toxic chemicals results in endogenous neurotoxins, ultimately contributing to neuronal toxicity (Cao et al., 2021). These endogenous neurotoxins work by interfering with neuronal transmissions, blocking ion channels, blocking receptors, and even inhibiting vital enzyme activity (Soddu et al., 2015). At the cellular level, these actions are exhibited by loss of membrane potential, increased oxidative stress, and upregulation of apoptotic proteins leading to an increase in the production of metabolites like aldehyde (Cao et al., 2021; Chen et al., 2022). A population study based in Canada observed the fumigants to be one of the major contributors to Alzheimer's disease, with farmers being more susceptible (Tyas et al., 2001). An association between pesticides and Parkinson's disease was noted after 10 years of follow-up, which might establish the risk between the both (Baldi et al., 2003). Though the mechanism between chemical neurotoxicity and neurodegenerative diseases might be similar, the exact downstream action is not yet known. At the cellular level, neurotoxicity can be diagnosed using various parameters. Under toxic and stressful conditions, neurons show elevated levels of ROS, protein damage, DNA damage, and lipid peroxidation which can be studied to know the toxicity in neurons. The products of these may be assessed along with other neurological-based tests to understand and diagnose neurotoxicity. The products of lipid peroxidation, for example, oxidized product of docosahexanoic acid, F-4 neuroprostanes, can be detected by mass spectrometry and flow cytometry. The reduced antioxidant mechanism in the neurons can be used as a marker by the detection of enzymes like catalase, glutathione reductase, peroxidase, and other antioxidants like vitamins, flavonoids, albumin protein, metallothionein, etc. Neurotoxicity is expressed as inflammation in the brain tissue, caused by an increase in various pro-inflammatory cytokines and activation of brain resident macrophages, called microglia. These microglia further activate the resident T-cells and other phagocytic cells to release pro-inflammatory cytokines, and these also serve as biomarkers (Shabab et al., 2017). In animal models, the markers for activated macrophages like Cluster of Differentiation 40 (CD40), CD86, and CD206 can be interpreted as neurotoxicity markers, and these can be analyzed using Flow Cytometry methods (Hellström Erkenstam et al., 2016). At a functional level, mitochondrial dysfunction is related to neurotoxicity, hence measuring the mitochondrial membrane potential using flow cytometry might give insights into the toxicity. Assessing these parameters on a routine basis might be useful in the detection of neurotoxicity, which will aid in the early detection of neurodegenerative disease indications before their clinical manifestations.

Cardiotoxicity The National Cancer Institute (NCI) defines cardiotoxicity as damage inflicted upon the heart valves, blood vessels, and the overall cardiovascular system. Cardiotoxic, which in turn is linked to cancer treatment, has the potential to compromise the left ventricular ejection fraction, ultimately resulting in heart failure (Tian et al., 2014). For heart disease, the diagnosis is usually done by echocardiography, but the consistency of the detection is skill-dependent. In this regard, biomarkers can serve as a simple and effective detection method to monitor cardiotoxicity. Several chemotherapeutic agents like anthracyclines, anti-microtubule agents, and angiogenesis inhibitors that are used in cancer treatment are also known to cause cardiotoxicity in a dose-dependent manner (Salvatici et al., 2015). Therefore, quantitative estimation of biomarkers such as cardiac troponins and natriuretic peptides present in serum can be a useful indicator for routine detection of cardiotoxicity in a non-invasive manner. Cardiac troponins are present in the cardiac muscle and have three protein subunits- troponin I, troponin T, and troponin C (Salvatici et al., 2015). These units are involved in the actin-myosin interaction and contraction of the heart muscle. However, troponin C which is commonly found across all types of muscle tissues, and hence cannot be used as a suitable biomarker for cardiotoxicity detection. On the contrary, some studies conducted in rat models have demonstrated that the level of Troponin T increases in response to doxorubicin treatment, suggesting that serum levels of troponin can be used as an effective biomarker. (Herman et al., 1999). Akin to the troponins, hormones are also released in the bloodstream by the myocardium tissue. For example, the natriuretic peptides, particularly B-type natriuretic peptides (BNP), a correlation was established between the BNP values and anthracycline toxicity as reported by Dodos et al. in patients receiving chemotherapy (Dodos et al., 2008). Considering the presence of fragments of these peptides in circulation, this can be eligible to be called as biomarkers to assess cardiotoxicity. Also, various immunoassays are now being employed to assess these biomarkers, providing valuable indications of both cardiotoxicity and disease progression.

Nephrotoxicity Numerous chemicals, including drugs, have been known to cause harm to the renal parts of the kidney, leading to nephrotoxicity. As the kidneys play a crucial role in blood purification through their renal tubules, they encounter various drugs in circulation and are consequently susceptible to drug-induced toxicity with symptoms ranging from inflammation, angiopathy, nephropathy, breakdown of muscle tissue, and buildup of toxins in the glomerular cells (Sharma et al., 2023). The diagnosis of nephrotoxicity involves the assessment of various blood parameters such as blood urea nitrogen content, glomerular filtration rate, serum creatinine and changes in any of these parameters might indicate potential nephrotoxicity. Various nephrotoxicity-inducing factors such as heavy metals, bacterial and fungal toxins and infections and a variety of medications contribute to the toxicity through diverse biochemical and physiological mechanisms. For example, Anti-angiotensin drugs, which are prescribed to elevate the blood pressure, affect the glomerular hemodynamics (Palmer, 2002; Schoolwerth et al., 2001). Similarly, nephrotoxic agents induce inflammation, leading to renal and intestinal nephritis (Perazella, 2005; Sharma et al., 2023). Crystal nephropathy is another type of toxicity characterized by the precipitation of drugs (Perazella, 2005; Sharma et al., 2023). It is also observed that increased toxicity may result in the release of intracellular content from muscle fibers like myoglobin, obstructing the filtration unit of the nephron (Coco & Klasner, 2004). A number of biochemical assays have been developed under which the Terminal deoxynucleotidyl transferase dUTP nick-end labeling (TUNEL) assay serves as a well-established method to detect DNA fragmentation caused by cell death, and the analysis of TUNEL patterns provides insights from toxic or hypoxic injury in *in vitro* and *in vivo* models (Moore et al., 2021). The presence of released constituents in the blood circulation due to nephrotoxicity also serves as non-invasive indicators and hence can be labelled as biomarkers. Additionally, metabolites excreted in the urine, including creatinine, albuminuria, calprotectin, osteopontin, clusterin, and macroglobulins, are commonly analyzed as biomarkers to detect nephrotoxicity in a non-invasive manner.

Reproductive Toxicity Biomarker utilization in predicting the potential adverse side effects has proven to be a convenient method to evaluate and monitor reproductive health. Reproductive toxicity alters various aspects of reproductive health like gametogenesis, hormonal cycle, development and maturation of gonads, and even gestational changes. To assess toxicity in the male reproductive system, sperm morphology and characteristics related to sperm maturation are commonly utilized. In females, biomarkers are associated with hormonal changes occurring during follicle development and endocrine function. A number of parameters have been evaluated to monitor the toxicity of the male reproductive system induced by different chemicals such as alkaloids, glycosides, etc., and these include counting sperm number, morphology, and motility. In addition, the metabolites of these toxic compounds are also indicative of toxicity. A study analyzed the effects of environmental endocrine-disrupting chemicals on sperms by examining the germ cell viability and acrosomal integrity using *Pisum sativum* agglutinin (PSA) as a marker (Tavares et al., 2013). Testosterone is one of the most widely analyzed biomarkers in male reproductive health and is often carried out using methods like ELISA or LC-MS (Brehm et al., 2019). In male mice, the effects of phthalate on the development of gonads were tested by hormonal levels and apoptosis assay, where it was found that phthalate affects reproductive development (Barakat et al., 2017). In females, it has been noted that the toxicants disrupt the endocrine levels, which are mirrored at the physiological levels in terms of developmental progression. Unlike male gametes, the availability of female gametes is restricted due to obvious limitations. Hence, biomarkers that are released into the circulation are preferred. Fluctuations in levels of chorionic gonadotrophin, follicle-stimulating hormone, estrogen, and progesterone, are yet another important indicator of reproductive health and development. New to this list, microplastics have been increasingly reported to cause oxidative stress, decrease antral follicles, and affect the maturation rate of oocytes in mouse models (Zhang et al., 2023). While ELISA and mass spectrometry have been extensively utilized for assessing biomarkers, there remains an ongoing need to discover more sensitive biomarkers related to reproductive toxicity. These advancements are crucial for better examining and maintaining overall reproductive health.

Hepatotoxicity The liver is involved in maintaining homeostasis along with drug metabolism. However, due to human exposure to various chemicals and toxic substances, the liver is highly susceptible to toxicity, further necessitating its proper periodic monitoring. In the case of the liver, the mechanism of hepatotoxicity can be intrinsic or idiosyncratic (Francis & Navarro, 2023). In chronic liver damage cases, invasive techniques like liver biopsy can be done to know the extent of liver damage. Manifestation of hepatocellular toxicity generally happens with fluctuating levels of various enzymes, such as alanine transaminase (ALT), aspartate aminotransaminase (AST), alkaline phosphatase (ALP), gamma-glutamyl transpeptidase (GGT), and metabolites like bilirubin (Francis & Navarro, 2023). While their estimation serves as biomarkers, ongoing research explores novel biomarkers for the identification of specific hepatotoxicity. A study conducted by Wang and Antonie, respectively, reported that during the acetaminophen-induced hepatotoxicity, a non-coding RNA, miR-122, which is generally involved in post-translational modifications, showed elevated levels in both blood and tissues and patients (Antoine et al., 2013). In addition to these, various other proteins like high-mobility group box-1 protein (HMGB1) and keratin-18 protein have also been shown to be present at elevated levels in the serum of similar animal models (Antoine et al., 2009) and at clinical levels (Antoine et al., 2013). Given that liver, hepatotoxicity involves extensive damage to the mitochondria, enzymes like glutamate dehydrogenase (GDH) and fragments of mitochondrial DNA could aid in investigating the extent of mitochondrial damage and, therefore, might have caused elevated levels in plasma samples of patients with acetaminophen overdose. In this regard, the identification of induced hepatotoxicity using biomarkers holds promise for patient-centric treatment of liver damage.

4 Use of Flow Cytometry in *In Vitro* Toxicity Testing

Flow Cytometry for Measurement of Oxidative Stress Oxidative stress in cells can be triggered by overproduction of reactive oxygen species (ROS) and free radicals, which have crucial roles in aging, senescence, and various genetic and acquired conditions. Free radicals such as superoxide (O_2^-), hydroxyl (OH^-), and hydrogen peroxide (H_2O_2) are some of the ROS present in biological systems that cause oxidative stress. Superoxide is one type of ROS that is unreactive initially but can generate H_2O_2 upon reacting with other molecules. H_2O_2 is not a free radicle but can diffuse easily through the membrane and facilitate metal-catalyzed ROS generation. Detection of ROS poses significant challenges owing to its low concentration and short half-life (Jávega et al., 2022). Several dyes are employed for the analysis of oxidative stress, and these include carboxy-H_2DFCDA, dihydroethidium (Hydroethidine), 2′,7′-dichlorodihydrofluorescin diacetate (H_2DCFH-DA), and dihydroergotamine 123. These dyes are membrane-permeable and non-fluorescent in nature but, upon mixing with cells, get converted into fluorescent form via oxidation of its key moieties. For instance, esterases convert H_2DCFH-DA to a fluorescent intermediate, 2′,7′-dichlorofluorescin (DCFH) that is retained inside the cell. During oxidation, the intermediate is converted into fluorescent 2′,7′-dichlorofluorescein (DCF) in the presence of both peroxide and superoxide, which is also retained in the cell. Ethidium bromide (EB), upon addition to sodium borohydride, gets reduced to hydroethidine. Hydroethidine is a cell-permeable compound that, upon oxidation, forms 2-hydroxy ethidium that intercalates into DNA and emits red fluorescence. Detection of ROS is a complex task due to limited concentration, short half-life, ROS interactions and intrinsic constraints of probe and experimental conditions (Jávega et al., 2022).

Detection of fluorescent products DCF and EB can be accomplished through fluorescence-based techniques such as confocal microscopy or flow cytometry. Flow cytometry is advantageous in terms of simplicity, comparatively high sensitivity, resolution, and high throughput nature in comparison with other techniques (Jávega et al., 2023). The capabilities of flow cytometry have been extensively utilized in numerous studies to delve deeper into the complexities of analyzing oxidative stress in diverse biological systems. A notable example of this was a study published in the journal “Biology Research” in 2022. Here, the researchers used flow cytometry to specifically assess oxidative stress in *E. coli* cells, highlighting the wide-ranging applications of this technique. Fluorescent dyes are generally auto-oxidative and may result in artefactual detection and, in high concentrations, can be toxic for cells. Oxidative stress was quantified by ROS-sensitive probes such as H_2DCF-DA upon interaction with significant peroxidative xenobiotics; hydrogen peroxide, tert-butyl hydroperoxide, and cumene hydroperoxide at different titer (Jávega et al., 2022).

Chronic granulomatous disease (CGD), inherited through genes, serves as a prime example highlighting the role of respiratory burst in our immune system. The respiratory burst involves the production of reactive oxygen intermediates (such as hydrogen peroxide, superoxide, and hypochlorous acid) in response to bacterial or fungal invaders. However, defects in the reduced nicotinamide dinucleotide phosphate (NADPH) oxidase system found in phagocytes can lead to a limited production of these powerful antibacterial agents, resulting in CGD. This genetic abnormality leads to the formation of granulomas and frequent, severe, and often fatal infections caused by pyogenic bacteria. Diagnostic techniques for CGD include dihydrorodamine (DHR) fluorescence assay (Segal et al., 2000). DHR assay is a flow cytometry-based technique to measure DHR oxidation to fluorescent rhodamine products. After stimulating leukocytes loaded with DHR with a potent activator (PMA) for cell activation, the conversion of DHR to rhodamine reflects the induction of intracellular oxidation (Vowells et al., 1995). DHR assay also provides genotype screening information to differentiate between normal and oxidase-defective neutrophil populations. Thalassemia is another genetic disease that is characterized by impairment in haemoglobin, and flow cytometry can be utilized for the diagnosis. In thalassemia, the initial alterations take place in globin genes. However, the formation of ROS is believed to be the prime suspect in damaging the cell components. For instance, chronic anemia is associated with increased iron content that involves different chemical reactions that generate ROS (Amer et al., 2004). A dye, namely C11 BODIPY 581/591, specifically binds with lipid ROS in the plasma membrane and is detected by flow cytometry (Pei et al., 2022). MitoSox Red is another dye that detects mitochondrial ROS generation involved induction of ferroptosis (Chen et al., 2023a, b).

Flow Cytometry to Assess Different Modes of Cell Death Cell death plays a fundamental role in both development and homeostasis. The elimination of damaged or transformed cells or the clearing of surplus cells is very crucial during development for proper organogenesis and morphogenesis, and homeostasis. Therefore, complications can arise if there is any imbalance between two processes that oversee too much cell death, as in the case of neurodegenerative disorders or the clearance of transformed cells (Yuan & Ofengeim, 2023). Various mechanisms and types of cell death have been described for apoptosis, necrosis, pyroptosis, autophagy, etc. (Fig. 3). Therefore, targeting a particular cell death pathway has gained a lot of attention for therapeutic intervention. To understand the various cellular events such as identification of activated protease, cytochrome c release by mitochondria, endonucleolytic cleavage of nuclear DNA, and cellular DNA content

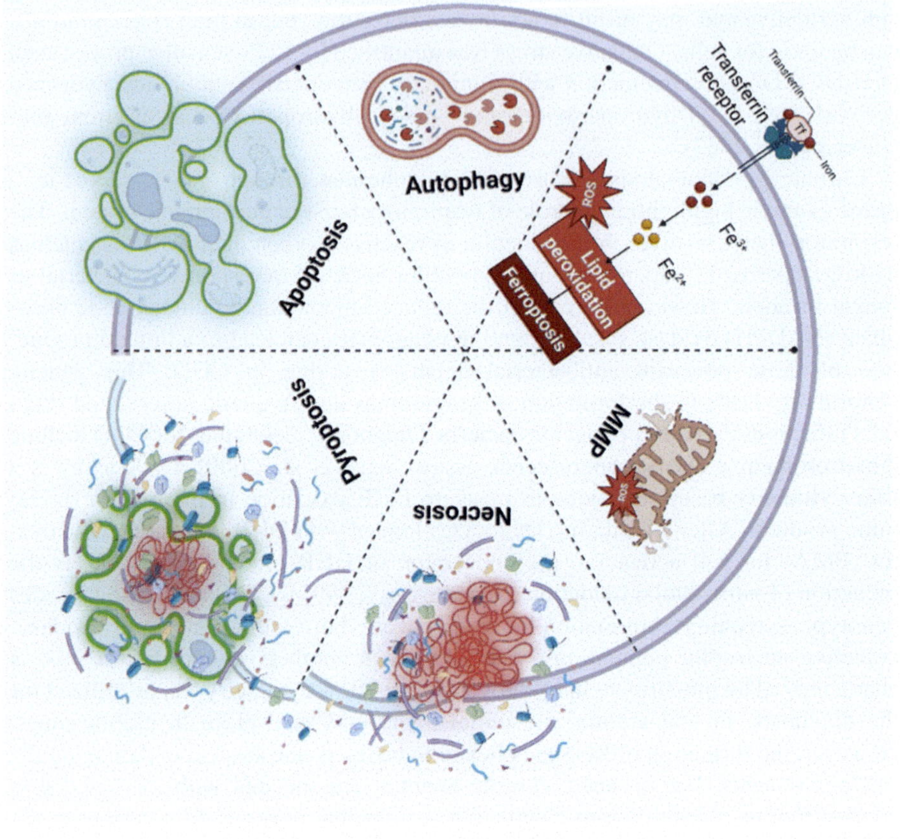

Fig. 3 Different forms of cell death frequently studied by flow cytometry

quantification, several techniques have been described. Among many, flow cytometry-based identification is the most sought-after for identifying and quantifying apoptotic cells and differentiating the different cell death pathways (George et al., 2004). Necrosis, pyroptosis, and apoptosis are the three main pathways through which a cell executes its death. Apoptosis is marked by nucleus fragmentation, cell atrophy, and the formation of apoptotic bodies, whereas necrosis results in the rapid release of enterocytes and membrane bursting. Pyroptosis is marked by increased cell membrane permeability, cytokine release, and the presence of pyroptotic bodies resembling features of both necrosis and apoptosis. Furthermore, pore-like structures and various light- and dark-colored vesicles are seen under a microscope due to the activation of gastrin (GSDM). However, these events are still being discussed for evaluating pyroptosis.

Based on the staining patterns observed during the type of cell death, it is possible to assign a few gold-standard techniques, such as Annexin V-FITC, which is helpful in detecting apoptotic cell death. In this experiment, healthy cells are negatively stained as the cell membrane is impermeable to both PI and Annexin V-FITC, indicating the intact nature of the cell membrane, particularly during the early apoptosis. However, during the apoptotic process of cell death, phosphatidylserine (PS) translocates from the inner leaflet to the outer leaflet of the plasma membrane, making it accessible for Annexin V-FITC to label and making the apoptotic fraction of cells as the single most dominant population during the flow cytometry acquisition. However, both PI and Annexin V are free to enter the cell to stain DNA and PS during pyroptosis as these populations of cells show greater membrane permeability. Pyroptotic cells, with increased membrane permeability, permit both PI and Annexin V to enter the cell, resulting in double-positive staining for both in flow cytometry. Notably, due to membrane rupture, necrotic cells display double staining with PI and Annexin V (Feng & Huang, 2021).

Apoptosis is an intricate process where cells are self-destructed in response to stimuli to protect surrounding cells from damage. The key features of apoptosis include diminished size, condensed chromatin, and fragmented nuclei. During apoptosis, the plasma membrane is initially preserved, but in later stages, it gets compromised, and cells shred organelle containing apoptotic bodies. On the other hand, necrosis is highly unregulated and has intact nuclei and limited chromatin condensation (George et al., 2004). In the flow cytometer, the apoptosis-associated events, such as changes in cell size and granularity, can be identified by analyzing the side scatter light. PI is another fluorescent DNA intercalating dye that is commonly employed in detecting cell death using a flow cytometer. PI can discriminate between intact and disrupted cell membranes and enter cells that do not have intact cell membranes, indicating necrosis or ferroptosis (Crowley et al., 2016). Pyroptosis is an inflammatory programmed cell death and has gained much attention lately due to its involvement in innate immunity and capability to affect tumor development. In pyroptosis, cell membrane integrity is lost due to the formation of pores that result in cell lysis. Mitochondrial membrane permeabilization (MMP) is a crucial event during cell death, affecting various cell types like lymphocytes, cardiomyocytes, and neurons. To analyze the MMP via flow cytometry, dyes, namely MitoTracker Deer Red FM and DIOC6, are used, in the case of studying the effects of various investigation drugs on cell death during cancer treatment (Ma et al., 2016). Tetramethylrhodamine methyl ester (TMRM)/JC-1 is another common dye that is used to detect the MMP via flow cytometry (Creed & McKenzie, 2019).

Ferroptosis, described in 2012, is another type of cell death that is driven by iron-dependent lipid peroxidation. Ferroptosis varies from apoptosis in various aspects, such as cellular morphology, genetics, and biochemistry. One such clear demarcation is the reduced or the absence of mitochondrial cristae, which is a kind of folds that allow the accumulation of a greater amount of electron transport enzymes and ATP synthase to synthesize more ATP. Ever since its description, ferroptosis has gained much attention due to its critical role in disease biology in relation to cancer especially metastasis cancers (Chen et al., 2023a, b). Flow cytometry technique can be used to study ferroptosis by considering various events and reactions taking place within cellular organelles, such as mitochondria, to identify cell death mechanisms. Cellular senescence or stable growth arrest is another metric that is physiologically significant for tissue balance and proper development and crucial for normal growth and tissue equilibrium. Various cellular events can be distinguished by morphological observations which contribute to the induction of cellular senescence. For instance, metabolic irregularities, changes in gene expression, activation of pro-inflammatory secretome pathway, and finally, age-induced telomere degradation have been documented in cellular senescence activation. 5-dodecanoylaminofluorescein di-β-D-galactopyranoside (C12FDG) is the dye that is commonly used to measure cellular senescence. C12FDG is a non-fluorescent, membrane-permeable dye that gets converted into fluorescent dye upon hydrolysis by a cellular enzyme β-galactosidase. Upon hydrolysis by β-galactosidase, a green fluorescence is emitted and remains confined into the cells and thus can be observed by cytochemical methods. Similarly, β-D-galactopyranoside with fluorophore bonded with galactosyl moieties also has superior sensitivity and permeability (Plovins et al., 1994). DDAOG is a similar substrate that in the presence of β-galactosidase gets hydrolyzed into DDAO and generates far-red fluorescent signals (Adewoye et al., 2020).

Autophagy is another type of cell death where cell digests their cellular contents. During this event, a protein, namely LC3, undergoes degradation by lysosomes within a particular cellular compartment called autophagosomes. By tagging LC3 protein with fluorescent protein GFP and measuring fluorescent intensity, it's possible to detect the autophagic activity and hence can serve as a good marker. Lipid-conjugated ATG8 protein is another autophagy marker, and analysis are done by measuring AT8 protein degradation. The fluorescent-tagged ATG8 protein, such as GFP-LC3, is used as an autophagy marker (Demishtein et al., 2015). The overall amount of LC3 degradation can be quantified by decreasing the concentration of GFP-LC3 fluorescence. This can serve as a good indicator of autophagic activity.

Intracellular Cytokine Assays Intracellular cytokine assays are usually carried out to investigate the inflammation caused by T-cells in response to external stimuli, and flow cytometry has become an ideal instrument to measure inflammation. The process begins with the isolation of T-cells from the peripheral blood mononuclear cells of a patient using the density gradient centrifugation method. This step is followed by the culturing of cells and, finally, passaging and cryopreservation (De Biasi et al., 2020). The measurement of cell proliferation is the method of choice to detect the T-cell response. However, several methods have been designed, which include the estimation of cytokine levels, and measurement of surface antigens, as conventional assays alone cannot distinguish between cell subsets. To identify T-cell populations from the isolated cells, staining with antibodies against T-cell receptors is performed (TCR) followed by identification of T-cell subsets using specific antibodies against $CD4^+$ and $CD8^+$. For instance, levels of $CD57^+$ T-cells showed a positive correlation with chronic infectious diseases *in vitro* and are a marker for replicative cell senescence. Another cytokine, namely, interferon-γ (IFN-γ), can be detected using membrane-based ELISPOT assay, which recognizes the CD8+ T-cells (Leehan & Koelsch, 2015), and this can be paired with the detection of other functional markers like CD107 (Lovelace & Maecker, 2018). The flow cytometer can also be used in the detection of intracellular cytokines like IL-4, IL-10, and TNF-α by employing beads tagged to specific antibodies against these cytokines (Rodrigues et al., 2017). Cell sorting using a fluorescence-activated cell sorter (FACS) instrument is another popular technique used to separate the T-cell subsets. For example, CD154, a co-stimulatory marker upregulated during the $CD4^{+}$ driven T-cells, can then be used to distinguish and sort other T-cell subsets and their levels are indicative of response against invaders or vaccines (Chattopadhyay et al., 2005). To avoid non-specific binding, a combination of amine-reactive dyes to detect live cells and surface markers like CD14 and CD19 can be used (Lamoreaux et al., 2006). Specific markers of degranulated cells like CD107a and CD107b can be examined using flow cytometry (Betts & Koup, 2004). Employing multichannel flow cytometry to analyze intracellular markers indicative of antigen-specific responses of cells under inflammatory conditions is a useful and beneficial approach for *in vitro* studies.

Flow Cytometry to Assess Cell Cycle Modulation The capacity of flow cytometers to detect up to 40 colors simultaneously has made this an instrument of choice to deeply analyze different aspects of the cell cycle (Park et al., 2020). The sensitivity of this can be utilized for single-cell cytometry, by including a multitude of markers. The simultaneous detection of diverse parameters such as cell cycle, apoptosis, and necrosis has made flow cytometric detection a reliable tool for screening toxicants *in vitro*. To assess cell cycle modulation, different dyes against biomarkers like Ki67 are used to differentiate between the quiescent G0 phase and active dividing phases of the cell cycle (Graefe et al., 2019). Another approach is to differentially stain DNA and RNA by using Hoechst 33342 and Pyronin Y dyes which help in the identification of G0 and G1 phases (Kim & Sederstrom, 2015). The interpretation of flow cytometry results is of utmost importance, wherein cell size and the intensity of fluorescence measured affect the cell cycle status, with cells in the G0/G1 phase showing low background fluorescence signal and cells in the G2/M phase showing higher fluorescence (Fedr et al., 2023). One of the critical uses of a flow cytometer is to analyze the functionality of biosimilars in comparison to the reference products, like analyzing cell cycle, apoptosis, receptor binding, etc. (Kirchhoff et al., 2017). A study validated the use of cell cycle analysis to test six toxic chemicals in HepG2, AAH-1, lymphocytes, and YAC-1 cells using flow cytometry for short and long durations (Tuschl & Schwab, 2005). A micronucleus assay on *in vitro* disease models showed 3-dimensional spheroids to have higher sensitivity than 2-dimensional cultures to identify multiple genotoxicants that act indirectly (Bryce et al., 2007).

5 Use of Flow Cytometry in Preclinical Toxicology and Safety Assessment

Hypersensitivity Testing Preclinical hypersensitivity testing by flow cytometry can be done by evaluating the immune response and efficacy of proposed therapies. Flow cytometry has been used in numerous studies for preclinical studies of drug delivery and novel medicines, including CAR T cell therapy for osteosarcoma and acute megakaryoblastic leukaemia treatment. In these investigations, flow cytometry was used for the identification of cytotoxicity, examination of effector-target cell contact, and evaluation of the anti-cancer activity of CAR T cells against specific cell types (Choe et al., 2023). Also, flow cytometry has enabled the characterization of tumor immunological environment in combined chemo-immunotherapy approaches to provide insights for integrating immune checkpoint inhibition with chemotherapy. Integration of flow cytometry with an automated drug screening platform facilitates the assessment of patients' samples and the identification of biomarkers for personalized medicines and preclinical drug delivery (Lassahn et al., 2023). Use of fluorescent markers and drug-primed dendritic cells to measure the proliferative response of diverse lymphocytic subpopulations and effector cells. The approach has demonstrated encouraging results in evaluating non-immediate drug hypersensitivity reactions (NIDHRs). Studies have shown that employing flow cytometric-based tests to target specific subpopulations of cells enhances the sensitivity of in vitro assessment of drug-specific proliferative responses in NIDHRs, providing a distinctive cellular *in vitro* test for diagnosing hypersensitivity reactions (Fernandez-Santamaria et al., 2021). Moreover, efforts are ongoing to standardize flow cytometric protocols and technical aspects for another flow-based assay, the basophil activation test (BAT), for the identification of acute drug hypersensitivity reactions (Mayorga et al., 2023). These results highlight diverse applications of flow cytometry in preclinical hypersensitivity testing and encompassing the effectiveness of novel treatments, prediction of patient response, and support in making informed treatment decisions.

Natural Killer cell's Flow Cytometry Assays Natural killer (NK) cells are a subset of lymphocytes and play a crucial role in innate defense against neoplastic and infected cells. Also, NK cells engage in adaptive immunity through physical interactions with T cells and dendritic cells and by releasing specific cytokines. Upon activation, NK cells eliminate suspected target cells by releasing secretory lysosomes containing perforin, also known as cytotoxic granules. Other than these, NK cells also secrete cytokines and chemokines. NK cells release soluble proteins such as MIP-1β, INF-γ, and TNF-α and contribute to the regulation of adaptive immunity and enhancement of cellular resistance to infections. Study of NK cell responses is necessary to assess clinical situations after stem cell transplantation to evaluate infection risk and diagnosis of immunodeficiency syndromes (Bryceson et al., 2010). Flow cytometry is a sensitive and quantitative method to characterize the phenotypic and functional properties of lymphocytes.

Various monoclonal antibodies target receptors and other surface molecules to assist in the evaluation of NK cell response and differentiating between the subsets. Specifically, the release study of lysosomes can be assessed by enhancing the expression of transmembrane proteins on the surface, typically located on secretory lysosomes, such as LAMP-3, Fas ligand/ 9CD178, LAMP-2, and CD107a (LAMP-1) present in lysosomes within inactivated cytotoxic cells. CD107a colocalizes with perforin in NK cells derived from healthy donors, and it is visible on the NK cell surface when target cells are lysed (Bryceson et al., 2005). Defective CD107a surface expression induction is associated with hyperinflammatory immunodeficiency disorders (Marcenaro et al., 2006). Some of the hyper-inflammatory forms of immunodeficiency disorders are linked to inadequate triggering of CD107a surface expression. Upon combination of degranulation and assessment of newly initiated transcriptional response, including the chemokine and cytokine synthesis, it allows the identification of novel signaling components associated with human immunological deficiencies that affect transcriptional response and diminish the functioning of lymphocyte effectors. These are expected to offer insights into NK cell response regulation in therapeutic and pathological scenarios (Lorenzo-Herrero et al., 2019).

Genotoxicity Assays Toxic agents can cause damage at genetic levels, which are inheritable and might lead to fatal consequences. New drugs are hence analyzed for their genotoxic properties at preclinical levels to assess risks associated with them. Flow cytometry technique is now being applied to test genotoxicity. In a study performed by Ueoka et al., the authors tested the genotoxic effect of various cosmetic ingredients using flow cytometry (Ueoka et al., 2023). The CHO cells served as preclinical models, and the micronuclei formation after treatment was analyzed. For this, a gating strategy specific to exclude any doublets and then counting micronuclei using sytox green dye was used (Bryce et al., 2007). Various studies have also analyzed the effects of nitrosamines on HepaRG cells as preclinical models by similar micronuclear staining with sytox green and ethidium monoazide dyes (Guo et al., 2020). In addition to micronucleus detection, genotypic stress can be interpreted through cell cycle analysis and apoptosis assay to study DNA damage (Miranda et al., 2023). Mutations resulting from any genotoxicity, for example, PIG-A mutations in patients with paroxysmal nocturnal hemoglobinuria (PNH), can be identified by looking at CD55 expression using flow cytometry (Lawrence et al., 2023). The phosphorylated γH2AX, which is a sensitive histone marker, is an indicator of DNA breaks and is identified using microscopy or flow cytometry (Křížkovská et al., 2023). A study upgraded the conventional flow cytometry-based assays, by combining TGx-DDI transcriptomic biomarker, micronucleus test, and multiflow assay for a new approach methodology strategy (Fortin et al., 2023). The use of flow cytometry to assess the genotoxic potential of toxicants can be a useful indicator of future risks in preclinical models.

Hematopoietic Toxicity Assays and Other Immunoassays in Preclinical Models Haematological toxicity, observed by toxicity in blood cells from bone marrow, is an effect of chemicals, drugs, and radiation. The toxicity in the hematopoietic lineage is reflected in the cells and traditionally is examined by peripheral complete blood count, blood smears, and microscopy, though a flow cytometer is now being utilized for looking at additional parameters. Examining the granulocyte nature of cells using a simple side scatter plot can give initial insights into the nature of cells. In preclinical models, the marrow makeup differs on various factors, majorly dietary intake (Levin et al., 1993), wherein it was observed that reduction in nutrition downregulates cyclin D1 and inhibits the cell cycle progression of hematopoietic cells (Nakajima et al., 2014). Another factor influencing the hematopoietic cellularity is the age of the preclinical models, as observed in Fisher 344/N rats across various age groups (Cline & Maronpot, 1985). A study used flow cytometry to explore the effects of an antimalarial drug on acute myeloid leukemia (AML) in cells of AML mice models (Liu et al., 2023b). The different hematopoietic cell lineages can be identified using surface-specific markers like Sca-1, CD48, CD150, CD-56, and c-kit, which can be analyzed and sorted using flow cytometry (Mata Forsberg et al., 2022). The whole marrow-derived cell populations can be analyzed using flow cytometry, enabling their classification into major categories like erythrocytic, neutrophilic, and lymphocytic based on their scatter plots (Biddle et al., 2022). An *ex vivo* analysis of T-reg cells to study the effects of a drug used flow cytometry for identification of T-cells (Kazemi et al., 2022). Another instance is examining the clonality of B-cells from the marrow by looking at the light chain kappa or lambda as given in module 6 of the quality and standards committee of the International Clinical Cytometry Society (www.cytometry.org). Given the substantial differences between the cells in diseased and normal conditions, flow cytometry provides a useful approach to evaluate toxicity.

6 Flow Cytometry to Assess Toxicity in Organs at the Clinical Level

Use of Flow Cytometry in a Biomarker-Based Assay for Various Diagnostic Purposes In various diagnoses, flowcytometry serves as an invaluable tool for biomarker-based tests. It is capable of counting and sorting cells easily and swiftly. It enables the characterization of diverse cell phenotypes and assessment of the efficiency of treatment on the intended target. Also, coupling with a multiparametric nature provides specific information regarding individual cells and makes it helpful for biomarker research in both preclinical as well as clinical settings (Barnard, 2012). Flow cytometry plays a vital role in influencing decisions related to cancer therapy, selecting doses for clinical trials, and donor-recipient-specific immunocompatibility. Additionally, flow cytometric assays can be tailored and validated for diverse study objectives, such as receptor occupancy assessments and characterization of cell growth and activation. These are being more and more employed in clinical trials to evaluate immunotherapy candidates (Millán & Brunet, 2015). Flow assays are used to track immune response, measure cellular activation, exhaustion, proliferation, and apoptosis, characterize proteins, and examine cell cycle and drug occupancy of cell target markers. Guidelines to validate standardized biomarker tests using flow cytometry are necessary during drug development, especially for clinical applications.

To maintain data quality in clinical research, it is essential to validate and develop methodologies for pharmacodynamic clinical indicators. The creation of predictive biomarkers demands documentation, specialized clinical expertise, and rigorous reagent selection and qualification (Wu et al., 2010). The continuous discussion between the clinical team and biomarker laboratory for the use of biomarker data is necessary in all stages of clinical trials for drug development.

Flow Cytometry in the Assessment of Inflammation The inflammatory burden caused by exposure to environmental toxicants is manifested, ranging from allergies and hypersensitivity reactions to more serious impairment in organ functions. Under inflammatory conditions, an imbalance in various immune cells, along with mitochondrial ROS production, has been reported (Mills et al., 2016). Nanoparticle-induced inflammation caused by environmental toxins like silica nanoparticles is known to cause allergic reactions, bronchitis, and asthma (Han et al., 2016), and the intracellular rise in cytokines like interleukins due to inflammation can be detected using methods like qPCR, ELISA, and flow cytometry. In addition to environmental factors, bacterial toxins also have a significant impact on the body's inflammatory response. The endotoxins and exotoxins released by the bacteria are known to impede the synthesis of protein, damage blood vessels, cause toxic shock syndrome, and impact the nervous system as well (Ghazaei., 2022). UroPathogenic *E. coli* has been known to secrete a protein toxin called alpha-hemolysin, which fluctuates calcium levels in renal epithelial cells (Söderblom et al., 2003). The effect of exercise in reducing inflammation was studied in *vitro* cultures of chondrocytes where flow cytometer analysis aided in deciphering the inhibition of PI3K/Akt/NF-kB and NLRP3/caspase-1/GSDMD signalling (Liu et al., 2023a). Investigation into various anti-inflammatory signalling pathways and anti-toxicants like *Ganoderma lucidum* in mitigating liver toxicity involved analysis of cell cycle, apoptosis, and cell interactions using flow cytometry (Chen et al., 2023a). In polycystic ovary syndrome mouse models, the relationship between PCOS and cardiovascular diseases was established by performing inflammation studies using different techniques, along with flow cytometry, to recognize specific immune cells in peripheral blood (Gao et al., 2023). In *vivo* models of placental inflammation, the impact on embryo development was studied in which flow cytometer was used for cell sorting by tagging the cells with viability dye (Ward et al., 2023). In recent reports, flow cytometry has been a useful tool in studying the cellular mechanisms and participating cells in various inflammatory diseases like ischemia (Coppin et al., 2023), rheumatoid arthritis (Vidal-Pedrola et al., 2023), Grave's orbitopathy (Lu et al., 2023), lupus (Fan et al., 2023). The capability of flow cytometry can aid in the detection of inflammation by looking at levels of immune cells, and cytokine levels and correlate them with the disease diagnosis.

Analysis of Blood for Clonality of B-Cells Examining the immune repertoire of blood majorly focuses on T- and B-cells. As a response to toxins, the immune cells respond by recognizing the specific antigen and producing antibodies against it. Hence, the clonality of cells can be evaluated. The rearrangement of the immunoglobin-heavy chain of B-cells has served as an important indicator in many lymphomas at a clinical level, and the detection of the same is usually done using polymerase chain reaction for complementarity determining region (CDR3), but smaller amounts of DNA might yield inaccurate results (Hoeve et al., 2000). The most predominant clonal families among memory B-cells are IgM, IgA, and IG2, with IG1 families with reduced levels of IgG1 families as seen by single-cell sequencing, with consistent stability over the years (Phad et al., 2022). Flow cytometry revealed that a minor B-cell subset of $CD19^{-}$, $CD38^{++}$, and $CD45^{-}$ can be used to distinguish healthy from myeloid phenotypes in multiple myeloma patients since other markers like CD19, CD20, CD24, CD138, and CD22 showed no significant difference (Luque et al., 1998). A study investigated the interplay between T and B cells using immortalized B-cells from IgG-related disease patients and found that there were anomalies in both B-cell receptors and T-cell receptors using multi-color flow cytometry and RNA sequencing (Cai et al., 2022). In various autoimmune diseases like Systemic Lupus Erythematous (SLE) and Schnitzler Syndrome, the B-cell clonality significantly influences the disease pathology. In Schnitzler syndrome, it was found that CDR3 and VDJ recombination and IgM reaction to nuclear proteins played an important role (Pathak et al., 2020). In lymphoma diagnosis, multiparametric flow cytometry has contributed to the detection of biclonal B-cells. Thus, implementing the flow cytometry technique can be valuable for in-depth analysis of blood, including assessment of B-cell clonality in various diseases, including toxicity.

Immunophenotyping of Markers Associated with Different Diseases Flow cytometry-based immunophenotyping is a robust diagnostic method for identifying a range of immunodeficiency disorders. The approach is to tag proteins with their specific antibodies with a fluorescent dye, and their expression would be an indication of the presence of the protein. In this context, broad targets like CD56 can be used to identify natural killer cells in a cell population. A popular application of immunophenotyping is the use of 17 markers as an indication of cytopenia in patients, which is detected using multiparameter flow cytometry (Porwit et al., 2023). For multiomics research, immunophenotyping of whole blood to identify and differentiate between myeloid lineages, T-cells, B-cells, and natural killer cells used 64 cell surface markers which gave a better resolution (Heubeck et al., 2023). Given the limited availability of a higher volume of samples, the use of immunophenotyping using such multicolor and multiparameter is of paramount importance. With immunophenotyping, accurate detection of cells, even with a frequency less than 0.001, can be done (Bocsi et al., 2011). In the case of viral toxicity caused by human immunodeficiency virus (HIV), flow cytometry has been used to determine the CD4/CD8 cell ratio (Masel et al., 2000). It was noticed that flow cytometry provided a high throughput and high content analysis when compared to slide-based imaging cytometry (Bocsi et al., 2011). The acute and sub-acute toxicity of novel nutraceuticals has been studied by looking at T-cell subpopulations using immunophenotyping markers $CD3^+CD8^+$ for cytotoxic suppressor T lymphocytes, $CD3^+CD4^+$ for helper T-lymphocytes, $CD3^-CD19^+$ for B-cells and CD3-NK1.1^+ for natural killer cells (Onisei et al., 2023). The toxicity caused by chemotherapy needs to be evaluated periodically, and for this, immunophenotyping using a 27-color panel has been optimized to analyze peripheral blood mononuclear cells (Edwards et al., 2023). In clinical settings, the use of flow cytometry remains an essential tool, and immunophenotyping panels can be developed specifically for disease indications.

7 Conclusion

In the growing area of drug development and modernization, the assessment of toxic components is of utmost importance. Flow cytometry gives a multi-parametric approach to detect different components related to toxicity like immunophenotyping, cell viability, and cell death, characterization, and many more. The area of personalized medicine can utilize this technique for tailoring the interventions according to the disease and the individual. The ever-expanding options of fluorescent dyes and probes will aid in utilizing cytometry to develop panels specific to diseases. The incorporation of other methods like microfluidics along with flow cytometry can be useful in developing strategies for efficient processing and detection. Coupled with other techniques, flow cytometry is an indispensable tool in the diagnosis of different types of toxicity, be it organ-specific or within laboratory conditions, at cellular and molecular levels, providing insights for diagnosis and prevention related to toxicants.

Acknowledgments The authors acknowledge the support provided by the Manipal Academy of Higher Education, Manipal, Karnataka, India.

References

Adewoye, A. B., Tampakis, D., Follenzi, A., & Stolzing, A. (2020). Multiparameter flow cytometric detection and quantification of senescent cells in vitro. *Biogerontology, 21*(6), 773–786. https://doi.org/10.1007/s10522-020-09893-9

Amer, J., Goldfarb, A., & Fibach, E. (2004). Flow cytometric analysis of the oxidative status of normal and thalassemic red blood cells. *Cytometry. Part A: The Journal of the International Society for Analytical Cytology, 60*(1), 73–80. https://doi.org/10.1002/cyto.a.20017

Antoine, D. J., Williams, D. P., Kipar, A., Jenkins, R. E., Regan, S. L., Sathish, J. G., Kitteringham, N. R., & Park, B. K. (2009). High-Mobility Group Box-1 protein and Keratin-18, circulating serum proteins informative of acetaminophen-induced necrosis and apoptosis in vivo. *Toxicological Sciences, 112*(2), 521–531. https://doi.org/10.1093/toxsci/kfp235

Antoine, D. J., Dear, J. W., Lewis, P. S., Platt, V., Coyle, J., Masson, M., Thanacoody, R. H., Gray, A. J., Webb, D. J., Moggs, J. G., Bateman, D. N., Goldring, C. E., & Park, B. K. (2013).

Mechanistic biomarkers provide early and sensitive detection of acetaminophen-induced acute liver injury at first presentation to hospital. *Hepatology (Baltimore, Md.), 58*(2), 777–787. https://doi.org/10.1002/hep.26294

Baldi, I., Lebailly, P., Mohammed-Brahim, B., Letenneur, L., Dartigues, J.-F., & Brochard, P. (2003). Neurodegenerative diseases and exposure to pesticides in the elderly. *American Journal of Epidemiology, 157*(5), 409–414. https://doi.org/10.1093/aje/kwf216

Barakat, R., Lin, P.-C. P., Rattan, S., Brehm, E., Canisso, I. F., Abosalum, M. E., Flaws, J. A., Hess, R., & Ko, C. (2017). Prenatal exposure to DEHP induces premature reproductive senescence in male mice. *Toxicological Sciences, 156*(1), 96–108. https://doi.org/10.1093/toxsci/kfw248

Barnard, R. M. (2012). Flow cytometry: A flexible tool for biomarker research. *Bioanalysis, 4*(20), 2471–2483. https://doi.org/10.4155/bio.12.225

Betts, M. R., & Koup, R. A. (2004). Detection of T-cell degranulation: CD107a and b. In *Methods in cell biology* (Vol. 75, pp. 497–512). Academic. https://doi.org/10.1016/S0091-679X(04)75020-7

Biddle, K. E. (2022). Opinion on the optimal histologic evaluation of the bone marrow in nonclinical toxicity studies. *Toxicologic Pathology, 50*(2), 266–273. https://doi.org/10.1177/01926233211061712

Bocsi, J., Mittag, A., & Tárnok, A. (2011). Comparison of Immunophenotyping and rare cell detection by slide-based imaging cytometry and flow cytometry. In *Advanced optical flow cytometry* (pp. 239–271). Wiley. https://doi.org/10.1002/9783527634286.ch9

Brehm, E., Rattan, S., Chiang, C., Warner, G. R., & Flaws, J. A. (2019). Chapter 17—Reproductive toxicity biomarkers. In R. C. Gupta (Ed.), *Biomarkers in toxicology* (2nd ed., pp. 287–301). Academic. https://doi.org/10.1016/B978-0-12-814655-2.00017-7

Bryce, S. M., Bemis, J. C., Avlasevich, S. L., & Dertinger, S. D. (2007). In vitro micronucleus assay scored by flow cytometry provides a comprehensive evaluation of cytogenetic damage and cytotoxicity. *Mutation Research/ Genetic Toxicology and Environmental Mutagenesis, 630*(1), 78–91. https://doi.org/10.1016/j.mrgentox.2007.03.002

Bryceson, Y. T., March, M. E., Barber, D. F., Ljunggren, H.-G., & Long, E. O. (2005). Cytolytic granule polarization and degranulation controlled by different receptors in resting NK cells. *The Journal of Experimental Medicine, 202*(7), 1001–1012. https://doi.org/10.1084/jem.20051143

Bryceson, Y. T., Fauriat, C., Nunes, J. M., Wood, S. M., Björkström, N. K., Long, E. O., & Ljunggren, H.-G. (2010). Functional analysis of human NK cells by flow cytometry. *Methods in Molecular Biology (Clifton, N.J.), 612*, 335–352. https://doi.org/10.1007/978-1-60761-362-6_23

Cai, S., Chen, Y., Hu, Z., Zhou, T., Huang, Y., Lin, S., Gao, R., Zhong, J., & Dong, L. (2022). The landscape of T and B lymphocytes interaction and synergistic effects of Th1 and Th2 type response in the involved tissue of IgG4-RD revealed by single cell transcriptome analysis. *Journal of Autoimmunity, 133*, 102944. https://doi.org/10.1016/j.jaut.2022.102944

Cao, Y., Li, B., Ismail, N., Smith, K., Li, T., Dai, R., & Deng, Y. (2021). Neurotoxicity and underlying mechanisms of endogenous neurotoxins. *International Journal of Molecular Sciences, 22*(23), 12805. https://doi.org/10.3390/ijms222312805

Chattopadhyay, P. K., Yu, J., & Roederer, M. (2005). A live-cell assay to detect antigen-specific CD4+ T cells with diverse cytokine profiles. *Nature Medicine, 11*(10), Article 10–Articl1117. https://doi.org/10.1038/nm1293

Chen, Z., Rasheed, M., & Deng, Y. (2022). The epigenetic mechanisms involved in mitochondrial dysfunction: Implication for Parkinson's disease. *Brain Pathology (Zurich, Switzerland), 32*(3), e13012. https://doi.org/10.1111/bpa.13012

Chen, C., Chen, J., Wang, Y., Fang, L., Guo, C., Sang, T., Peng, H., Zhao, Q., Chen, S., Lin, X., & Wang, X. (2023a). Ganoderma lucidum polysaccharide inhibits HSC activation and liver fibrosis via targeting inflammation, apoptosis, cell cycle, and ECM-receptor interaction mediated by TGF-β/Smad signaling. *Phytomedicine, 110*, 154626. https://doi.org/10.1016/j.phymed.2022.154626

Chen, Z., Lin, H., Wang, X., Li, G., Liu, N., Zhang, M., & Shen, Y. (2023b). The application of approaches in detecting ferroptosis. *Heliyon, 10*(1), e23507. https://doi.org/10.1016/j.heliyon.2023.e23507

Chinedu, E., Arome, D., & Ameh, F. S. (2013). A new method for determining acute toxicity in animal models. *Toxicology International, 20*(3), 224–226. https://doi.org/10.4103/0971-6580.121674

Choe, M., Kirkey, D., Ries, R. E., Hylkema, T., Menashe, S. J., Davis, L. E., Wagner, M. J., Cranmer, L. D., Malhotra, S., Gorlick, R. G., Meshinchi, S., & Loggers, E. T. (2023). Preclinical testing of FOLR1-CAR T cells against osteosarcoma (OS). *Journal of Clinical Oncology, 41*(16_suppl), 2555–2555. https://doi.org/10.1200/JCO.2023.41.16_suppl.2555

Cline, J. M., & Maronpot, R. R. (1985). Variations in the histologic distribution of rat bone marrow cells with respect to age and anatomic site. *Toxicologic Pathology, 13*(4), 349–355. https://doi.org/10.1177/019262338501300411

Coco, T. J., & Klasner, A. E. (2004). Drug-induced rhabdomyolysis. *Current Opinion in Pediatrics, 16*(2), 206–210.

Coppin, E., Zhang, X., Ohayon, L., Johny, E., Dasari, A., Zheng, K. H., Stiekema, L., Cifuentes-Pagano, E., Pagano, P. J., Chaparala, S., Stroes, E. S., & Dutta, P. (2023). Peripheral ischemia imprints epigenetic changes in hematopoietic stem cells to propagate inflammation and atherosclerosis. *Arteriosclerosis, Thrombosis, and Vascular Biology, 43*(6), 889–906. https://doi.org/10.1161/ATVBAHA.123.318956

Coulter, W. H. (1953). *Means for counting particles suspended in a fluid* (United States Patent US2656508A). https://patents.google.com/patent/US2656508A/en

Creed, S., & McKenzie, M. (2019). Measurement of mitochondrial membrane potential with the fluorescent dye Tetramethylrhodamine methyl Ester (TMRM). In M. Haznadar (Ed.), *Cancer metabolism: Methods and protocols* (pp. 69–76). Springer. https://doi.org/10.1007/978-1-4939-9027-6_5

Crosland-Taylor, P. J. (1953). A device for counting small particles suspended in a fluid through a tube. *Nature, 171*(4340), Article 4340–Article 4338. https://doi.org/10.1038/171037b0

Crowley, L. C., Scott, A. P., Marfell, B. J., Boughaba, J. A., Chojnowski, G., & Waterhouse, N. J. (2016). Measuring cell death by Propidium iodide uptake and flow cytometry. *Cold Spring Harbor Protocols, 2016*(7), pdb.prot087163. https://doi.org/10.1101/pdb.prot087163

De Biasi, S., Meschiari, M., Gibellini, L., Bellinazzi, C., Borella, R., Fidanza, L., Gozzi, L., Iannone, A., Lo Tartaro, D., Mattioli, M., Paolini, A., Menozzi, M., Milić, J., Franceschi, G., Fantini, R., Tonelli, R., Sita, M., Sarti, M., Trenti, T., et al. (2020). Marked T cell activation, senescence, exhaustion and skewing towards TH17 in patients with COVID-19 pneumonia. *Nature Communications, 11*(1), Article 1. https://doi.org/10.1038/s41467-020-17292-4

Demishtein, A., Porat, Z., Elazar, Z., & Shvets, E. (2015). Applications of flow cytometry for measurement of autophagy. *Methods, 75*, 87–95. https://doi.org/10.1016/j.ymeth.2014.12.020

Dodos, F., Halbsguth, T., Erdmann, E., & Hoppe, U. C. (2008). Usefulness of myocardial performance index and biochemical markers for early detection of anthracycline-induced cardiotoxicity in adults. *Clinical Research in Cardiology: Official Journal of the German Cardiac Society, 97*(5), 318–326. https://doi.org/10.1007/s00392-007-0633-6

Edwards, J. M., Andrews, M. C., Burridge, H., Smith, R., Owens, C., Edinger, M., Pilkington, K., Desfrancois, J., Shackleton, M., Senthi, S., & van Zelm, M. C. (2023). Design, optimisation and standardisation of a high-dimensional spectral flow cytometry workflow assessing T-cell immunophenotype in patients with melanoma. *Clinical & Translational Immunology, 12*(9), e1466. https://doi.org/10.1002/cti2.1466

Fan, X., Ng, C. T., Guo, D., Lim, F., Tan, J. C., Law, A., Goh, L. H., Poon, Z. Y., Cheung, A., Kong, S. L., Tan, M., Li, S., Loh, A., James, A., Lim, T., Chen, J., Thumboo, J., Hwang, W., & Low, A. (2023). Dampened inflammation and improved survival after CXCL5 Administration in Murine Lupus via myeloid and neutrophil pathways. *Arthritis & Rheumatology (Hoboken, N.J.), 75*(4), 553–566. https://doi.org/10.1002/art.42383

Fedr, R., Kahounová, Z., Remšík, J., Reiterová, M., Kalina, T., & Souček, K. (2023). Variability of fluorescence intensity distribution measured by flow cytometry is influenced by cell size and cell cycle progression. *Scientific Reports, 13*(1), Article 1. https://doi.org/10.1038/s41598-023-31990-1

Feng, Y., & Huang, X. (2021). Methodology for comprehensive detection of Pyroptosis. In A. B. Alvero & G. G. Mor (Eds.), *Detection of cell death mechanisms: Methods and protocols* (pp. 149–157). Springer US. https://doi.org/10.1007/978-1-0716-1162-3_13

Fernandez-Santamaria, R., Bogas, G., Palomares, F., Salas, M., Fernandez, T. D., Jimenez, I., Barrionuevo, E., Doña, I., Torres, M. J., & Mayorga, C. (2021). Dendritic cells inclusion and cell-subset assessment improve flow-cytometry-based proliferation test in non-immediate drug hypersensitivity reactions. *Allergy, 76*(7), 2123–2134. https://doi.org/10.1111/all.14755

Fortin, A.-M. V., Long, A. S., Williams, A., Meier, M. J., Cox, J., Pinsonnault, C., Yauk, C. L., & White, P. A. (2023). Application of a new approach methodology (NAM)-based strategy for genotoxicity assessment of data-poor compounds. *Frontiers in Toxicology, 5*. https://www.frontiersin.org/articles/10.3389/ftox.2023.1098432

Francis, P., & Navarro, V. J. (2023). Drug-induced hepatotoxicity. In *StatPearls*. StatPearls Publishing. http://www.ncbi.nlm.nih.gov/books/NBK557535/

Gao, L., Zhao, Y., Wu, H., Lin, X., Guo, F., Li, J., Long, Y., Zhou, B., She, J., Zhang, C., Sheng, J., Jin, L., Wu, Y., & Huang, H. (2023). Polycystic ovary syndrome fuels cardiovascular inflammation and aggravates ischemic cardiac injury. *Circulation, 148*(24), 1958–1973. https://doi.org/10.1161/CIRCULATIONAHA.123.065827

George, T. C., Basiji, D. A., Hall, B. E., Lynch, D. H., Ortyn, W. E., Perry, D. J., Seo, M. J., Zimmerman, C. A., & Morrissey, P. J. (2004). Distinguishing modes of cell death using the ImageStream multispectral imaging flow cytometer. *Cytometry. Part A: The Journal of the International Society for Analytical Cytology, 59*(2), 237–245. https://doi.org/10.1002/cyto.a.20048

Ghazaei, C. (2022). Advances in the study of bacterial toxins, their roles and mechanisms in pathogenesis. *The Malaysian Journal of Medical Sciences, 29*(1), 4–17. https://doi.org/10.21315/mjms2022.29.1.2

Graefe, C., Eichhorn, L., Wurst, P., Kleiner, J., Heine, A., Panetas, I., Abdulla, Z., Hoeft, A., Frede, S., Kurts, C., Endl, E., & Weisheit, C. K. (2019). Optimized Ki-67 staining in murine cells: A tool to determine cell proliferation. *Molecular Biology Reports, 46*(4), 4631–4643. https://doi.org/10.1007/s11033-019-04851-2

Guo, X., Seo, J.-E., Petibone, D., Tryndyak, V., Lee, U. J., Zhou, T., Robison, T. W., & Mei, N. (2020). Performance of HepaRG and HepG2 cells in the high-throughput micronucleus assay for in vitro genotoxicity assessment. *Journal of Toxicology and Environmental Health. Part A, 83*(21–22), 702–717. https://doi.org/10.1080/15287394.2020.1822972

Gwaltney-Brant, S. (2014). Chapter 21—Blood and bone marrow toxicity biomarkers. In R. C. Gupta (Ed.), *Biomarkers in toxicology* (pp. 361–371). Academic. https://doi.org/10.1016/B978-0-12-404630-6.00021-X

Han, H., Park, Y. H., Park, H. J., Lee, K., Um, K., Park, J.-W., & Lee, J.-H. (2016). Toxic and adjuvant effects of silica nanoparticles on ovalbumin-induced allergic airway inflammation in mice. *Respiratory Research, 17*(1), 60. https://doi.org/10.1186/s12931-016-0376-x

Heck, D. E., Joseph, L. B., & Kim, H. D. (2014). Chapter 18—Dermal toxicity biomarkers. In R. C. Gupta (Ed.), *Biomarkers in toxicology* (pp. 309–316). Academic. https://doi.org/10.1016/B978-0-12-404630-6.00018-X

Heimann, K. G., & Doughty, K. (2014). Examination of acute and chronic toxicity. In F.-X. Reichl & M. Schwenk (Eds.), *Regulatory toxicology* (pp. 81–88). Springer. https://doi.org/10.1007/978-3-642-35374-1_31

Hellström Erkenstam, N., Smith, P. L. P., Fleiss, B., Nair, S., Svedin, P., Wang, W., Boström, M., Gressens, P., Hagberg, H., Brown, K. L., Sävman, K., & Mallard, C. (2016). Temporal characterization of microglia/macrophage phenotypes in a mouse model of neonatal hypoxic-ischemic brain injury. *Frontiers in Cellular Neuroscience, 10*. https://www.frontiersin.org/articles/10.3389/fncel.2016.00286

Herman, E. H., Zhang, J., Lipshultz, S. E., Rifai, N., Chadwick, D., Takeda, K., Yu, Z. X., & Ferrans, V. J. (1999). Correlation between serum levels of cardiac troponin-T and the severity of the chronic cardiomyopathy induced by doxorubicin. *Journal of Clinical Oncology: Official Journal of the American Society of Clinical Oncology, 17*(7), 2237–2243. https://doi.org/10.1200/JCO.1999.17.7.2237

Heubeck, A., Savage, A., Henderson, K., Roll, C., Hernandez, V., Torgerson, T., Bumol, T., & Reading, J. (2023). Cross-platform immunophenotyping of human peripheral blood mononuclear cells with four high-dimensional flow cytometry panels. *Cytometry Part A, 103*(6), 500–517. https://doi.org/10.1002/cyto.a.24715

Hoeve, M. A., Krol, A. D. G., Philippo, K., Derksen, P. W. B., Veenendaal, R. A., Schuuring, E., Kluin, P. M., & van Krieken, J. H. J. M. (2000). Limitations of clonality analysis of B cell proliferations using CDR3 polymerase chain reaction. *Molecular Pathology, 53*(4), 194–200.

Jávega, B., Herrera, G., & O'Connor, J.-E. (2022). Flow Cytometric analysis of oxidative stress in Escherichia coli B strains deficient in genes of the antioxidant Defence. *International Journal of Molecular Sciences, 23*(12). https://doi.org/10.3390/ijms23126537

Jávega, B., Herrera, G., Martínez-Romero, A., & O'Connor, J.-E. (2023). Flow cytometry of oxygen and oxygen-related cellular stress. *Oxygen, 3*(2), Article 2–Articl255. https://doi.org/10.3390/oxygen3020016

Kaur, G. (2019). Chapter 14—Hepatic toxicity biomarkers. In R. C. Gupta (Ed.), *Biomarkers in toxicology* (2nd ed., pp. 251–266). Academic. https://doi.org/10.1016/B978-0-12-814655-2.00014-1

Kazemi, M. H., Shokrollahi Barough, M., Ghanavatinejad, A., Momeni-Varposhti, Z., Khorrami, S., Sadeghi, B., & Falak, R. (2022). Decrease of tumor-infiltrating regulatory T cells using Pentoxifylline: An ex vivo analysis in triple-negative breast cancer mouse model. *Iranian Journal of Allergy, Asthma, and Immunology, 21*(2), 167–177. https://doi.org/10.18502/ijaai.v21i2.9224

Kim, K. H., & Sederstrom, J. M. (2015). Assaying cell cycle status using flow cytometry. Current protocols in molecular biology/edited by Frederick M. Ausubel ... [et Al.], *111*, 28.6.1–28.6.11. https://doi.org/10.1002/0471142727.mb2806s111

Kirchhoff, C. F., Wang, X.-Z. M., Conlon, H. D., Anderson, S., Ryan, A. M., & Bose, A. (2017). Biosimilars: Key regulatory considerations and similarity assessment tools. *Biotechnology and Bioengineering, 114*(12), 2696–2705. https://doi.org/10.1002/bit.26438

Kodavanti, U. P. (2014). Chapter 12—Respiratory toxicity biomarkers. In R. C. Gupta (Ed.), *Biomarkers in toxicology* (pp. 217–239). Academic. https://doi.org/10.1016/B978-0-12-404630-6.00012-9

Kontadakis, G. A., Fragou, D., Plaka, A., Hernández, A. F., Kymionis, G. D., & Tsatsakis, A. M. (2019). Chapter 21—Ocular biomarkers in diseases and toxicities. In R. C. Gupta (Ed.), *Biomarkers in toxicology* (2nd ed., pp. 375–383). Academic. https://doi.org/10.1016/B978-0-12-814655-2.00021-9

Křížkovská, B., Schätz, M., Lipov, J., Viktorová, J., & Jablonská, E. (2023). In vitro high-throughput Genotoxicity testing using γH2AX biomarker, microscopy and reproducible automatic image analysis in ImageJ—A pilot study with Valinomycin. *Toxins, 15*(4), Article 4. https://doi.org/10.3390/toxins15040263

Lamoreaux, L., Roederer, M., & Koup, R. (2006). Intracellular cytokine optimization and standard operating procedure. *Nature Protocols, 1*(3), Article 3–Artic1516. https://doi.org/10.1038/nprot.2006.268

Lassahn, K. A., Elrod, A. K., Carlson, A. L., Dance, N. A., Millard, M., Wick, M. J., DesRochers, T. M., & Appleton, K. M. (2023). Abstract 2275: Preclinical testing of therapeutic biologics using patient-derived 3D spheroids. *Cancer Research, 83*(7_Supplement), 2275–2275. https://doi.org/10.1158/1538-7445.AM2023-2275

Lawrence, R., Munn, K., Naser, H., Thomas, L., Haboubi, H., Williams, L., Doak, S., & Jenkins, G. (2023). The PIG-A gene mutation assay in human biomonitoring and disease. *Environmental and Molecular Mutagenesis, 64*(8–9), 480–493. https://doi.org/10.1002/em.22577

Leehan, K. M., & Koelsch, K. A. (2015). T cell ELISPOT: For the identification of specific cytokine-secreting T cells. In B. T. Kurien & R. H. Scofield (Eds.), *Western blotting: Methods and protocols* (pp. 427–434). Springer. https://doi.org/10.1007/978-1-4939-2694-7_43

Levin, S., Semler, D., & Ruben, Z. (1993). Effects of two weeks of feed restriction on some common toxicologic parameters in Sprague-Dawley rats. *Toxicologic Pathology, 21*(1), 1–14. https://doi.org/10.1177/019262339302100101

Liu, J., Jia, S., Yang, Y., Piao, L., Wang, Z., Jin, Z., & Bai, L. (2023a). Exercise induced meteorin-like protects chondrocytes against inflammation and pyroptosis in osteoarthritis by inhibiting PI3K/Akt/NF-κB and NLRP3/caspase-1/GSDMD signaling. *Biomedicine & Pharmacotherapy, 158*, 114118. https://doi.org/10.1016/j.biopha.2022.114118

Liu, Y., Li, H., Luo, Z., Yu, Y., Yang, J., Zhang, M., Law, B. Y. K., Huang, Z., & Li, W. (2023b). Artesunate, a new antimalarial clinical drug, exhibits potent anti-AML activity by targeting the ROS/Bim and TFRC/Fe2+ pathways. *British Journal of Pharmacology, 180*(6), 701–720. https://doi.org/10.1111/bph.15986

Lorenzo-Herrero, S., Sordo-Bahamonde, C., González, S., & López-Soto, A. (2019). A flow Cytometric NK cell-mediated cytotoxicity assay to evaluate anticancer immune responses in vitro. *Methods in Molecular Biology (Clifton, N.J.), 1884*, 131–139. https://doi.org/10.1007/978-1-4939-8885-3_8

Lorke, D. (1983). A new approach to practical acute toxicity testing. *Archives of Toxicology, 54*(4), 275–287. https://doi.org/10.1007/BF01234480

Lovelace, P., & Maecker, H. T. (2018). Multiparameter intracellular cytokine staining. In T. S. Hawley & R. G. Hawley (Eds.), *Flow cytometry protocols* (pp. 151–166). Springer. https://doi.org/10.1007/978-1-4939-7346-0_9

Lu, Y., Wang, Y., Wang, Y., Wu, Y., Huang, Y., Liu, X., Zhang, S., Zhong, S., Li, Y., Li, B., Sun, J., Fang, S., & Zhou, H. (2023). M1-like macrophages modulate fibrosis and inflammation of orbital fibroblasts in graves' Orbitopathy: Potential relevance to soluble Interleukin-6 receptor. *Thyroid: Official Journal of the American Thyroid Association, 33*(3), 338–350. https://doi.org/10.1089/thy.2022.0254

Luque, R., Brieva, J. A., Moreno, A., Manzanal, A., Escribano, L., Villarrubia, J., Velasco, J. L., López-Jiménez, J., Cerveró, C., Otero, M. J., Martínez, J., Bellas, C., & Roldán, E. (1998). Normal and clonal B lineage cells can be distinguished by their differential expression of B cell antigens and adhesion molecules in peripheral blood from multiple myeloma (MM) patients—Diagnostic and clinical implications. *Clinical and Experimental Immunology, 112*(3), 410–418. https://doi.org/10.1046/j.1365-2249.1998.00600.x

Ma, S., Henson, E. S., Chen, Y., & Gibson, S. B. (2016). Ferroptosis is induced following siramesine and lapatinib treatment of breast cancer cells. *Cell Death & Disease, 7*(7), Article 7. https://doi.org/10.1038/cddis.2016.208

Maciejska, A., Skórkowska, A., Jurczyk, J., Pomierny, B., & Budziszewska, B. (2022). Biomarkers of neurotoxicity. In V. B. Patel, V. R. Preedy, & R. Rajendram (Eds.), *Biomarkers in toxicology* (pp. 1–30). Springer International Publishing. https://doi.org/10.1007/978-3-030-87225-0_17-1

Marcenaro, S., Gallo, F., Martini, S., Santoro, A., Griffiths, G. M., Arricó, M., Moretta, L., & Pende, D. (2006). Analysis of natural killer-cell function in familial hemophagocytic lymphohistiocytosis (FHL): Defective CD107a surface expression heralds Munc13-4 defect and discriminates between genetic subtypes of the disease. *Blood, 108*(7), 2316–2323. https://doi.org/10.1182/blood-2006-04-015693

Masel, J., Arnaout, R. A., O'Brien, T. R., Goedert, J. J., & Lloyd, A. L. (2000). Fluctuations in HIV-1 viral load are correlated to CD4+ T-lymphocyte count during the natural course of infection. *Journal of Acquired Immune Deficiency Syndromes (1999), 23*(5), 375–379. https://doi.org/10.1097/00126334-200004150-00003

Mata Forsberg, M., Arasa, C., van Zwol, W., Uzunçayır, S., Schönbichler, A., Regenthal, P., Schelin, J., Lindkvist-Petersson, K., Björkander, S., & Sverremark-Ekström, E. (2022). Activation of human γδ T cells and NK cells by Staphylococcal enterotoxins requires both monocytes and conventional T cells. *Journal of Leukocyte Biology, 111*(3), 597–609. https://doi.org/10.1002/JLB.3A1020-630RR

Mayorga, C., Çelik, G. E., Pascal, M., Hoffmann, H. J., Eberlein, B., Torres, M. J., Brockow, K., Garvey, L. H., Barbaud, A., Madrigal-Burgaleta, R., Caubet, J. C., & Ebo, D. G. (2023).

Flow-based basophil activation test in immediate drug hypersensitivity. An EAACI task force position paper. *Allergy, 79*, 580–600. https://doi.org/10.1111/all.15957

Millán, O., & Brunet, M. (2015). Flow cytometry as platform for biomarker discovery and clinical validation. In V. R. Preedy & V. B. Patel (Eds.), *General methods in biomarker research and their applications* (pp. 141–164). Springer Netherlands. https://doi.org/10.1007/978-94-007-7696-8_16

Mills, E. L., Kelly, B., Logan, A., Costa, A. S. H., Varma, M., Bryant, C. E., Tourlomousis, P., Däbritz, J. H. M., Gottlieb, E., Latorre, I., Corr, S. C., McManus, G., Ryan, D., Jacobs, H. T., Szibor, M., Xavier, R. J., Braun, T., Frezza, C., Murphy, M. P., & O'Neill, L. A. (2016). Succinate dehydrogenase supports metabolic repurposing of mitochondria to drive inflammatory macrophages. *Cell, 167*(2), 457–470.e13. https://doi.org/10.1016/j.cell.2016.08.064

Miranda, L. A., de Souza, V. V., Campos, R. A., de Campos, J. M. S., & da Silva Souza, T. (2023). Phytotoxicity and cytogenotoxicity of pesticide mixtures: Analysis of the effects of environmentally relevant concentrations on the aquatic environment. *Environmental Science and Pollution Research, 30*(52), 112117–112131. https://doi.org/10.1007/s11356-023-30100-y

Moldavan, A. (1934). Photo-electric technique for the counting of microscopical cells. *Science.* (New York, N.Y.), *80*(2069), 188–189. https://doi.org/10.1126/science.80.2069.188

Moore, C. L., Savenka, A. V., & Basnakian, A. G. (2021). TUNEL assay: A powerful tool for kidney injury evaluation. *International Journal of Molecular Sciences, 22*(1), 412. https://doi.org/10.3390/ijms22010412

Muller, P. Y., & Dieterle, F. (2009). Tissue-specific, non-invasive toxicity biomarkers: Translation from preclinical safety assessment to clinical safety monitoring. *Expert Opinion on Drug Metabolism & Toxicology, 5*(9), 1023–1038. https://doi.org/10.1517/17425250903114174

Nakajima, K., Crisma, A. R., Silva, G. B., Rogero, M. M., Fock, R. A., & Borelli, P. (2014). Malnutrition suppresses cell cycle progression of hematopoietic progenitor cells in mice via cyclin D1 down-regulation. *Nutrition (Burbank, Los Angeles County, Calif.), 30*(1), 82–89. https://doi.org/10.1016/j.nut.2013.05.029

Onisei, T., Tihăuan, B.-M., Dolete, G., Axinie (Bucos), M, Răscol, M., & Isvoranu, G. (2023). In vivo acute toxicity and immunomodulation assessment of a novel nutraceutical in mice. *Pharmaceutics, 15*(4), 1292. https://doi.org/10.3390/pharmaceutics15041292

Palmer, B. F. (2002). Renal dysfunction complicating the treatment of hypertension. *New England Journal of Medicine, 347*(16), 1256–1261. https://doi.org/10.1056/NEJMra020676

Park, L. M., Lannigan, J., & Jaimes, M. C. (2020). OMIP-069: Forty-color full spectrum flow cytometry panel for deep immunophenotyping of major cell subsets in human peripheral blood. *Cytometry Part A, 97*(10), 1044–1051. https://doi.org/10.1002/cyto.a.24213

Parker, J. C., & Horst, W. R. (1959). Method of simultaneously counting red and white blood cells (United States Patent US2875666A) https://patents.google.com/patent/US2875666A/en

Pathak, S., Rowczenio, D., Lara-Reyna, S., Kacar, M., Owen, R., Doody, G., Krause, K., Lachmann, H., Doffinger, R., Newton, D., & Savic, S. (2020). Evidence of B cell clonality and investigation into properties of the IgM in patients with Schnitzler syndrome. *Frontiers in Immunology, 11*. https://www.frontiersin.org/articles/10.3389/fimmu.2020.569006

Pei, Z., Qin, Y., Fu, X., Yang, F., Huo, F., Liang, X., Wang, S., Cui, H., Lin, P., Zhou, G., Yan, J., Wu, J., Chen, Z.-N., & Zhu, P. (2022). Inhibition of ferroptosis and iron accumulation alleviates pulmonary fibrosis in a bleomycin model. *Redox Biology, 57*, 102509. https://doi.org/10.1016/j.redox.2022.102509

Perazella, M. A. (2005). Drug-induced nephropathy: An update. *Expert Opinion on Drug Safety, 4*(4), 689–706. https://doi.org/10.1517/14740338.4.4.689

Phad, G. E., Pinto, D., Foglierini, M., Akhmedov, M., Rossi, R. L., Malvicini, E., Cassotta, A., Fregni, C. S., Bruno, L., Sallusto, F., & Lanzavecchia, A. (2022). Clonal structure, stability and dynamics of human memory B cells and circulating plasmablasts. *Nature Immunology, 23*(7), Article 7–Artic1085. https://doi.org/10.1038/s41590-022-01230-1

Plovins, A., Alvarez, A. M., Ibañez, M., Molina, M., & Nombela, C. (1994). Use of fluorescein-di-beta-D-galactopyranoside (FDG) and C12-FDG as substrates for beta-galactosidase detection by flow cytometry in animal, bacterial, and yeast cells. *Applied and Environmental Microbiology, 60*(12), 4638–4641. https://doi.org/10.1128/aem.60.12.4638-4641.1994

Porwit, A., Béné, M. C., Duetz, C., Matarraz, S., Oelschlaegel, U., Westers, T. M., Wagner-Ballon, O., Kordasti, S., Valent, P., Preijers, F., Alhan, C., Bellos, F., Bettelheim, P., Burbury, K., Chapuis, N., Cremers, E., Della Porta, M. G., Dunlop, A., Eidenschink-Brodersen, L., et al. (2023). Multiparameter flow cytometry in the evaluation of myelodysplasia: Analytical issues. *Cytometry Part B: Clinical Cytometry, 104*(1), 27–50. https://doi.org/10.1002/cyto.b.22108

Rodrigues, V., Baudier, J. B., & Chantal, I. (2017). Development of a bead-based multiplexed assay for simultaneous quantification of five bovine cytokines by flow cytometry. *Cytometry Part A, 91*(9), 901–907. https://doi.org/10.1002/cyto.a.23170

Salvatici, M., Cardinale, D., Colombo, A., & Sandri, M. T. (2015). Cancer cardiotoxicity and cardiac biomarkers. In V. R. Preedy & V. B. Patel (Eds.), *Biomarkers in cancer* (pp. 73–105). Springer Netherlands. https://doi.org/10.1007/978-94-007-7681-4_20

Schoolwerth, A. C., Sica, D. A., Ballermann, B. J., & Wilcox, C. S. (2001). Renal considerations in angiotensin converting enzyme inhibitor therapy. *Circulation, 104*(16), 1985–1991. https://doi.org/10.1161/hc4101.096153

Segal, B. H., Leto, T. L., Gallin, J. I., Malech, H. L., & Holland, S. M. (2000). Genetic, biochemical, and clinical features of chronic granulomatous disease. *Medicine, 79*(3), 170–200. https://doi.org/10.1097/00005792-200005000-00004

Shabab, T., Khanabdali, R., Moghadamtousi, S. Z., Kadir, H. A., & Mohan, G. (2017). Neuroinflammation pathways: A general review. *International Journal of Neuroscience, 127*(7), 624–633. https://doi.org/10.1080/00207454.2016.1212854

Sharma, R., Sinha, R., Kaur, R., & Rani, S. (2023). Drug-induced nephrotoxicity and use of biomarkers. In V. B. Patel, V. R. Preedy, & R. Rajendram (Eds.), *Biomarkers in toxicology* (pp. 797–829). Springer International Publishing. https://doi.org/10.1007/978-3-031-07392-2_50

Soddu, E., Rassu, G., Giunchedi, P., Sarmento, B., & Gavini, E. (2015). From naturally-occurring neurotoxic agents to CNS shuttles for drug delivery. *European Journal of Pharmaceutical Sciences: Official Journal of the European Federation for Pharmaceutical Sciences, 74*, 63–76. https://doi.org/10.1016/j.ejps.2015.04.005

Söderblom, T., Oxhamre, C., Torstensson, E., & Richter-Dahlfors, A. (2003). Bacterial protein toxins and inflammation. *Scandinavian Journal of Infectious Diseases, 35*(9), 628–631. https://doi.org/10.1080/00365540310016303

Tavares, R. S., Mansell, S., Barratt, C. L. R., Wilson, S. M., Publicover, S. J., & Ramalho-Santos, J. (2013). P,p′-DDE activates CatSper and compromises human sperm function at environmentally relevant concentrations. *Human Reproduction (Oxford, England), 28*(12), 3167–3177. https://doi.org/10.1093/humrep/det372

Tian, S., Hirshfield, K. M., Jabbour, S. K., Toppmeyer, D., Haffty, B. G., Khan, A. J., & Goyal, S. (2014). Serum biomarkers for the detection of cardiac toxicity after chemotherapy and radiation therapy in breast cancer patients. *Frontiers in Oncology, 4*. https://www.frontiersin.org/articles/10.3389/fonc.2014.00277

Tu, P., Chi, L., Bodnar, W., Zhang, Z., Gao, B., Bian, X., Stewart, J., Fry, R., & Lu, K. (2020). Gut microbiome toxicity: Connecting the environment and gut microbiome-associated diseases. *Toxics, 8*(1), 19. https://doi.org/10.3390/toxics8010019

Tuschl, H., & Schwab, C. E. (2005). The use of flow cytometric methods in acute and long-term in vitro testing. *Toxicology In Vitro, 19*(7), 845–852. https://doi.org/10.1016/j.tiv.2005.06.026

Tyas, S. L., Manfreda, J., Strain, L. A., & Montgomery, P. R. (2001). Risk factors for Alzheimer's disease: A population-based, longitudinal study in Manitoba, Canada. *International Journal of Epidemiology, 30*(3), 590–597. https://doi.org/10.1093/ije/30.3.590

Ueoka, A. R., Sufi, B. S., Magalhães, W. V., Fernandes, L., Andreo-Filho, N., Leite-Silva, V. R., & Lopes, P. S. (2023). Flow cytometry as an alternative method to evaluate genotoxicity of natural cosmetic actives. *Journal of Cosmetic Dermatology, 22*(3), 958–968. https://doi.org/10.1111/jocd.15473

Vaidya, V. S., Ferguson, M. A., & Bonventre, J. V. (2008). Biomarkers of acute kidney injury. *Annual Review of Pharmacology and Toxicology, 48*, 463–493. https://doi.org/10.1146/annurev.pharmtox.48.113006.094615

Vidal-Pedrola, G., Naamane, N., Cameron, J. A., Pratt, A. G., Mellor, A. L., Isaacs, J. D., Scheel-Toellner, D., & Anderson, A. E. (2023). Characterization of age-associated B cells in early drug-naïve rheumatoid arthritis patients. *Immunology, 168*(4), 640–653. https://doi.org/10.1111/imm.13598

Vowells, S. J., Sekhsaria, S., Malech, H. L., Shalit, M., & Fleisher, T. A. (1995). Flow cytometric analysis of the granulocyte respiratory burst: A comparison study of fluorescent probes. *Journal of Immunological Methods, 178*(1), 89–97. https://doi.org/10.1016/0022-1759(94)00247-t

Wahid, B., Bashir, H., Bilal, M., Wahid, K., & Sumrin, A. (2017). Developing a deeper insight into reproductive biomarkers. *Clinical and Experimental Reproductive Medicine, 44*(4), 159–170. https://doi.org/10.5653/cerm.2017.44.4.159

Ward, E. J., Bert, S., Fanti, S., Malone, K. M., Maughan, R. T., Gkantsinikoudi, C., Prin, F., Volpato, L. K., Piovezan, A. P., Graham, G. J., Dufton, N. P., Perretti, M., Marelli-Berg, F. M., & Nadkarni, S. (2023). Placental inflammation leads to abnormal embryonic heart development. *Circulation, 147*(12), 956–972. https://doi.org/10.1161/Circulationaha.122.061934

Wu, D. Y., Patti-Diaz, L., & Hill, C. G. (2010). Development and validation of flow cytometry methods for pharmacodynamic clinical biomarkers. *Bioanalysis, 2*(9), 1617–1626. https://doi.org/10.4155/bio.10.33

Wysowski, D. K., & Swartz, L. (2005). Adverse drug event surveillance and drug withdrawals in the United States, 1969-2002: The importance of reporting suspected reactions. *Archives of Internal Medicine, 165*(12), 1363–1369. https://doi.org/10.1001/archinte.165.12.1363

Yuan, J., & Ofengeim, D. (2023). A guide to cell death pathways. *Nature Reviews. Molecular Cell Biology, 25*, 379–395. https://doi.org/10.1038/s41580-023-00689-6

Zhang, Y., Wang, X., Zhao, Y., Zhao, J., Yu, T., Yao, Y., Zhao, R., Yu, R., Liu, J., & Su, J. (2023). Reproductive toxicity of microplastics in female mice and their offspring from induction of oxidative stress. *Environmental Pollution, 327*, 121482. https://doi.org/10.1016/j.envpol.2023.121482

Flowcytometry-Based Genotoxicity Assays

Sneha Verma, Priti Prabha, Shambhavi Jha, Priyanka Singh, and Alok Kumar Pandey

Abstract

Flow cytometry is an advanced technology that relies on the idea that certain fluorescent probe-labeled cells would scatter light and emit fluorescence when they move across a laser beam. Given that it enables quick, comparatively quantitative, multiparametric examination of cell populations on a single-cell basis, it has a number of special benefits. It also makes it possible to physically sort the cells in order to divide the subpopulation according to various criteria. Due to the great sensitivity and precision of this technology, research in the subject of flow cytogenetics, which is the analysis and sorting of chromosomes using flow cytometric techniques, has increased during the past 40 years. Nucleic acid-specific fluorescent dyes have reduced the subjectivity of previous slide-based approaches by enabling partially automated quantitative chromosomal analysis. Cell cycle analysis is still a developing area among those who use flow cytometry due to the discovery of new technologies and more cell cycle phase markers. Chromosomal aberrations can be detected by flow cytometric categorization of chromosomes (flow karyotyping), and large-scale isolation of individual chromosome types is possible for chromosome genomics through flow sorting. This chapter discusses the uses, limitations, and potential future developments of flow cytometry in genotoxicity assessment.

Keywords

Flow cytometry · Genotoxicity · Fluorescent probe-labeled cells · Chromosomal aberrations

---

S. Verma · P. Prabha · S. Jha · P. Singh · A. K. Pandey (✉)
Nanomaterial Toxicology Laboratory, Drug & Chemical Toxicology Group (FEST Division), CSIR-Indian Institute of Toxicology Research, Lucknow, India
e-mail: alokpandey@iitr.res.in

AB Pant et al. (eds.), *Flow Cytometry: Applications in Cellular and Molecular Toxicology*, https://doi.org/10.1007/978-981-97-9758-5_6

1 Introduction

Initially developed in the late 1960s, flow cytometry is a very effective and adaptable instrument for the quantitative and qualitative examination of individual cells (Schmit et al., 2021). The cells travel over several lasers in a continuous stream while staying in a buffered solution. Every particle that passes is counted and examined for several fluorescence metrics as well as the visible light scatter. Though information can be obtained in any direction from visible light dispersion, a standard cytometer produces two signals: a forward light scatter signal making use of scatter angles close to the direction of the laser beam and a wide-angle side scatter signal using angles about 90 degrees. A particle's size is indicated by the forward scatter (FSC), and the cell's granularity is shown by the side scatter (SSC) as shown in Fig. 1. Furthermore, depending on the accessibility of dyes and fluorochromes, individual samples can be labeled with antibody combinations and dual fluorescent dyes or their fluorescence can be assessed according to the fluorescence protein expression (Hulett et al., 1969).

The advancement of equipment and dyes in recent years has resulted in an astonishing expansion in the number of potential parameters that may be employed in analysis and experimentation with flow cytometry. The significant advancements in polymer dyes, fluorochromes, and fluorescent proteins make it feasible to work with multiple parameters at once. Furthermore, developments in machine technology have made it possible to employ many laser systems on a daily basis for a system

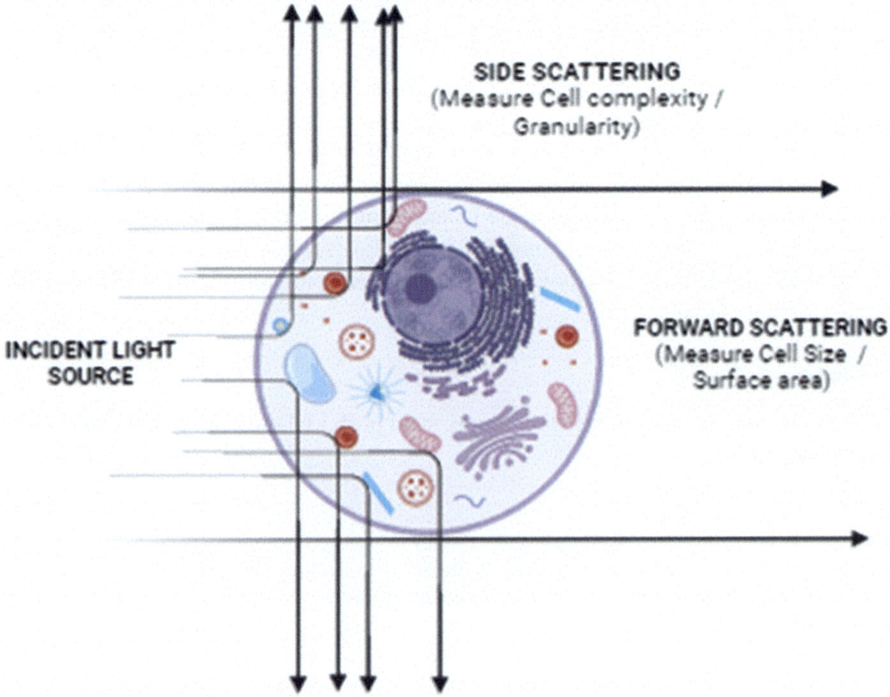

Fig. 1 Scattering of light. (Created by Bio Render)

that combines flow cytometry and microscopy devices with the goal of sorting individual cells. Protocols for flow cytometry can be categorized into different classes: droplet fluorescent flow cytometry, high throughput screening applications, isolation of certain cell types or subpopulations, and cell surface display (Vitelli et al., 2021).

Originally designed as an automated blood cell counting instrument to replace microscopy, flow cytometry's advancement was primarily motivated by its clinical applications. Flow cytometry has demonstrated significant utility in clinical settings and has been used increasingly in immunophenotyping over the past 20 years. In conjunction with multiparametric flow cytometric techniques, immunophenotyping has been shown to be a prognostic as well as a disease-specific diagnostic tool.

The principles of chromosomal sorting and analysis using flow cytometrics, or flow cytogenetics, have been studied more over the past 40 years because of the technique's great sensitivity and precision. The independent histories of genetics and cytology were combined to become cytogenetics, a new field of study, with Walter Sutton's (1903) proposal of the chromosomal hypothesis of inheritance (Sutton, 1903). The advancement in this subject was mostly dependent on the creation of increasingly sophisticated methods for chromosomal isolation, labeling, and differentiation. The chromosomes mounted on a flat surface were analyzed by all these techniques until Grey et al. used flow cytometry to classify and sort the Chinese hamster cell chromosomes according to dye concentration, ushering in a new era of flow cytogenetics (Gray et al., 1975b).

Flow cytometry's adaptability is its most important attribute, which accounts for its wide range of applications in the present and the future. A suspension of particles labeled with dye or probe that can be detected optically is the only strict requirement needed for a successful flow cytometric study. It offers a ton of data on cell cycle status, apoptosis, ploidy, genomic structure, and viability of cells. As a result, in many other biological research domains such as immunology, pathology, and oncology, it has become an indispensable tool. The present book chapter outlines the utility of flow cytometry in genotoxicity assays, recent advances, limitations, and future perspectives.

2 Genotoxicity Assays Based on Flow Cytometry

The genotoxicity profile of pharmaceuticals and biological/chemical agents is ascertained through the examination of genomic instability and mutation frequency, which is helpful in the safety assessment of novel medications and goods (da Rocha et al., 2019; Hassani et al., 2019; Paz et al., 2020; Shokrzadeh et al., 2021; Shokrzadeh et al., 2020). Furthermore, a multitude of breakdown products, chemical synthesis intermediates and contaminants may necessitate testing; hence, the creation of novel technologies is imperative for the dependable and precise evaluation of genotoxic potential (Custer & Sweder, 2008; Gollapudi & Krishna, 2000; McKim & James, 2010). Consequently, a number of high-throughput, quick tests have been created that promise to anticipate the results of the essential battery of tests for genetic toxicology (Knight et al., 2009; Westerink et al., 2009). Table 1 summarizes the Genotoxic Potential of biological/chemical agents based on Flow Cytometric Assays.

There are various applications of flow cytometry in genotoxicity assays, outlined in Fig. 2.

2.1 Cell Cycle Analysis

Cell cycle is the repeated cell division to perform life activities for growth. It is divided into mitosis (M) phase & interphase. The interphase is divided into G1, where many proteins get ready for the preparation of DNA synthesis; S phase, where the DNA gets replicated; the so-called synthesis phase & last G2 phase, where the cells get prepared for cell division. But when the cells immediately halt or do not move towards their next division, they remain in the quiescent phase, the G0 phase. During the M (mitosis) phase, the cells complete their cell division, where they get divided into two daughter cells. Each daughter cell contains the same genetic material as like parent cells & equal amount of other cellular components. Flow cytometry is the most used method to analyze cell cycle, which is based on DNA content analysis.

DNA Content & Ploidy

Flow cytometry works effectively for analyzing DNA due to the availability of dyes that bind DNA linearly and proportionately. This makes it possible to quantify the amount of DNA present, which in turn makes it possible to distinguish between diploid normal cells that are at rest and those that are actively synthesizing DNA, as well as between premitotic and mitotic cells. Once these cell cycle phases have been discovered, protein expression and cell cycle stages may be connected by combining DNA content analysis with antibodies that detect proteins associated to the cell cycle. It will also be possible to identify the cells that are actively synthesizing DNA by adding bromodeoxyuridine (BrdU), a thymidine analog, throughout the culture process. DNA content analysis can also be used to assess chromosomal numbers, whether they are abnormal or not. So, it is possible to evaluate ploidy (for example, in megakaryocytes) or aneuploidy (for example, in neoplastic illness).

Each autosomal chromosome as well as each sex chromosome (XX in females and XY in men) is found in two copies in mammalian cells. Ploidy is the term for multiples of whole pairs of homologous chromosomes which are represented by the numbers 4n, 8n, 16n, 32n, etc. Aneuploidy is the term for an abnormally high number of individual chromosomal pairs. This can result from a variety of conditions, such as monosomy (having one chromosome for each pair), trisomy (having triplets of a certain pair), or polysomal copies of certain chromosomes (for instance, chromosomes 1 and 7 in breast cancer) (Davies & Allen, 2007). It is, therefore, possible to determine the type of cell by analyzing its DNA content. For e.g., a cell that has undergone endoreduplication (replication of DNA without dividing cells) exhibits abnormal chromosomes, which is a result of aberrant mitosis, a condition observed in cancerous neoplastic cells and a number of human disorders, including Down's syndrome (trisomy of chromosome no. 21).

Table 1 Genotoxic potential of biological/chemical agents based on flow cytometric assays

S.No.	Genotoxic agent	Cell Type	Parameter analyzed by flow cytometer	Results	References
1	CuO,ZnO, TiO_2, $Fe_3O_4$and Ag nanoparticles	Chinese hamster ovary(CHO)-K1 cells	γ-H_2AX	Significant γ-H2AX generation was generated by CuO and ZnO particles, and this production was dose and time-dependent. Small increase in generation of γ-H_2AX in case of TiO_2, Ag, and Fe_3O_4 nanoparticles.	Toduka et al. (2012)
2.	17 known *in vivo* genotoxins and 31 known *in vivo* non genotoxins	TK6 cells	Micronucleus assay	It accurately forecasts the *in vivo* genotoxicity of a broad variety of substances, including commercially available medications.	Franz et al. (2020)
3	Dimethylbenz[a]anthrace (DMBA), 2acetylaminofluorene (2-AAF), benzo[a]pyrene (B[a]P), methyl nitrosourea, (MNU) methyl methane sulphonate (MMS) and 4-nitroquinoline oxide (4NQO)	Mouse lymphoma L5178Y cells	γ- H_2AX	γ-H_2AX increased significantly (≥1.5 times) in response to prototypical genotoxins concentration-dependently.	Smart et al. (2011)
4	TiO_2 nanomaterials(anatase, 50–150 nm, anatase, 5–8 nm, rutile, 20–28 nm)	BEAS-2B cells	Micronucleus assay	All three of the studied TiO2 compounds had a modest genotoxic effect. Anatase (50–150 nm) was most potent	Di Bucchianico et al. (2017)
5.	Nickel oxide (NiO) NPs	Human bronchial epithelial cells (HBEC-3kt) and THP-1 derived macrophages	Micronucleus assay	When the HBEC were exposed indirectly, no MN production was seen either. Nonetheless, in THP-1 monoculture, the exposure to NiO NPs clearly induces MN.	Vallabani and Karlsson (2022)

(continued)

Table 1 (continued)

S.No.	Genotoxic agent	Cell Type	Parameter analyzed by flow cytometer	Results	References
6.	64 chosen substances with established genotoxic characteristics	HepG2 cells	γ- H_2AX	In contrast to previous combinations, the γH2AX test produced higher accuracy (62–70%) when it was included in the standard battery instead of the MLA assay.	Tsamou et al. (2012)
7.	31 compounds with well proven mode of action	TK6 cells	γ- H_2AX	19/20 (95%) genotoxicants and 10/11(91%) nongenotoxicants were categorized accurately.	Wilde et al. (2017)
8.	8 prototypical clastogens, 8 aneugens, and 15 nongenotoxicants	CHO-K1 cells	Micronucleus assay	The resultant flow cytometric MN data showed a strong correlation with the anticipated in vitro cytogenetics: Specificity = 14/15, sensitivity = 16/16	Bryce et al. (2011)
9.	Industrial waste waters	RTG-2 (rainbow trout gonads)	Micronucleus assay	In fourteen of them, the frequency of micronuclei significantly increased.	Kohlpoth et al. (1999)
10.	Mitomycin C (MMC) and Nitrofurazone (NFZ)	*E.coli* strain, MG1655 and HME63	Tunnel Assay	To identify free 3-OH DNA ends in the E. coli genome, terminal deoxyribonucleotide transferase (TdT)-mediateddUTP nick end labelling (TUNEL) can enzymatically label ssDNA ends with bromo-deoxy uridine triphosphate (BrdU). Fluorescence microscopy and flow cytometry are two methods for identifying and measuring labelled DNA ends.	Behrmann and Trakselis (2022)
11.	Thiophanate-methyl	Human Blood Pheripheral Lymphocytes	Fluorescence *in situ* hybridization (FISH)	Cellular proliferation was delayed and the frequency of apoptotic cells was increased	Fimognari et al. (1999)
12.	S9-generated metabolites	TK6 cells	GADD45a-GFP Assay	The majority of pro-genotoxins tested was detected using Flow Cytometry	Jagger et al. (2008)

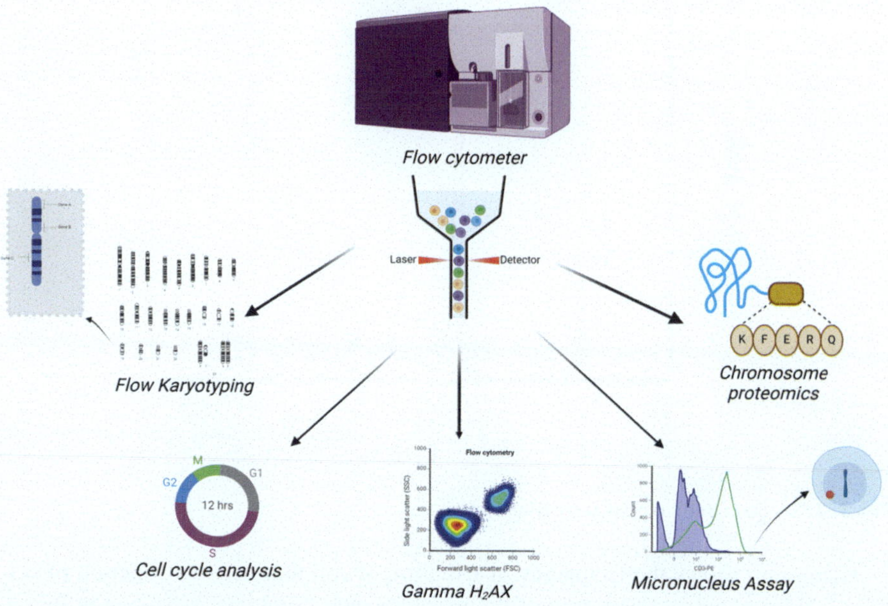

Fig. 2 Application of flow cytometry in genotoxicity assays. (Created by BioRender)

There are some commonly used methods for flow cytometry cell cycle analysis. The first two are based on the study of the cellular DNA content using a single variable and the deconvolution of the cellular DNA content frequency histograms as well as differentiation between noncyclic & cyclic proteins connected to proliferation. The third one is based on the identification of mitotic cells by histone H3 phosphorylation. The bivariate analysis of proteins connected to proliferation and DNA content forms the basis of the fourth strategy. In the fifth strategy, the DNA-replicating cells are identified by looking for the incorporation of 5′-bromo2′-deoxyuridine (BrdU). These techniques make it possible to identify apoptotic cells containing fractions of DNA and show the distribution of cells over the three primary cycle stages (G1, S, G2/M) (Darzynkiewicz et al., 2001).

2.1.1 The Analysis of Cellular DNA Content in Univariate Form to Distinguish Between Cells in Go, S, and G2/M Phase

DNA content during the cell cycle serves as an indicator of cellular maturity. DNA is present in cells in the Go/G1 phase set to one unit, whereas cells in the S phase have phase replicated DNA, which increases its content proportionately as it moves via S phase; upon going into the G2 phase, and subsequently M phase, the amount of Go/G1 phase DNA is doubled. Univariate study of the amount of DNA in cells is the most widely used technique to identify the frequency of cells in Go/1, S, and G2/M phase cells. Propidium iodide (PI) or 4′6′-diamidino-2-phenylindole (DAPI), both of which are specific fluorochromes for nucleic acids, are used in most tests to fix or permeabilize the cells.

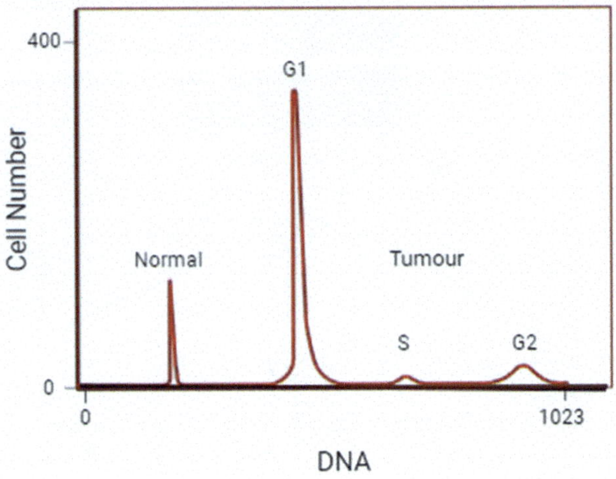

Fig. 3 DNA histogram. (Created by BioRender)

Histogram shows the frequency distribution of cell number as a function of cellular DNA content as the output of fluorescence measurements, as shown in Fig. 3. The histograms present the proportion of the cells in each phase of the cycle according to variations in fluorescence intensity. In certain tumor forms, the increased cell frequency in the S and G2/M phases is thought to suggest enhanced proliferative capability, which is associated with a negative prognostic characteristic. When asynchronously growing cells are exposed to an agent that disrupts the cell cycle either *in vitro* or *in vivo*, the cells gather in the phase of the cycle where the disruption to their progression is greatest. This is represented by a rise in the equivalent proportion of cells in that stage on histograms of DNA content frequency (Crissman & Tobey, 1990). In contrast, frequency histograms cannot provide information regarding cell kinetics or the pace at which the cell cycle progresses, although the duration of the cell cycle (Tc) or the time it takes for cells to double in culture is necessary for these analyses. With this information, the length of a specific cell cycle phase (S; Ts) may be calculated using the following equation:

Ts / TcInFs1 / In2

where Fs is an estimated proportion of S phase cells derived from the frequency histogram of DNA content, Ts is the amount of time of the S phase, and Tc is the length of the cell cycle or the period during which cells double (Darzynkiewicz et al., 2001).

The DNA histogram's Quality

Width of DNA peak from cells throughout the G1 phase of the cell cycle indicates the DNA histogram's quality. This is determined by calculating the coefficient of variation (CV) over the peak by taking the standard deviation using the formula:

$$\mathbf{CV = 100 \times SD / (peak\ channel)\%}$$

The mean channel number of the peak is represented by the peak channel.

The size of the genome, or the amount of nuclear DNA, is a crucial aspect of biodiversity, and research on it provides a solid biological unifier with analytical and practical value. The haploid nuclear DNA's content that is not duplicated chromosomal complement is represented by the C-value. Flow cytometry may be utilized to measure the quantity of DNA or DNA content present in cells to display the distribution of cells during the main cell cycle phases.

2.1.2 Using Proliferation-Associated Proteins to Distinguish Between Proliferating Cells and Non Cycling (Go; Quiescent) Cells

Prognostic implications may arise from the number of quiescent or G0 cells, as they are frequently indicative of tumor proliferative capacity. Nevertheless, in numerous cell types, characteristics apart from DNA content—such as low RNA levels that illustrate Go cells' low ribosome count, differentiate G0 from G1 cells (Darzynkiewicz et al., 1976). Therefore, depending on this variation in the RNA composition, differentiation between G0 and G1 cells is possible using bivariate DNA vs RNA content analysis. There are commercially available antibodies against several "proliferation-associated proteins," which are expressed differently in cycling and noncycling cells. Hence, noncycling cells may be identified immunocytochemically from cells based on the existence or lack of these proteins. The immunocytochemical marker is frequently employed in conjunction with DNA content measurement during cytometric analysis to show the specificity of marker expression with respect to the cell cycle phase. The antibody Ki-67 is the most widely utilized immunocytochemical indicator for cycling cells. The role of the protein Ki-67 is still unclear, although it has been shown that this marker is widespread in cycling cells but commonly lacking in noncyclic normal and malignant cells (Endl et al., 2001). The cell cycle status can be determined by using flow cytometric measurement of DNA content and Ki-67. According to earlier research, a tiny percentage of G2/M cells had a markedly elevated Ki-67 level (Landberg et al., 1990). Although it is still up for debate, these cells were thought to be the earliest mitotic cells. Other markers, such as Cyclins and phospho-Ser10-histone H3 (which detects M phase), must be coupled in order to further distinguish M phase cells (Juan et al., 2001; Landberg et al., 1990; Vignon et al., 2013).

Protocol for Cell Cycle Analysis Using Ki-67 Antibody and DNA Content

This protocol for cell cycle analysis based on Ki-67 and DNA content has been taken from (Kim & Sederstrom, 2015).

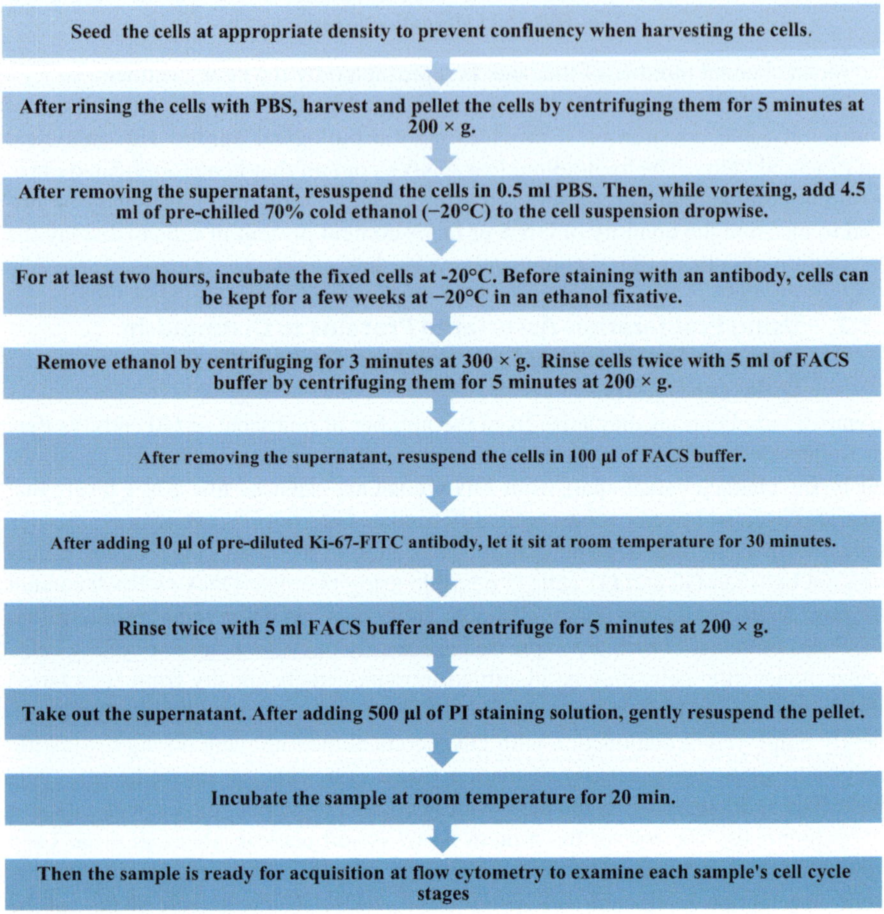

2.1.3 Phosphorylation of Histone (H3) Allows for the Identification of Mitotic Cells

Univariate cellular DNA content analysis cannot distinguish between cells in the G2 and M phase of the cell cycle since their DNA contents are identical. For this purpose, a variety of mitotic cell markers have been used. One of the most focused methods relies on a bivariate comparison between the amount of cellular DNA and the phosphorylation of serine-10 residue. This occurs on almost all of the chromatin's histone H3 molecules from prophase to telophase. Recently, commercially accessible antibodies that selectively identify the histone H3 phosphorylated epitope (anti-H3Pt) have been developed; these antibodies serve as a useful indicator of the mitotic cell and may be employed to distinguish between M and G phases. (Juan et al., 2001) as shown in Fig. 4.

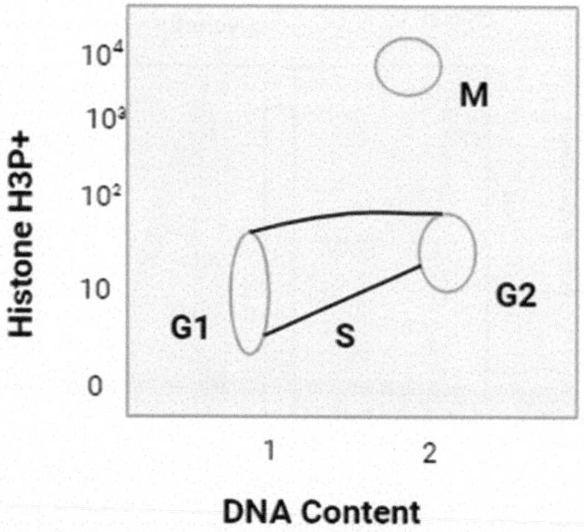

Fig. 4 Determination of mitotic cells. (Created by BioRender)

2.1.4 Bivariate Comparison of DNA Quantity and Expression of Cyclin D, E, A, or B1

The essential parts of the machinery that moves the cell cycle progression are cyclins. Numerous cyclins, notably D, E, A, and B cyclins, exhibit irregular, distinct expression timing during the undisturbed growth of normal cells (Darzynkiewicz et al., 1996). Moreover, to be divided into four main phases, the cell cycle may also be divided into several sub-compartments using the landmarks provided by the periodicity of cyclin expression. Additionally, it is feasible to distinguish among cells with the same amount of DNA existing throughout various stages of the cycle using a bivariate comparison of DNA content and cyclin expression. Figure 5 shows a diagrammatic representation of the distinct pattern of corresponding cyclins based on cell cycle position in normal, non-tumor cells ascertained by the DNA content of cells. Cyclin D was found only in G1 and a tiny percentage of G2/M phase cells; Cyclin D-negative cells are those that enter the S-phase, move through the S-phase, and spend the majority of G2/M in this state. During G1, there is variability in the expression of cyclin E; some cells express maximum levels of the protein, while others are cyclin E-negative. The highest level of cyclin E expression over the threshold is seen in cells that enter the S phase. Cyclin A is not expressed during G1, but it begins in S-phase, increases progressively during S, and culminates in G2. M phase cells do not express cyclin A. During the S phase, cyclin E decreases. Cyclic B1 is identified in late S & maximally expressed in the G2/M phase in cells.

2.1.5 Investigation of Cell Kinetics and Identification of DNA Replicating Cells Using the Bivariate Distributions of 5 Bromo-2′-Deoxyuridine (BrdU) Incorporation Versus DNA Content

A popular cytometric test for analyzing the cell cycle is one that compares DNA content and the bivariate incorporation of 5-bromo-2′-deoxyuridine (BrdU) (Gratzner, 1982). This analog of thymidine is introduced either *in vivo* or in *vitro to* living cells,

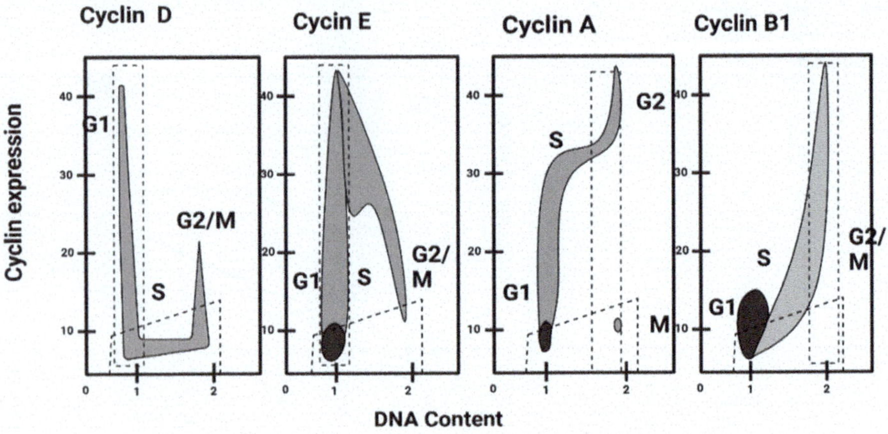

Fig. 5 Bivariate distribution of DNA content vs Cyclin D, E, A, and B1 expression

where it is integrated into DNA during replication and identified immunocytochemically as the incorporated precursor. The fluorescein-tagged anti-BrdU antibodies can be applied directly or indirectly, and the DNA is counterstained using an intercalating fluorochrome that fluoresces red, such as PI or 7-amino actinomycin D (7-AAD). The differentiation of S phase cells from G1 or G2/M phase is based on a bivariate analysis of BrdU antibody immunofluorescence vs DNA content. Additionally, S phase cells are identified by having integrated the green fluorescence and the precursor (BrdU).

Time-lapse analysis and the incorporation of BrdU can be used together for kinetic investigations of cell cycle progression. Following BrdU pulse labelling, cells are cultured without the precursor for a certain amount of time before fixation, in which the group of cells bearing the BrdU label transitions from S to G2, splits, and then returns to G1. The bivariate DNA content vs BrdU incorporation plots were used to determine the fraction of BrdU-labeled cells that entered G2, mitosis, or returned to G1, offering data for determining the cell cycle's kinetic characteristics (Begg et al., 1985; Terry & Whiter, 2001). With a DNA content of between 1 and 2 (arbitrary units), S-phase cells exposed to the precursor throughout the whole pulse are maximally and fairly evenly labelled. Cells exposed to a precursor for varying periods of time and displaying varying degrees of labeling were those that entered or exited the S phase (into G2) during the pulse. They have DNA contents of about 1 or 2, respectively (Fig. 6).

Flow Cytometry-Based Cell Progression Analysis Steps

The steps for cell cycle progression analysis from flow cytometry given below are according to Dubey et al. (2022).

Fig. 6 Diagrammatic representation of the subpopulations of cells identified by using bivariate distributions of DNA content vs BrdU immunofluorescence after BrdU incorporation or labelling

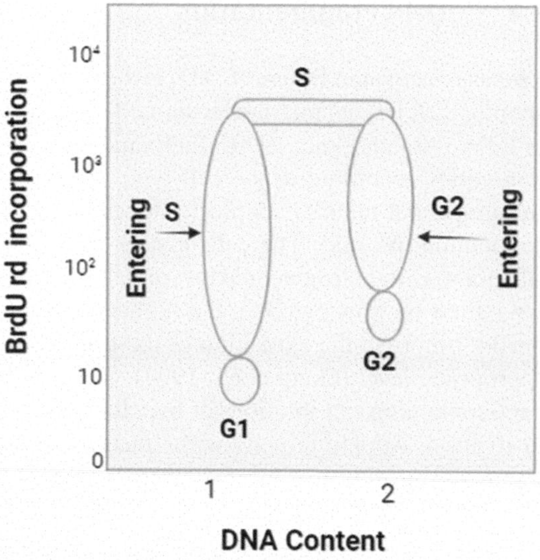

Seed 5x10^5 cells /well in a cell culture plate & incubate.

↓

Add respective treatment. Then add fresh complete media after removing treatment & incubate for next complete cycle.

↓

After harvesting of cells, complete media should be added to quench cells & make single cell suspension.

↓

Then centrifuge cell suspension and resuspend the pellet in 1 ml 70% ethyl alcohol and kept for 30 min at -20° to fix the cells.

↓

After centrifugation, resuspend the pellet in lysis buffer (1xPBS along with 0.2% triton-X-100) for 30 minutes

↓

Resuspend pellet in 1 ml of 1x PBS containing RNAse (10 mg/ml) & incubate.

↓

Resuspend in PI dye (1 mg/ml) & store at 4°C. Then the sample is ready for acquisition at flow cytometry.

2.2 DNA Fragmentation

Caspases-activated DNase (CAD) is the primary endonuclease that functions during apoptosis. It targets internucleosomal DNA and produces substantial fragmentation of DNA (Arends et al., 1990; Oberhammer et al., 1993). The level of DNA degradation varies according to the cell type, apoptosis stage, and frequently the type of substance that induces apoptosis. As a result, DNA's extractability changes during the staining process. The cells can be treated with precipitating fixatives such as ethanol to extract fragmented low molecular weight DNA. To accomplish their ideal separation by flow cytometry, a high-molarity phosphate-citrate buffer can be utilized to improve the extraction of the fragmented DNA from apoptotic cells to the appropriate level (Gong et al., 1994). Some techniques call for lysis cells in detergent-containing salt solutions or hypotonic buffers before DNA staining, as opposed to fixation. Apoptotic cells with fractured nuclei might release many chromatin fragments during lysis; hence, the sub-G1 signals on the DNA histogram might not represent individual apoptotic cells but rather chromatin fragments, isolated chromosomes, or cell debris. A quantitative estimation of the frequency of apoptosis cannot be obtained from cell lysis.

A reduction in the amount of DNA in apoptotic cells can be used to identify them due to the process of apoptosis-induced internucleosomal DNA fragmentation. One of the apoptotic enzymes that cleave DNA, caspase-activated DNase (CAD), is in an inhibitory complex (ICAD) until it is activated by effector caspase-3 or -7. During the staining process, tiny DNA fragments are removed, as well as apoptotic cells (APs) are subsequently identified as cells with a fractional DNA content (sub-G1) on DNA content frequency histograms (Figs. 7 and 8).

DNA Strand Break Detection

DNA fragmentation associated with apoptosis causes many nucleus DNA strand splits, especially as it advances to internucleosomal regions (Oberhammer et al., 1993). A deoxynucleotide that has been fluorochrome-tagged can be attached to the 3′ OH ends of the breaks to determine DNA strand break. TUNEL, which denotes TDT-mediated deoxyuridine triphosphate-biotin Nick-End Labelling, is the reaction's popular name. When it comes to reaction simplicity, affordability, and sensitivity, BrdUTP seems to be the best label for DNA breaks (Darzynkiewicz et al., 1992; Gold et al., 1994). 5′-bromo-2′-deoxyuridine (BrdU) incorporation identifies the cells that replicate DNA. During the S phase of DNA synthesis, an analog BrdU of thymine may inserted into the replicating DNA. In BrdU assay, DNA partially gets denatured by strong acid or heat. To get the best possible equilibrium between DNA sections that are non denatured (stainable with PI) and denatured (single-stranded, accessible to BrdU Ab), DNA denaturation by acid or heat is the crucial step (Dolbeare et al., 1983; Li & Darzynkiewicz, 1995).

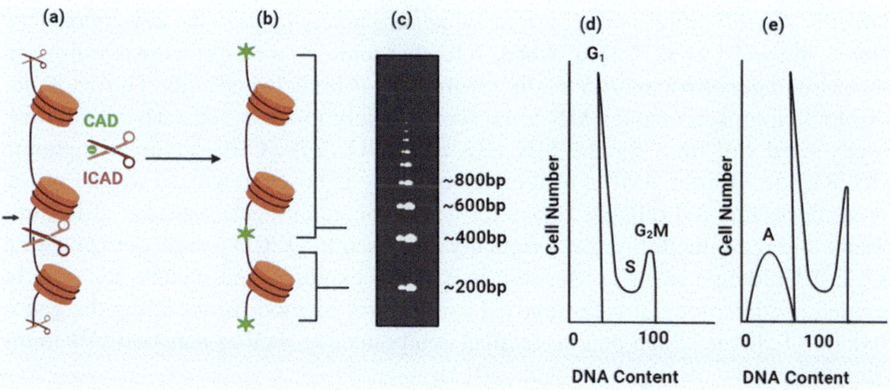

Fig. 7 DNA fragmentation during apoptosis via internucleosomal breaks. (Created by Biorender)

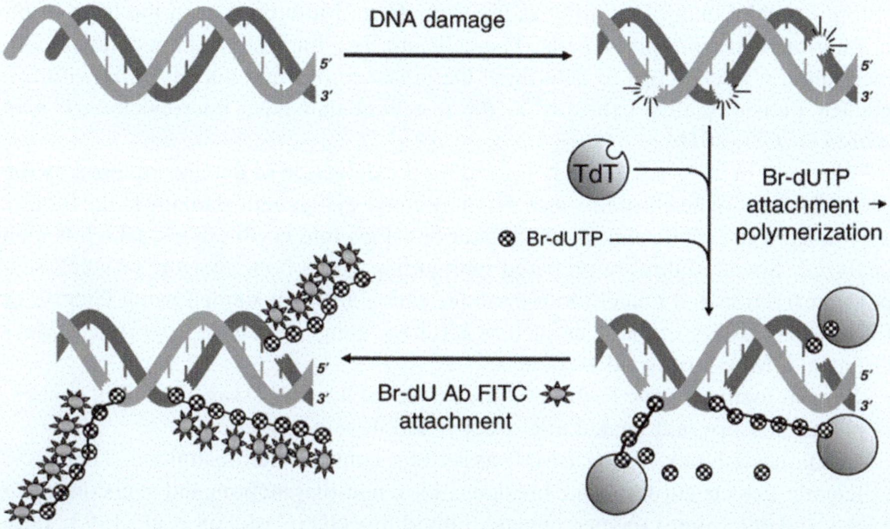

Fig. 8 Diagram illustrating the polymerization, immunocytochemical (FITC) detection of BrdU in DNA strand-break labelling, and TdT-mediated attachment of Br-dUTP to 3′OH ends. (Created by Biorender)

2.3 Micronucleus Formation

Micronuclei (MNs) are little extra nuclear entities that arise from the lagging chromosomes that appear during the anaphase of the cell cycle. They are tiny bodies made of chromatin that are about 1/5th to 1/20th the size of a nucleus (Fenech et al., 2003). When chromosome lesions fail to be reincorporated into a major nucleus after mitosis or meiosis, they extend into interphase and cause chromosome missegregation (numerical change, antigenicity), spindle failure, kinetochore damage, defect in the cell cycle control system, which ultimately leads to the development of MN (Landsiedel et al., 2022). Since, cells that undergo a single mitosis following exposure to genotoxicants are easily identifiable as binucleated cells. The cytokinesis-block approach enables MN to be scored solely in these cells. The most commonly used cell lines for the MN tests are CHO, V79, CHL, mouse lymphoma L5178Y, etc. (Carter, 1967). Micronucleus assay is now recognized as one of the most successful and reliable assays for genotoxic carcinogens because of its simplicity and reproducibility, i.e., originally proposed in OECD test recommendation (OECD Guideline no.487). Because it is being explored and widely accepted in regulatory toxicology, it is currently a well-known method for detecting the genotoxicity of chemical and pharmaceutical compounds as well as nanoparticles in *in-vitro* and *in-vivo* testing (Kim et al., 2013).

The *in-vitro* MN assay is based on exposing cells to the test substance with and without metabolic activity. After exposure to the test substance, cytochalasin B is added to prevent cytokinesis and the cells are cultured for another cycle to detect chromosomal damage, resulting in the formation of micronuclei in the form of bi- or multinucleate interphase cells. The cells are then harvested, stained, and examined under a microscope to determine the presence of micronuclei. Cells with full nuclear divisions after exposure to the test item that have micronuclei are then scored OECD (2016).

Whereas, in *in-vivo* MN assay is used for the detection of damage induced by the genotoxicants to the chromosomes that may cause cytogenetic damage to the mitotic apparatus of the erythroblast, which later develops into erythrocytes, which is then analyzed. Smear slides are prepared with either whole bone marrow or a cell suspension fractionated on cellulose column, which are then stained using Giemsa or fluorescent DNA-specific dyes such as acridine orange, Hoechst + pyronin Y, May-Grunwald, May-Grunwald + Giemsa.

Several methods have been developed to score the MN assay, including conventional microscopy, automated microscopy, and flow cytometry.

Cytokinesis-block MN (CBMN) assay is a conventional technique. This assay efficiently detects chromosome breakage, loss, non-disjunction, and excision repair events and determines the percentage of dividing cells (Castelain et al., 1993; Tates et al., 1990). In this assay, Cytochalsin-B is utilized to impede spindle assembly and stop the nuclear division-induced cytokinesis. In this method, the evaluation of binucleated cells is simple but tedious to observe under a microscope (George, 2010). A reliable, inexpensive automated image analysis scoring system is needed to eliminate methodological factors contributing to the high efficiency of variation observed with MN assays (George, 2010).

Automating the analysis can be done with high throughput techniques like flow cytometric MN analysis. Flow cytometry techniques distinguish between MN populations (released from cells following lysis) and nuclei using fluorescence and scatter intensities (Avlasevich et al., 2006; Bryce et al., 2013; Nüsse & Marx, 1997). Because the differences in DNA content between micronuclei and nuclei can be as much as a factor of 100, flow cytometric studies of micronuclei's DNA content are carried out using log mode. Lysing the cell membrane allows for rapid data collection from thousands of cells, but it also produces a debris-filled solution that makes scoring event verification difficult (Fenech, 2006). A gate should be established for data analysis based on the frequency and DNA distribution of micronuclei labeled with ethidium bromide. The Multispectral Imaging Flow Cytometry (MIFC) setup is also an optimal tool for conducting this work as it integrates both traditional flow cytometry and high-resolution microscope imaging (Basiji, 2016).

Protocol for Flow Cytometry-Based Micronucleus Assay Steps

The steps for the Micronucleus assay given below are according to Mourya et al. (2024).

Seed cells in 6 well plate (0.5×10^6 cells/well). Incubate for 18 hr.

↓

Extract medium and add respective treatment to cells along with control.

↓

Remove a treatment and add fresh complete medium and incubate for about 1.5 -2.0 normal cell cycle length.

↓

Transfer the medium into centrifuge tubes after incubation, then add trypsin-EDTA to harvest cells.

↓

To make a single cell suspension and halt the trypsin's enzymatic activity, add complete media.

↓

Centrifuge suspension for 10 minutes at 1200 rpm.

↓

To wash, discard the supernatant and resuspend the pellet in 1X PBS.

↓

Using 1200 rpm, centrifuge this suspension once more for 10 min, then discard the supernatant

↓

Re-suspend the pellet in solution 1, vortex for 2-3 sec, and then incubate at room temperature for one hour.

↓

Add Solution 2 into the centrifuge tube, whirl for 2-3 seconds, and then store at 4°C.

↓

Utilizing a flow cytometer, analyze.

Advantages and Limitations of Different MN Scoring Methods

In visual microscopy, scoring can be done manually by direct visualization of thousands of mononucleated, binucleated, and polynucleated cells for micronucleus frequency on a regular basis at any time, and slides can be stored. But it is a bit laborious task and low throughput as a limited number of cells can be scored manually. Microscope and slide quality can also vary according to laboratory conditions.

In Automated microscopy, scanning of cells in slides is done automatically, images can be stored and slides can be re-scanned. When compared with visual microscopy, we get improved throughput gain because 1000 to 2000 cells can be detected in 2–3 min (Rodrigues et al., 2018). A significant limitation for researchers who require more customization and flexibility in their experiments is that they need high-quality microscopic slides that can be varied.

MN and nuclei can be scored by conventional flow cytometry. In a matter of minutes, high throughput data can be collected and scored. It is simple to distinguish between visible nuclei from necrotic or apoptotic cells. The disadvantage of this method is the cell lysis that produces a debris-filled solution. Therefore, it cannot clearly distinguish between MN and DNA-positive detritus. Sometimes, it is impossible to quantify BN and POLY cells. One of the major drawbacks of automated flow cytometry is the lack of flexibility because automated systems are efficient at processing large volumes of samples, but they are limited in their ability to adapt to different experimental setups or protocols.

Imaging flow cytometry is a cutting-edge method that enables the potential for fully automated scoring of cells, including MONO, BN, POLY cells, and MN. This innovative technology allows for direct visualization of the required cell types, revolutionizing the way samples are analyzed. Through bright field imagery, multiple lasers, and image channels, this method allows cytoplasm to be visualized. This technology enables the simultaneous imaging of multiple cellular structures, including nuclei, centromeres, and telomeres. This method not only saves time but also improves the consistency and reproducibility of results. In minutes, high-speed data acquisition can be achieved for cellular imaging, allowing for multiple scoring. 2-D projection image of 3-D cell can hide some MN behind a main nucleus due to a analyses using stored algorithms (Rodrigues et al., 2018). They also help us to score BN and MN which can be implemented. Imaging flow cytometry enables researchers to directly visualize cells in real-time, providing valuable insights into their morphology and characteristics.

2.4 Gamma H2AX Foci

Repairing DSBs presents greater difficulty compared to many other lesions, and if they are repaired incorrectly (for example, by rejoining broken strands of DNA from different chromosomes incorrectly), it may cause cytotoxic effects or changes to the genome. DNA double-strand breaks are considered to be the hazardous type of DNA damage (DSBs), which are followed by phosphorylation of histone H2AX, which results in the production of γ-H2AX, close to broken DNA ends (Rogakou et al., 1998). γ-H2AX, a recently phosphorylated protein, is the starting point for recruiting and localising DNA repair proteins. The γ-H2AX test has been used for the past 10 years to establish a connection between γ-H2AX levels and DNA damage and repair in various cell and tissue types (Pilch et al., 2003; Redon et al., 2009;

Redon et al., 2011; Turner et al., 2011; Valdiglesias et al., 2013). Research has demonstrated that the numbers of DSBs are associated with the quantity of γ-H2AX focal locations in cell nuclei (Rothkamm & Löbrich, 2003). Since methods for detecting DNA double-strand breaks (DSBs) are typically time-consuming, advancements in these areas are crucial for scientists and medical professionals. Following DNA damage, the two primary detection techniques used to measure γ-H2AX are flow cytometry and immunofluorescence (Hernández et al., 2013). Immunofluorescent staining is often followed by a manual count of each γ-H2AX foci under a fluorescent microscope. The most efficient technique for detecting DSBs right now is flow cytometry because γ-H2AX foci may be directly quantified using fluorescence intensity measurements and the automated count of γ-H2AX positive cells. By utilizing an antibody against γ-H2AX, it is possible to assess DNA damage in a vast number of cells and track the distribution of cell cycles within cell populations. To evaluate flow cytometry's effectiveness in identifying γH2AX, Smart et al. (2011) tested certain medications, including genotoxic compounds with variable mechanisms of operation. To confirm the sensitivity of the procedure, the L5178Y cell line was utilized to compare the γH2AX results with the mouse lymphoma assay, the chromosomal aberration assay, or the Ames bacterial assay results. The validation set results showed 82% concordance between Ames positive compounds and flow cytometry-based measurement of γH2AX (Smart et al., 2011). In a different investigation, Tsamou et al. (2012) reached the same conclusion: one possible genotoxicity assay is the measurement of γH2AX using flow cytometry. Throughout the investigation, HepG2 cells were exposed to 64 carefully selected compounds having known genotoxic properties. The γH2AX assay was shown to be more accurate in measuring chemicals than other combinations of *in vivo* and *in vitro* assays, according to their findings (Tsamou et al., 2012).

Fundamentals of the γ-H2AX Flow Cytometric Test

Lately, flow cytometry γ-H2AX (FC γ-H2AX) was created as a high-throughput variant of the γ-H2AX staining technique for the identification and quantification of DSBs. A key feature of FC is its high-speed analysis; for example, 10 thousand of cells can be examined in a single second. Following treatment, a single-cell suspension needs to be prepared prior to fixation. Subsequently, cells are stained with either a γ-H2AX primary antibody with fluorescence labelling or γ-H2AX primary antibody combined with fluorescence labeling secondary antibody. Using numerous fluorescent antibodies or fluorescent stains, the FC γ-H2AX test offers the advantage of providing more information on other cellular processes. For instance, cell cycle information can be obtained through the use of propidium iodide (PI) staining. This is crucial for determining how the DNA damage response is timed. It is recommended to examine the side-scattered (SS) light and fluorescence signal simultaneously when using the FC γ-H2AX test system on particle-exposed cells. Finding SSC and γ-H2AX together can reveal details regarding the connection between particle incorporation and DSBs generation (Fig. 9).

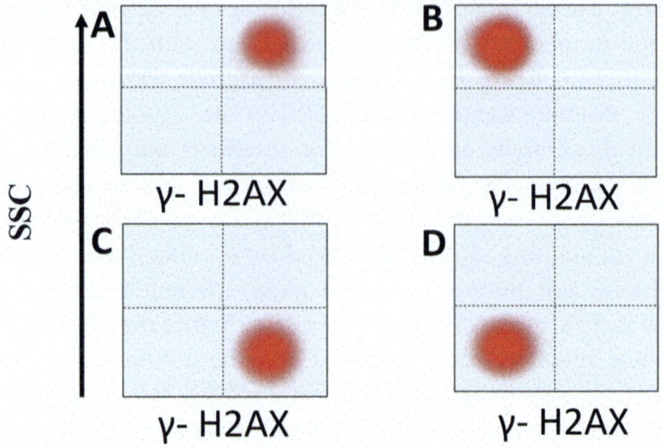

Fig. 9 Hypothetical dot plot patterns for SSC versus γ-H2AX responses following particle treatment. Four types of dot plot patterns can be identified: (**a**) high genotoxicity (γ-H2AX) and high uptake of particles; (**b**) low genotoxicity and high uptake; (**c**) high genotoxicity and low uptake; and (**d**) low genotoxicity and low uptake

2.5 Flow Karyotyping and Sorting

Flow cytometry was first used to analyze chromosomes in the mid of 1970s (Gray et al., 1975a, b). Although it is not commonly utilized in human disease, it gave us rapid and high-throughput analysis, as well as the ability to flow sort chromosomes. Subsequent advancements in flow cytometric instrumentation, particularly by incorporation of digital cameras and visual image processing, have now contributed to a resurgence of flow cytometry for the purpose of karyotyping.

Flow cytometry can accurately analyze chromosome numbers and structure in suspension (Gray et al., 1975a, b) and used for karyotyping, FISH can investigate telomeres, centromeres, and chromosomal sorting for further research. Although flow karyotyping and flow FISH are not commonly used for human chromosome research, Stepanov et al. (1996) provide an overview of each technique to provide context for the subject.

Karyotyping

Flow karyotyping detects metaphase chromosomes in suspension using any of the two DNA stains: Hoechst 33342 (HO) and Chromomycin A3 (CA3), which binds with the A-T and G-C rich sections of the DNA, respectively (Stanley et al., 2021). The relative distribution of the genome's complete chromosomal DNA content is depicted by the flow karyotype. This made it possible to purify various chromosomes into different forms in vast quantities for additional analysis, creating a new field of cytogenetic research known as flow cytogenetics. But for chromosomal sorting and flow karyotyping, sample quality is a critical factor (Halder et al., 2017). The goal should be to achieve pure chromosomes with stable morphology and structure, free of all types of contaminants and without clumping. Bivariate flow karyotyping did not significantly advance flow cytogenetics for plants, in contrast to animals and humans, where it did give a response (Lee et al., 2000; Lucretti & Doležel, 1997). Current advances in research are flow karyotyping and chromosome genomics, which will transform methods for examining genome structure and chromosome structure, respectively (Halder et al., 2017). It is a quick and precise technique for quantitatively identifying structural and numerical chromosome abnormalities in addition to describing a species' chromosomal complement (Otto & Oldiges, 1980). Flow karyotyping may identify genetic aberrations that cause chromosome variations (deletions, insertions, tranlocations) as many as 3–5 Mbp (Trask et al., 1989), although numerical chromosome aberrations are still unknown but can be found using DNA content analysis. Nevertheless, flow karyotyping is unable to identify translocations involving the same DNA content or inversions. An essential method for examining chromosomal rearrangements connected to cancer and inherited diseases is aberration analysis. Breakpoints on aberrant chromosomes can be found using FCM (Flowcytometry) studies and sorting, which involves hybridization with a wide range of probes or spot blots (Halder et al., 2017).

Reverse chromosome painting, which shows the structure and breakpoint locations of an aberrant chromosome by developing a probe from a flow-sorted aberrant chromosome of interest and hybridizing it into a normal karyotype, is made possible by chromosomal DNA amplification (Blennow, 2004).

The development of the method of fluorescent in situ hybridization in suspension (FISHIS) has enabled high-resolution mapping on sorted chromosomes. Sorted chromosome DNA can be utilized for a number of processes, including positional cloning, in situ hybridization, marker development, PCR-based physical mapping, and the creation of chromosome-specific libraries. Flow sorting by flow cytometry method has changed genome sequencing by reducing nuclear genomes to individual chromosomes. This is especially useful for animals with big or complicated genomes, since it can identify structural and numerical chromosome abnormalities (Halder et al., 2017). Currently, flow cytometry is the method of choice for examining chromosomes because it enables quick, quantitative analysis that is statistically appropriate. The range of parameters that may be investigated with this method was expanded by the development of nucleic acid-specific fluorochromes and probes, as well as the sensitivity and accuracy of the flow cytometer.

3 Recent Advances in Flow Cytometry

Though there has been considerable advancement in research depending on flow cytometry, there are still many areas in which the technique has to be improved. The cell cycle may now be dynamically analyzed with small samples, high throughput, multiparametric analysis, and current technologies. New concepts and avenues will also arise as image analysis, flow cytometry, and microfluidics technologies progress and become more integrated.

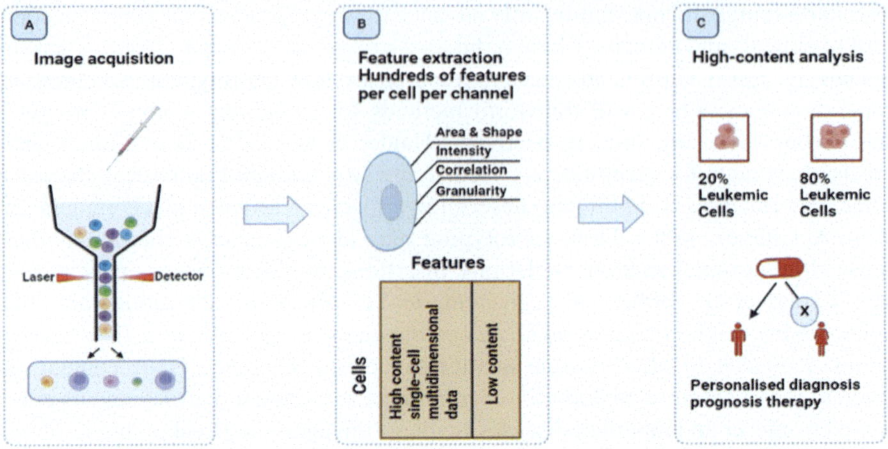

Fig. 10 Imaging flow cytometry obtains high throughput pictures of individual cells. Charge-coupled device detectors record the pictures of each cell as it passes by the light sources. For individual diagnosis, prognosis, and treatment, high-content analysis may be utilized. (Created by Biorender)

3.1 Imaging Flow Cytometry

Due to the conventional flow cytometry's lack of examination of the morphological characteristics of cells (such as their area and shape), imaging flow cytometry (IFC) has been introduced, which can examine cell morphology. It includes the benefits of image analysis (integrating cell signals into cell morphology) and flow cytometry, eliminating the need for subsequent segmentation of cells, which greatly enhances the use of cytometry in applications. In addition to minimizing fluorescent dye interference, the approach offers a non-destructive way to monitor cells. Additionally, the imaging flow cytometry system offers a way to analyze cellular heterogeneity (Blasi et al., 2016; Doan et al., 2018; Filby et al., 2011, 2016). Filby et al. combined fluorescent DNA labelling using tracking dyes for proliferation and mitotic cell antigen staining to analyze asymmetric cell division (Fig. 10).

3.2 Mass Flow Cytometry

Mass spectrometry and conventional flow cytometry are connected via mass flow cytometry. The idea is that precisely the proteins that need to be identified are bound by metal-labeled antibodies. Mass spectrometry is capable of analyzing more than 40 antigens at once and is used to quantitatively and multiparametrically identify the antibody-antigen complexes labeled with metals. Mass flow cytometry, in contrast to conventional flow cytometry techniques, makes use of metal element markers to significantly increase the number of detection channels and prevent signal overlap, which is typically present along with fluorescent signals. This lowers the need for laborious compensation calculations and boosts data reliability (Behbehani, 2018; Spitzer & Nolan, 2016).

3.3 Droplet Flow Cytometry (for Single-Cell Detection)

Flow cytometry, which is among the most adaptable and high-throughput examinations of a single cell, works by identifying various fluorescence characteristics of individual cells present in an aqueous solution passing across an excitation laser focus. This includes fluorescence-activated cell sorting. However, this method is dependent on cell surface and intracellular biomarker expressions, which are bound to be devoid of cell activities and phenotypes (spatial and temporal), including secreted proteins, generation of extracellular metabolites, and proliferation.

Manipulation and encapsulation of millions of individual cells have been made possible by the recent development of droplet microfluidics as a potent technique. Connecting single-cell genotypes and cell phenotypes through new cellular assays is made possible by combining flow cytometry with microdroplet structures surrounded by aqueous solutions, similar to hydrogel droplets and a double emulsion of water-in-oil-in-water (W/O/W).

Using this method, light energy from fluorescence molecules is emitted or scattered when floating individual cells in a fluid flow pass across a laser beam one at a time. Fluorescence-activated cell sorting (FACS), a specialized kind of flow cytometer, adds the ability to sort cells. This allows the isolation of certain cells of interest from heterogeneous populations and their subsequent placement in other containers (tubes, other collection containers, or well plates) for additional study on the selected cells. Significant advancements in clinical diagnosis, drug discovery, and the life and biomedical sciences have been made possible using single-cell analysis via FC, including FACS (Li et al., 2021) (Fig. 11).

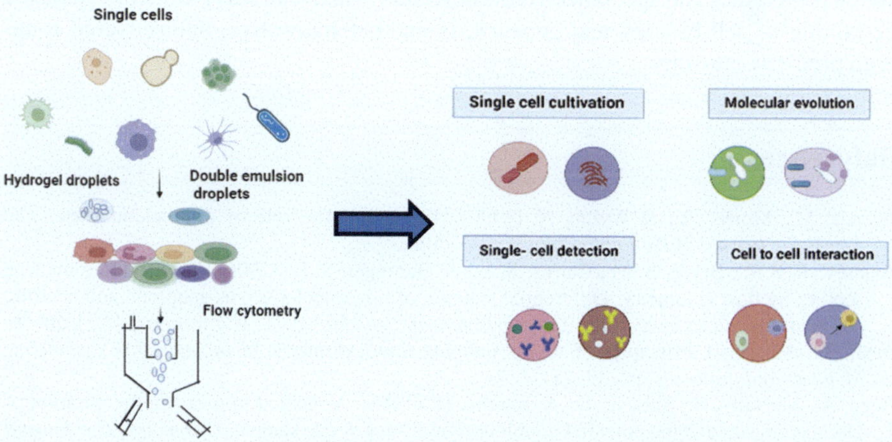

Fig. 11 Droplet flow cytometry allows for a wide range of high-throughput single-cell analysis by combining commercially accessible emulsion microdroplet, including hydrogel droplets and water-in-oil-water double emulsions

4 Limitations and Future Prospects

Since flow cytometry can perform large-scale, accurate, and fast investigation of extensive cellular or molecular features, it is currently the method of choice for both scientific and clinical research. There are limitations, too, such as cells must be in a single-cell solution in order to perform a flow cytometry-based study. While the analysis of solid tumor and tissue samples using flow cytometry has expanded due to recent advancements in cell dissociation methods, researchers frequently run into challenges such as decreased cell viability and modification of function after tissue structural disruption. The proper clinical context must be taken into consideration while analyzing flow cytometry data. Handling human information, in particular, requires caution because typical patterns are not static, and homeostatic conditions are not always well-defined. Furthermore, the expression of antigens can be readily altered by various treatment modalities. Therefore, factors like dosage, time after therapy, and fundamental patient characteristics must be carefully taken into account. This makes assessing and making decisions based on facts in a therapeutic or clinical situation very difficult. Flow cytometers are highly complex and costly devices. They need to be warmed up, have their lasers calibrated, and cleaned after each usage because they are vulnerable to issues like fluidic system obstructions. As a result, skilled workers and stringent regular quality control procedures must be followed in order to maintain the equipment.

There are substantial future implications for the application of flowcytometry in plant biology, which is currently understudied. Chromosome genomics and flow karyotyping are current frontiers in research that will transform methods for examining genome structure and chromosomes, respectively. The analysis of chromosomal proteomics and specialized chromosomes, which can solve numerous puzzles pertaining to cell biology and genetics, is another innovative, undiscovered route with predictive novelty.

References

Arends, M., Morris, R., & Wyllie, A. (1990). Apoptosis. The role of the endonuclease. *The American Journal of Pathology, 136*(3), 593–608.

Avlasevich, S. L., Bryce, S. M., Cairns, S. E., & Dertinger, S. D. (2006). In vitro micronucleus scoring by flow cytometry: Differential staining of micronuclei versus apoptotic and necrotic chromatin enhances assay reliability. *Environmental and Molecular Mutagenesis, 47*(1), 56–66.

Basiji, D. A. (2016). Principles of Amnis imaging flow cytometry. In *Imaging flow cytometry: Methods and protocols* (pp. 13–21).

Begg, A., McNally, N., Shrieve, D., & Kärche, H. (1985). A method to measure the duration of DNA syntheses and the potential doubling time from a single sample. *Cytometry: The Journal of the International Society for Analytical Cytology, 6*(6), 620–626.

Behbehani, G. K. (2018). Cell cycle analysis by mass cytometry. In *Cellular quiescence: Methods and protocols* (pp. 105–124).

Behrmann, M. S., & Trakselis, M. A. (2022). In vivo fluorescent TUNEL detection of single stranded DNA gaps and breaks induced by dnaB helicase mutants in Escherichia coli. In *Methods in enzymology* (Vol. 672, pp. 125–142). Elsevier.

Blasi, T., Hennig, H., Summers, H. D., Theis, F. J., Cerveira, J., Patterson, J. O., Davies, D., Filby, A., Carpenter, A. E., & Rees, P. (2016). Label-free cell cycle analysis for high-throughput imaging flow cytometry. *Nature Communications, 7*(1), 10256.

Blennow, E. (2004). Reverse painting highlights the origin of chromosome aberrations. *Chromosome Research, 12*, 25–33.

Bryce, S. M., Avlasevich, S. L., Bemis, J. C., & Dertinger, S. D. (2011). Miniaturized flow cytometry-based CHO-K1 micronucleus assay discriminates aneugenic and clastogenic modes of action. *Environmental and Molecular Mutagenesis, 52*(4), 280–286.

Bryce, S. M., Avlasevich, S. L., Bemis, J. C., Tate, M., Walmsley, R. M., Saad, F., Van Dijck, K., De Boeck, M., Van Goethem, F., & Lukamowicz-Rajska, M. (2013). Flow cytometric 96-well microplate-based in vitro micronucleus assay with human TK6 cells: Protocol optimization and transferability assessment. *Environmental and Molecular Mutagenesis, 54*(3), 180–194.

Carter, S. (1967). Effects of cytochalasins on mammalian cells. *Nature, 213*(5073), 261–264.

Castelain, P., Van Hummelen, P., Deleener, A., & Kirsch-Volders, M. (1993). Automated detection of cytochalasin-B blocked binucleated lymphocytes for scoring micronuclei. *Mutagenesis, 8*(4), 285–293.

Crissman, H. A., & Tobey, R. A. (1990). Specific staining of DNA with the fluorescent antibiotics, mithramycin, chromomycin, and olivomycin. In *Methods in cell biology* (Vol. 33, pp. 97–103). Elsevier.

Custer, L., & Sweder, K. (2008). The role of genetic toxicology in drug discovery and optimization. *Current Drug Metabolism, 9*(9), 978–985.

da Rocha, S. M. M., dos Santos Cardoso, P. C., de Oliveira Bahia, M., do Pessoa, Ó. C., Soares, P. C., da Rocha, S. M., Burbano, R. M. R., & da Rocha, C. A. M. (2019). Effect of the kaurenoic acid on genotoxicity and cell cycle progression in cervical cancer cells lines. *Toxicology In Vitro, 57*, 126–131.

Darzynkiewicz, Z., Traganos, F., Sharpless, T., & Melamed, M. (1976). Lymphocyte stimulation: A rapid multiparameter analysis. *Proceedings of the National Academy of Sciences, 73*(8), 2881–2884.

Darzynkiewicz, Z., Bruno, S., Del Bino, G., Gorczyca, W., Hotz, M., Lassota, P., & Traganos, F. (1992). Features of apoptotic cells measured by flow cytometry. *Cytometry, 13*(8), 795–808.

Darzynkiewicz, Z., Gong, J., Juan, G., Ardelt, B., & Traganos, F. (1996). Cytometry of cyclin proteins. *Cytometry: The Journal of the International Society for Analytical Cytology, 25*(1), 1–13.

Darzynkiewicz, Z., Bedner, E., & Smolewski, P. (2001). Flow cytometry in analysis of cell cycle and apoptosis. *Seminars in Hematology, 38*, 179–193.

Davies, D., & Allen, P. (2007). DNA analysis by flow cytometry. In *Flow cytometry: Principles and applications* (pp. 165–179).

Di Bucchianico, S., Cappellini, F., Le Bihanic, F., Zhang, Y., Dreij, K., & Karlsson, H. L. (2017). Genotoxicity of TiO2 nanoparticles assessed by mini-gel comet assay and micronucleus scoring with flow cytometry. *Mutagenesis, 32*(1), 127–137.

Doan, M., Vorobjev, I., Rees, P., Filby, A., Wolkenhauer, O., Goldfeld, A. E., Lieberman, J., Barteneva, N., Carpenter, A. E., & Hennig, H. (2018). Diagnostic potential of imaging flow cytometry. *Trends in Biotechnology, 36*(7), 649–652.

Dolbeare, F., Gratzner, H., Pallavicini, M., & Gray, J. (1983). Flow cytometric measurement of total DNA content and incorporated bromodeoxyuridine. *Proceedings of the National Academy of Sciences, 80*(18), 5573–5577.

Dubey, K., Maurya, R., Mourya, D., & Pandey, A. K. (2022). Physicochemical characterization and oxidative potential of size fractionated particulate matter: Uptake, genotoxicity and mutagenicity in V-79 cells. *Ecotoxicology and Environmental Safety, 247*, 114205.

Endl, E., Hollmann, C., & Gerdes, J. (2001). Antibodies against the Ki-67 protein: Assessment of the growth fraction and tools for cell cycle analysis. *Methods in Cell Biology, 63*, 399–418.

Fenech, M. (2006). Commentary on the SFTG international collaborative study on the in vitro micronucleus test: To Cyt-B or not to Cyt-B? *Mutation Research, 607*(1), 9–12.

Fenech, M., Bonassi, S., Turner, J., Lando, C., Ceppi, M., Chang, W., Holland, N., Kirsch-Volders, M., Zeiger, E., & Bigatti, M. (2003). HUman MicroNucleus project. Intra-and inter-laboratory variation in the scoring of micronuclei and nucleoplasmic bridges in binucleated human lymphocytes. Results of an international slide-scoring exercise by the HUMN project. *Mutation Research, 534*(1–2), 45–64.

Filby, A., Perucha, E., Summers, H., Rees, P., Chana, P., Heck, S., Lord, G. M., & Davies, D. (2011). An imaging flow cytometric method for measuring cell division history and molecular symmetry during mitosis. *Cytometry Part A, 79*(7), 496–506.

Filby, A., Day, W., Purewal, S., & Martinez-Martin, N. (2016). The analysis of cell cycle, proliferation, and asymmetric cell division by imaging flow cytometry. In *Imaging flow cytometry: Methods and protocols* (pp. 71–95).

Fimognari, C., Nüsse, M., & Hrelia, P. (1999). Flow cytometric analysis of genetic damage, effect on cell cycle progression, and apoptosis by thiophanate-methyl in human lymphocytes. *Environmental and Molecular Mutagenesis, 33*(2), 173–176.

Franz, P., Bürkle, A., Wick, P., & Hirsch, C. (2020). Exploring flow cytometry-based micronucleus scoring for reliable nanomaterial genotoxicity assessment. *Chemical Research in Toxicology, 33*(10), 2538–2549.

George, E. (2010). Occupational hazard for pathologists: Microscope use and musculoskeletal disorders. *American Journal of Clinical Pathology, 133*(4), 543–548.

Gold, R., Schmied, M., Giegerich, G., Breitschopf, H., Hartung, H., Toyka, K., & Lassmann, H. (1994). Differentiation between cellular apoptosis and necrosis by the combined use of in situ tailing and nick translation techniques. *Laboratory Investigation; a Journal of Technical Methods and Pathology, 71*(2), 219–225.

Gollapudi, B. B., & Krishna, G. (2000). Practical aspects of mutagenicity testing strategy: An industrial perspective. *Mutation Research/Fundamental and Molecular Mechanisms of Mutagenesis, 455*(1–2), 21–28.

Gong, J., Traganos, F., & Darzynkiewicz, Z. (1994). A selective procedure for DNA extraction from apoptotic cells applicable for gel electrophoresis and flow cytometry. *Analytical Biochemistry, 218*(2), 314–319.

Gratzner, H. G. (1982). Monoclonal antibody to 5-bromo-and 5-iododeoxyuridine: A new reagent for detection of DNA replication. *Science, 218*(4571), 474–475.

Gray, J., Carrano, A., Moore, D., Steinmetz, L., Minkler, J., Mayall, B., Mendelsohn, M., & Van Dilla, M. (1975a). High-speed quantitative karyotyping by flow microfluorometry. *Clinical Chemistry, 21*(9), 1258–1262.

Gray, J., Carrano, A., Steinmetz, L., Van Dilla, M., Moore, D., 2nd, Mayall, B., & Mendelsohn, M. (1975b). Chromosome measurement and sorting by flow systems. *Proceedings of the National Academy of Sciences, 72*(4), 1231–1234.

Halder, M., Nath, S., & Jha, S. (2017). Flow cytometry and its utility. In *Chromosome structure and aberrations* (pp. 109–126).

Hassani, M., Ghassemi-Barghi, N., Modanloo, M., Mohammadpour, A., & Shokrzadeh, M. (2019). Cytotoxic effects of duloxetine on MKN45 and NIH3T3 cell lines and genotoxic effects on human peripheral blood lymphocytes. *Arquivos de Gastroenterologia, 56*, 372–376.

Hernández, L., Terradas, M., Martín, M., Tusell, L., & Genescà, A. (2013). Highly sensitive automated method for DNA damage assessment: Gamma-H2AX foci counting and cell cycle sorting. *International Journal of Molecular Sciences, 14*(8), 15810–15826.

Hulett, H. R., Bonner, W. A., Barrett, J., & Herzenberg, L. A. (1969). Cell sorting: Automated separation of mammalian cells as a function of intracellular fluorescence. *Science, 166*(3906), 747–749.

Jagger, C., Tate, M., Cahill, P. A., Hughes, C., Knight, A. W., Billinton, N., & Walmsley, R. M. (2008). Assessment of the genotoxicity of S9-generated metabolites using the GreenScreen HC GADD45a–GFP assay. *Mutagenesis, 24*(1), 35–50.

Juan, G., Traganos, F., & Darzynkiewicz, Z. (2001). Methods to identify mitotic cells by flow cytometry. *Methods in Cell Biology, 63*, 343–354.

Kim, K. H., & Sederstrom, J. M. (2015). Assaying cell cycle status using flow cytometry. *Current Protocols in Molecular Biology, 111*(1), 28.26. 21–28.26. 11.

Kim, H. R., Park, Y. J., Oh, S. M., & Chung, K. H. (2013). Appropriate in vitro methods for genotoxicity testing of silver nanoparticles. *Environmental Health and Toxicology, 28*, e2013003.

Knight, A. W., Little, S., Houck, K., Dix, D., Judson, R., Richard, A., McCarroll, N., Akerman, G., Yang, C., & Birrell, L. (2009). Evaluation of high-throughput genotoxicity assays used in profiling the US EPA ToxCast™ chemicals. *Regulatory Toxicology and Pharmacology, 55*(2), 188–199.

Kohlpoth, M., Rusche, B., & Nüsse, M. (1999). Flow cytometric measurement of micronuclei induced in a permanent fish cell line as a possible screening test for the genotoxicity of industrial waste waters. *Mutagenesis, 14*(4), 397–402.

Landberg, G., Tan, E. M., & Roos, G. (1990). Flow cytometric multiparameter analysis of proliferating cell nuclear antigen/cyclin and Ki-67 antigen: A new view of the cell cycle. *Experimental Cell Research, 187*(1), 111–118.

Landsiedel, R., Honarvar, N., Seiffert, S. B., Oesch, B., & Oesch, F. (2022). Genotoxicity testing of nanomaterials. *Wiley Interdisciplinary Reviews: Nanomedicine and Nanobiotechnology, 14*(6), e1833.

Lee, J.-H., Arumuganathan, K., Chung, Y.-S., Kim, K.-Y., Chung, W.-B., Bae, K.-S., Kim, D.-H., Chung, D.-S., & Kwon, O.-C. (2000). Flow cytometric analysis and chromosome sorting of barley (Hordeum vulgare L.). *Molecules and Cells, 10*(6), 619–625.

Li, X., & Darzynkiewicz, Z. (1995). Labelling DNA strand breaks with BrdUTP. Detection of apoptosis and cell proliferation. *Cell Proliferation, 28*(11), 571–579.

Li, M., Liu, H., Zhuang, S., & Goda, K. (2021). Droplet flow cytometry for single-cell analysis. *RSC Advances, 11*(34), 20944–20960.

Lucretti, S., & Doležel, J. (1997). Bivariate flow karyotyping in broad bean (Vicia faba). *Cytometry: The Journal of the International Society for Analytical Cytology, 28*(3), 236–242.

McKim, J., & James, M. (2010). Building a tiered approach to in vitro predictive toxicity screening: A focus on assays with in vivo relevance. *Combinatorial Chemistry & High Throughput Screening, 13*(2), 188–206.

Mourya, D., Dubey, K., Jha, S., Maurya, R., & Pandey, A. K. (2024). In vitro effects of zirconia nanoparticles: Uptake, genotoxicity, and mutagenicity in V-79 cells. *Biological Trace Element Research, 202*(3), 927–940.

Nüsse, M., & Marx, K. (1997). Flow cytometric analysis of micronuclei in cell cultures and human lymphocytes: Advantages and disadvantages. *Mutation Research/Genetic Toxicology and Environmental Mutagenesis, 392*(1–2), 109–115.

Oberhammer, F., Wilson, J. W., Dive, C., Morris, I., Hickman, J. A., Wakeling, A. E., Walker, P. R., & Sikorska, M. (1993). Apoptotic death in epithelial cells: Cleavage of DNA to 300 and/or 50 kb fragments prior to or in the absence of internucleosomal fragmentation. *The EMBO Journal, 12*(9), 3679–3684.

OECD (2016). *Test No. 474: mammalian erythrocyte micronucleus test. OECD Publishing].*

Otto, F. J., & Oldiges, H. (1980). Flow cytogenetic studies in chromosomes and whole cells for the detection of clastogenic effects. *Cytometry: The Journal of the International Society for Analytical Cytology, 1*(1), 13–17.

Paz, M. F. C. J., de Alencar, M. V. O. B., de Lima, R. M. P., Sobral, A. L. P., do Nascimento, G. T. M., Dos Reis, C. A., de Sousa, M. D. P. S., do Nascimento, M. L. L. B., Júnior, A. L. G., & da Conceição Machado, K. (2020). Pharmacological effects and toxicogenetic impacts of omeprazole: Genomic instability and cancer. *Oxidative Medicine and Cellular Longevity, 2020*, 3457890–3457821.

Pilch, D. R., Sedelnikova, O. A., Redon, C., Celeste, A., Nussenzweig, A., & Bonner, W. M. (2003). Characteristics of γ-H2AX foci at DNA double-strand breaks sites. *Biochemistry and Cell Biology, 81*(3), 123–129.

Redon, C. E., Dickey, J. S., Bonner, W. M., & Sedelnikova, O. A. (2009). γ-H2AX as a biomarker of DNA damage induced by ionizing radiation in human peripheral blood lymphocytes and artificial skin. *Advances in Space Research, 43*(8), 1171–1178.

Redon, C. E., Nakamura, A. J., Sordet, O., Dickey, J. S., Gouliaeva, K., Tabb, B., Lawrence, S., Kinders, R. J., Bonner, W. M., & Sedelnikova, O. A. (2011). γ-H2AX detection in peripheral blood lymphocytes, splenocytes, bone marrow, xenografts, and skin. In *DNA damage detection in situ, ex vivo, and in vivo: Methods and protocols* (pp. 249–270).

Rodrigues, M. A., Beaton-Green, L. A., Wilkins, R. C., & Fenech, M. F. (2018). The potential for complete automated scoring of the cytokinesis block micronucleus cytome assay using imaging flow cytometry. *Mutation Research/Genetic Toxicology and Environmental Mutagenesis, 836*, 53–64.

Rogakou, E. P., Pilch, D. R., Orr, A. H., Ivanova, V. S., & Bonner, W. M. (1998). DNA double-stranded breaks induce histone H2AX phosphorylation on serine 139. *Journal of Biological Chemistry, 273*(10), 5858–5868.

Rothkamm, K., & Löbrich, M. (2003). Evidence for a lack of DNA double-strand break repair in human cells exposed to very low x-ray doses. *Proceedings of the National Academy of Sciences, 100*(9), 5057–5062.

Schmit, T., Klomp, M., & Khan, M. N. (2021). An overview of flow cytometry: Its principles and applications in allergic disease research. In *Animal models of allergic disease: Methods and protocols* (pp. 169–182).

Shokrzadeh, M., Etebari, M., & Ghassemi-Barghi, N. (2020). An engineered non-erythropoietic erythropoietin-derived peptide, ARA290, attenuates doxorubicin induced genotoxicity and oxidative stress. *Toxicology In Vitro, 66*, 104864.

Shokrzadeh, M., Bagheri, A., Ghassemi-Barghi, N., Rahmanian, N., & Eskandani, M. (2021). Doxorubicin and doxorubicin-loaded nanoliposome induce senescence by enhancing oxidative stress, hepatotoxicity, and in vivo genotoxicity in male Wistar rats. *Naunyn-Schmiedeberg's Archives of Pharmacology, 394*, 1803–1813.

Smart, D., Ahmedi, K., Harvey, J., & Lynch, A. (2011). Genotoxicity screening via the γH2AX by flow assay. *Mutation Research/Fundamental and Molecular Mechanisms of Mutagenesis, 715*(1–2), 25–31.

Spitzer, M. H., & Nolan, G. P. (2016). Mass cytometry: Single cells, many features. *Cell, 165*(4), 780–791.

Stanley, J., Hui, H., Erber, W., Clynick, B., & Fuller, K. (2021). Analysis of human chromosomes by imaging flow cytometry. *Cytometry Part B: Clinical Cytometry, 100*(5), 541–553.

Stepanov, S. I., Konyshev, V. N., Kotlovanova, L. V., & Roganov, A. P. (1996). Karyotyping of individual cells with flow cytometry. *Cytometry: The Journal of the International Society for Analytical Cytology, 23*(4), 279–283.

Sutton, W. S. (1903). The chromosomes in heredity. *The Biological Bulletin, 4*(5), 231–250.

Tates, A., Van Welie, M., & Ploem, J. (1990). The present state of the automated micronucleus test for lymphocytes. *International Journal of Radiation Biology, 58*(5), 813–825.

Terry, N. H., & Whiter, R. A. (2001). Cell cycle kinetics estimated by analysis of bromodeoxyuridine incorporation. *Methods in Cell Biology, 63*, 355–374.

Toduka, Y., Toyooka, T., & Ibuki, Y. (2012). Flow cytometric evaluation of nanoparticles using side-scattered light and reactive oxygen species-mediated fluorescence–correlation with genotoxicity. *Environmental Science & Technology, 46*(14), 7629–7636.

Trask, B., Van den Engh, G., & Gray, J. (1989). Inheritance of chromosome heteromorphisms analyzed by high-resolution bivariate flow karyotyping. *American Journal of Human Genetics, 45*(5), 753–760.

Tsamou, M., Jennen, D. G., Claessen, S. M., Magkoufopoulou, C., Kleinjans, J. C., & van Delft, J. H. (2012). Performance of in vitro γH2AX assay in HepG2 cells to predict in vivo genotoxicity. *Mutagenesis, 27*(6), 645–652.

Turner, H. C., Brenner, D. J., Chen, Y., Bertucci, A., Zhang, J., Wang, H., Lyulko, O. V., Xu, Y., Shuryak, I., & Schaefer, J. (2011). Adapting the γ-H2AX assay for automated processing in human lymphocytes. 1. Technological aspects. *Radiation Research, 175*(3), 282–290.

Valdiglesias, V., Giunta, S., Fenech, M., Neri, M., & Bonassi, S. (2013). γH2AX as a marker of DNA double strand breaks and genomic instability in human population studies. *Mutation Research/Reviews in Mutation Research, 753*(1), 24–40.

Vallabani, N. S., & Karlsson, H. L. (2022). Primary and secondary genotoxicity of nanoparticles: Establishing a co-culture protocol for assessing micronucleus using flow cytometry. *Frontiers in Toxicology, 4*, 845987.

Vignon, C., Debeissat, C., Georget, M.-T., Bouscary, D., Gyan, E., Rosset, P., & Herault, O. (2013). Flow cytometric quantification of all phases of the cell cycle and apoptosis in a two-color fluorescence plot. *PLoS One, 8*(7), e68425.

Vitelli, M., Budman, H., Pritzker, M., & Tamer, M. (2021). Applications of flow cytometry sorting in the pharmaceutical industry: A review. *Biotechnology Progress, 37*(4), e3146.

Westerink, W. M., Stevenson, J. C., Lauwers, A., Griffioen, G., Horbach, G. J., & Schoonen, W. G. (2009). Evaluation of the Vitotox™ and RadarScreen assays for the rapid assessment of genotoxicity in the early research phase of drug development. *Mutation Research/Genetic Toxicology and Environmental Mutagenesis, 676*(1–2), 113–130.

Wilde, S., Dambowsky, M., Hempt, C., Sutter, A., & Queisser, N. (2017). Classification of in vitro genotoxicants using a novel multiplexed biomarker assay compared to the flow cytometric micronucleus test. *Environmental and Molecular Mutagenesis, 58*(9), 662–677.

Application of Flow Cytometry in Elucidating the Intricate Mechanism of Neurotoxicity

Abhishek Pandeya and AB Pant

Abstract

Neurotoxicity, the detrimental impact of substances on the nervous system, necessitates a comprehensive exploration of its underlying cellular and molecular mechanisms. This chapter highlights the multifaceted applications of flow cytometry in advancing our understanding of neurotoxicity. Flow cytometry has become a pivotal tool in the realm of neurotoxicity research, enabling precise assessments across various dimensions. It serves as a linchpin for evaluating cell viability and apoptosis, offering insights into the effects of neurotoxic agents on neural cell populations. Moreover, flow cytometry facilitates the isolation and purification of specific neuronal subtypes, aiding in the meticulous analysis of cells afflicted by neurotoxic insults. The technique's utility extends to probing intracellular signalling pathways, allowing the examination of protein phosphorylation changes critical to neurotoxicity mechanisms. Additionally, it empowers the quantification of oxidative stress levels in neural cells, a hallmark of neurotoxicity. Neuroinflammation, a substantial contributor to neurotoxicity, is scrutinized through flow cytometry's ability to characterize immune responses in the central nervous system. The analysis of immune cell populations and cytokine secretion

A. Pandeya
Systems Toxicology Group, FEST Division, CSIR-Indian Institute of Toxicology Research, Lucknow, Uttar Pradesh, India

AB Pant (✉)
Systems Toxicology Group, FEST Division, CSIR-Indian Institute of Toxicology Research, Lucknow, Uttar Pradesh, India

Academy ofScientific and Innovative Research (AcSIR), Ghaziabad, India
e-mail: abpant@iitr.res.in

AB Pant et al. (eds.), *Flow Cytometry: Applications in Cellular and Molecular Toxicology*, https://doi.org/10.1007/978-981-97-9758-5_7

provides invaluable insights into the inflammatory processes accompanying neurotoxic insults. Flow cytometry further illuminates neural stem cell behaviour, facilitating the assessment of differentiation, proliferation, and survival in the presence of neurotoxic agents. It assists in quantifying alterations in neuronal protein expression, potentially identifying biomarkers or therapeutic targets. Additionally, it enables the evaluation of mitochondrial function, a critical facet of neurotoxicity, through the measurement of mitochondrial membrane potential and function. Lastly, flow cytometry plays a pivotal role in high-throughput drug screening, aiding in the identification of compounds with neuroprotective or neurotoxic properties. In conclusion, flow cytometer's diverse applications have established it as an indispensable tool for delving into the intricate facets of neurotoxicity, offering an encompassing perspective on cellular and molecular responses to neurotoxic agents. Its versatility in assessing cell viability, signalling pathways, immune responses, and stem cell behaviour positions it as an indispensable asset in neurotoxicity research and therapeutic development endeavours.

Keywords

Neurotoxicity · Neurodegenerative disorder · Flow cytometry · Neuroprotection

1 Introduction

The brain is the most sophisticated organ in the human body that comprises of billions of interconnected neuronal cells. Neuronal cells, together with glial cells providing structural and metabolic support, constitute the fundamental building blocks of the nervous system. The nervous system comprises the central nervous system, encompassing the brain and spinal cord, and the peripheral nervous system, which includes the autonomic and somatic nervous systems. While the central nervous system houses the majority of neurons in vertebrates, some are found in peripheral ganglia, and sensory neurons are distributed in sensory organs like the retina and cochlea. Due to certain factors, the brain gets sick and starts malfunctioning (Cunha-Oliveira et al., 2008). This can occur due to continuous exposure to chronic toxicants which might be environmental, chemicals, food, and biological factors such as viruses, bacteria or toxic venoms etc. This leads to cause Neurotoxicity which refers to the capacity of certain substances or conditions to cause damage to the nervous system, including the brain and peripheral nerves. This damage can result in various neurological symptoms and disorders (Sainio Sr., 2015). Neurotoxicity is a critical field of study within the broader realms of neuroscience and toxicology, focusing on the adverse effects of various agents on the nervous system. Most of the neurotoxicants inhibit GABA receptor leading to the closure of chloride channel and excess nerve excitation. GABA receptors shows dizziness, headache, nausea, vomiting, tremors, convolutions and some time may lead to death (Dai et al., 2019). Understanding neurotoxicity is crucial for public health, environmental protection, and the development of safe pharmaceuticals. Lots of techniques are available for assessing the neurotoxicity.

The assessment of neurotoxicity encompasses a variety of techniques, each tailored to specific aspects of nervous system impairment. *In vitro* models, including primary neuronal cultures and cell lines like SH-SY5Y, PC12, LUHMES or many more neurological cells offer initial insights into cytotoxic effects of substances at the cellular level or assessing the neurodegenerative disorders (Harry et al., 2014; Radio & Mundy, 2008; Slanzi et al., 2020). These models are instrumental in elucidating mechanisms of neurotoxicity and screening potential neurotoxic agents. Electrophysiological techniques, such as patch-clamp recordings, provide a direct measure of neuronal excitability and synaptic function, crucial for understanding the functional impact of neurotoxic substances (Maurice et al., 2006). Behavioural assessments in animal models, like the Morris water maze and open field test, are used to correlate neurotoxic exposure with changes in cognitive and motor functions (Bushnell et al., 2007). Neuroimaging techniques, including MRI and PET, offer non-invasive ways to observe neurotoxic effects *in vivo*, giving insights into structural and functional changes in the brain (Cecil et al., 2008). These diverse techniques, each with their own strengths and limitations, collectively contribute to a comprehensive understanding of neurotoxicity. There are numerous causes of neurotoxicity, including chemicals and environmental toxins such as heavy metals, pesticides, industrial chemicals, air pollution, continuous consumption of alcohol, and drug abuse; whereas biological materials such as infections of viruses, bacteria, or certain other kinds of microbes and can also due to neurotoxin venoms from the certain reptiles such as snakes etc. (Costa et al., 2008; Keifer & Firestone, 2007; Richardson et al., 2019) as shown in the Figure 1.

These neurotoxins can be categorized in following groups:

1. **Chemical and environmental toxins:** A significant concern in this field arises from the myriad of chemical and environmental toxins that humans are exposed to, either occupationally, environmentally, or through lifestyle choices. Such toxins can have profound implications for neural health, leading to functional impairments, neurodevelopmental disorders, and degenerative diseases. Chemical and environmental neurotoxins can emanate from various sources. Heavy metals like lead, mercury, and arsenic are classic examples of neurotoxicants found in contaminated water, food, and sometimes even in household items (Needleman, 2004). Pesticides, commonly used in agriculture, have also been implicated in neurotoxicity, with organophosphates being particularly notorious (Costa, 2006; Srivastava et al., 2023). Additionally, certain industrial chemicals and solvents, such as toluene and manganese, can exhibit neurotoxic properties when humans are exposed to them in occupational settings (Guilarte, 2013). The nervous system's complexity means that toxins can interfere at multiple levels. Some neurotoxins, such as heavy metals, can disrupt neurotransmitter function or alter ionic balances within neurons (Jomova & Valko, 2011). Others, like certain pesticides, inhibit enzymes crucial for neurotransmitter metabolism, leading to an accumulation of neurotransmitters and subsequent

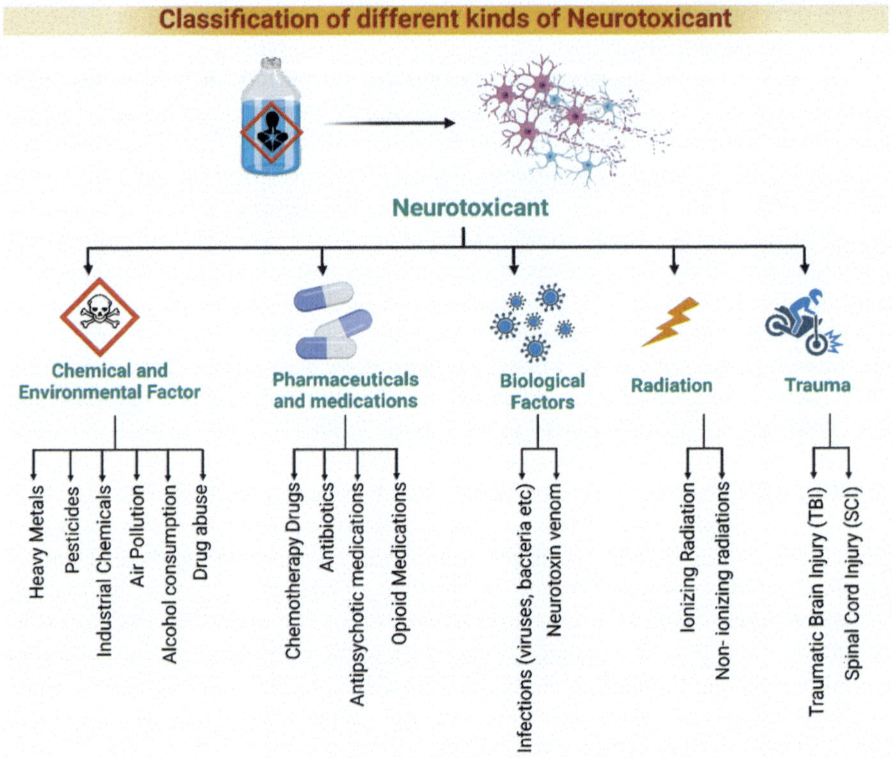

Fig. 1 Different kinds of neurotoxicants and their respective molecules. (Created through BioRender)

overstimulation of neurons (Costa et al., 2008). Furthermore, some toxins can induce oxidative stress, inflammation, or even trigger apoptotic pathways, leading to neuronal death (Bhat et al., 2015). These are the toxin that can induce neurotoxicity by prolonged exposure, these include:

(a) **Heavy Metals:** Lead, mercury, arsenic, and cadmium are examples of heavy metals that can cause neurotoxicity.
(b) **Pesticides:** Some pesticides and insecticides, such as organophosphates and carbamates, can be neurotoxic.
(c) **Industrial Chemicals:** Exposure to industrial chemicals like solvents, PCBs, and certain cleaning agents can lead to neurotoxic effects.
(d) **Air Pollution:** Airborne pollutants, including fine particulate matter and toxic gases, can have neurotoxic effects, especially in urban areas.
(e) **Alcohol:** Chronic alcohol abuse can result in alcohol-induced neurotoxicity, causing conditions like Wernicke-Korsakoff syndrome.
(f) **Drugs of Abuse:** Illicit drugs like cocaine, methamphetamine, and opioids can damage the nervous system.

2. **Pharmaceuticals and Medications:** Pharmaceuticals, while primarily designed for therapeutic benefits, can sometimes have unintended neurotoxic effects. Within the vast spectrum of medications available, a subset has been identified to potentially elicit adverse neurological outcomes. The neurotoxicological implications of such pharmaceuticals warrant careful consideration, both in clinical practice and in pharmaceutical research. Several classes of drugs have been identified with potential neurotoxic effects. Chemotherapy agents, such as cisplatin and methotrexate, can cause peripheral neuropathy and cognitive deficits (Argyriou et al., 2008). Antipsychotic medications, particularly the older typical antipsychotics, have been associated with tardive dyskinesia, a movement disorder (Carbon et al., 2017). Certain antiretroviral drugs used in the treatment of HIV can lead to HAND (HIV-associated neurocognitive disorders) (Clifford & Ances, 2013). Even some antibiotics, like metronidazole and isoniazid, can cause neuropathies and other neurotoxic effects when used for extended periods. The pathways leading to drug-induced neurotoxicity are diverse. Some drugs might interfere with neurotransmitter metabolism or receptor function (Olney et al., 1991). Others may induce oxidative stress, mitochondrial dysfunction, or directly damage neuronal structures (Lin & Beal, 2006). Additionally, certain drugs might have off-target effects, inadvertently affecting neural pathways and leading to adverse outcomes. Neurotoxic effects of pharmaceuticals can manifest in various ways, ranging from mild symptoms like dizziness or insomnia to severe conditions like seizures, movement disorders, or even cognitive decline (Lipton & Rosenberg, 1994). Often, the symptoms are dose-dependent and might resolve upon discontinuation of the medication. However, in some instances, they can be persistent or even irreversible.

These contains different kind of drugs molecules which are described below:

(a) **Chemotherapy Drugs:** Some chemotherapy agents can cause peripheral neuropathy, affecting the nerves outside the central nervous system. Vinca alkaloids, cisplatin, and taxanes rank prominently among agents known to induce peripheral neurotoxicity. These pharmacological compounds are extensively utilized in the treatment of diverse malignancies, including ovarian and breast cancers, as well as various haematological neoplasms (Verstappen et al., 2003; Was et al., 2022).

(b) **Antibiotics:** Certain antibiotics like fluoroquinolones can lead to peripheral neuropathy. Reports of neurotoxicity have been associated with cephalosporins across multiple generations, including first-generation agents like cefazolin, second-generation compounds such as cefuroxime, third-generation varieties exemplified by ceftazidime, and fourth-generation cephalosporins like cefepime. The spectrum of neurotoxic manifestations may extend from encephalopathy to non-convulsive status epilepticus (Grill & Maganti, 2011; Schwartz & Calvert, 1990).

(c) **Antipsychotic Medications:** Some antipsychotic drugs may cause movement disorders or other neurological side effects. Antipsychotic drugs, while pivotal in managing psychiatric disorders, are not devoid of neurotoxic effects, presenting a significant concern in their therapeutic use. These effects can manifest in a diverse spectrum, ranging from subtle cognitive impairments to severe conditions like tardive dyskinesia and neuroleptic malignant syndrome. The underlying mechanisms are multifaceted, often involving dopaminergic blockade, oxidative stress, and disruption of neural pathways. Chronic use of these medications, particularly the first-generation antipsychotics, is frequently associated with structural brain changes, evident in imaging studies. This neurotoxic potential necessitates a careful evaluation of the risk-benefit ratio in long-term therapy and underscores the importance of monitoring for early signs of neurological compromise. Additionally, it prompts ongoing research into the development of safer antipsychotics with a reduced propensity for neurotoxicity (Feder & Goverman, 2022; Zareifopoulos et al., 2021).

(d) **Opioid Medications:** Chronic use of opioids can result in opioid-induced hyperalgesia and other neurological problems. Opioid-induced neurotoxicity generally manifests within a temporal window ranging from several days to 1 week following the initiation of opioid therapy or upon achieving a dosage level that facilitates the accumulation of neurotoxic metabolites. Contributory factors, including dehydration, infectious processes, and the concomitant administration of central nervous system depressants, may precipitate a state of opioid toxicity, particularly in geriatric patients with preexisting fragility. The opioid neurotoxicity shows the symptoms like hypersomnolence, delirium, hallucinations, allodynia (pain from a stimulus that does not normally provoke pain), hyperalgesia (abnormally increased sensitivity to pain), myoclonus, tremor, and seizures (Gallagher, 2007; Godwin et al., 2022).

3. **Biological Factors:** Neurotoxicity, the adverse effect on the structure and function of the nervous system, can be instigated by a myriad of biological factors. Genetic predispositions can render some individuals more susceptible to neurotoxicants, as evidenced by polymorphisms in cytochrome P450 enzymes which affect toxin metabolism (Neafsey et al., 2009). Additionally, immune responses can contribute to neurotoxicity; for instance, the activation of microglia and astrocytes can lead to the release of pro-inflammatory cytokines and reactive oxygen species, exacerbating neural damage (Brown & Bal-Price, 2003). Age is another crucial biological factor, with both the developing brains of the young and the aging brains of the elderly showing heightened vulnerability to toxic insults (Grandjean & Landrigan, 2006). Furthermore, the interactions in relate with the miRNAs level have been implicated in modulating neurotoxic responses, though this area requires further exploration (Fu et al., 2023; Singh & Yadav, 2020).

   (a) **Infections:** Various pathogens have been identified as causative agents of neurotoxicity in humans. Viral pathogens, including Human Immunodeficiency Virus (HIV) and herpes simplex virus, bacterial agents such as *Borrelia burgdorferi* responsible for Lyme disease, and prion-related disorders like Creutzfeldt-Jakob disease, can all induce detrimental effects on the nervous system (Le Govic et al., 2022).

Table 1 Shows the biological components which leads to cause neurotoxic effect in human brain (Bilge, 2022; Spencer & Lein, 2014)

S No.	Biological form	Substances with neurotoxic potential
1	Bacterium	Deptheria A toxin
2	Fungus	3-Nitropropanoic acid
3	Plant	L-beta-N-oxalylamino-L-alanine
4	Insect	Apamin
5	Reptiles	Dendrotoxin
6	Bird	Bactrachotoxin

(b) **Neurotoxic Venoms:** Neurotoxic venoms, primarily produced by certain snakes, spiders, and marine animals, are specialized toxic secretions that target the nervous system. These venoms contain a complex mixture of proteins, peptides, and small molecules, which can exert their neurotoxic effects by disrupting neuronal signaling. Specifically, many of these toxins act by modulating ion channels, such as sodium, potassium, and calcium channels, or by interfering with neurotransmitter release, leading to paralysis, respiratory failure, or other neurological symptoms. For instance, the venom of the black mamba snake contains dendrotoxins that inhibit voltage-dependent potassium channels, leading to an increase in neuro-transmitter release and over-excitation of the nervous system. Similarly, the venom of the funnel-web spider contains δ-atracotoxins that slow the inactivation of voltage-gated sodium channels, causing prolonged neuronal depolarization. Understanding the mechanisms of action of neurotoxic venoms not only sheds light on their potent neurotoxic effects but also offers potential therapeutic avenues. For instance, the peptides found in these venoms can be utilized as pharmacological tools to study ion channel function or even as leads for drug development targeting neurological disorders.

The Table 1 shows the potential infection of biological life form which may leads to cause neurotoxication in human body.

4. **Radiation:** Ionizing radiation, which includes X-rays, gamma rays, and particle radiation, has the potential to induce neurotoxic effects when the central nervous system (CNS) is exposed. The extent and nature of the damage depend on factors such as the dose, rate of exposure, type of radiation, and the developmental stage of the CNS at the time of exposure. Radiation can damage cellular integrity through direct DNA damage, leading to mutations, and can also produce reactive oxygen species (ROS) that damage cellular components, including DNA, lipids, and proteins. Neurons, due to their limited regenerative capacity, are particularly vulnerable to such damage, which can lead to cell death or functional impairment. It also affects the cognitive impairment: Several studies have shown that exposure to ionizing radiation, even at doses considered safe for therapeutic procedures, can lead to cognitive deficits. This is particularly evident in patients who undergo radiation therapy for CNS tumours. The hippocampus, a region critical for memory formation, is especially sensitive to radiation-induced damage. Whereas neuroinflammation can be caused due to radiation, which can

induce an inflammatory response in the CNS. This involves the activation of microglia and astrocytes, which release pro-inflammatory cytokines and can exacerbate neuronal damage. Other than this this radiation can shows the delayed effects, which manifests years after exposure. This includes radiation-induced necrosis and the increased risk of developing neurodegenerative disorders. The major source of It's worth noting that while ionizing radiation poses a risk for neurotoxicity, non-ionizing radiation (e.g., from mobile phones, Wi-Fi) has been a topic of ongoing debate and research regarding its potential neurotoxic effects. Current evidence suggests that typical environmental exposure levels to non-ionizing radiation are unlikely to cause significant neurotoxic effects, but research in this area continues.

5. **Trauma:** Traumatic events, whether physical (e.g., traumatic brain injury) or psychological (e.g., severe stress or emotional trauma), can precipitate neurotoxic effects on the central nervous system (CNS). Physical trauma can disrupt the blood-brain barrier, leading to inflammation and the accumulation of potentially neurotoxic substances in the brain (Shlosberg et al., 2010). Concurrently, inflammatory mediators and oxidative stress can induce neuronal damage (Witcher et al., 2015). On the other hand, psychological trauma can trigger prolonged activation of the hypothalamic-pituitary-adrenal (HPA) axis, resulting in elevated cortisol levels. Chronic exposure to high cortisol can be neurotoxic, particularly in regions like the hippocampus, affecting memory and cognitive function (Sapolsky, 2000). Furthermore, both forms of trauma can elevate the risk of neurodegenerative diseases and contribute to synaptic dysfunction, emphasizing the multifaceted neurotoxic implications of trauma (McEwen et al., 2015).
   (a) **Traumatic Brain Injury (TBI):** Severe head injuries can cause neurotoxicity and lead to various neurological deficits.
   (b) **Spinal Cord Injury:** Damage to the spinal cord can result in permanent neurological impairment.

It's important to note that the severity and specific neurological effects of neurotoxicity can vary widely depending on the cause, duration, and individual susceptibility. Prevention and early detection are crucial in managing and mitigating the impact of neurotoxicity. If you suspect neurotoxicity due to exposure to any of these factors, seek medical evaluation and intervention promptly.

To understand the mechanisms underlying neurotoxicity and identifying potential risk factors is a complex and multidisciplinary endeavour. Researchers employ various techniques, including advanced methods like flow cytometer, to explore the intricate cellular and molecular processes involved in neurotoxicity. These investigations aim to elucidate the underlying mechanisms, assess the impact on neuronal and glial cells, and ultimately inform the development of strategies for prevention and treatment. The neurotoxicity is a field of growing importance, given the ubiquity of neurotoxic agents in our environment and the potential implications for human health. A comprehensive understanding of neurotoxicity and its mechanisms is not only crucial for safeguarding public health but also for advancing our knowledge of the nervous system and its vulnerabilities. Ongoing research in this field is essential to mitigate the risks posed by neurotoxic agents and to develop effective strategies for intervention and mitigation. In this field, the flow cytometry play a crucial role for the analysis and identification of the neurotoxic effect of any of the above mentioned agent(s) which can hinder the neuronal growth and development. As of the day, flow cytometry can be used to access the toxic effects, cell cycle arresting in the toxic substances exposed neuronal cells etc.

2 Know About Flow Cytometry

Flow cytometry is a technique which uses fluorescent probes/markers/dyes to identify, quantify, and even sorting the cells for further study (de Graaf et al., 2011; El-Hajjar et al., 2023). Cells can be identified by their uniquely expressed cell surface receptors or by cytoplasmic expression of cytokines, chemokine, or intracellular receptors. In a current scenario, this technique become a powerful analytical technique which is almost used in every field of biological sciences for the quantification and analysis of physical and chemical characteristics of particles as they flow through a fluidic system (Shapiro, 2003). This technique helps to reveals the mystery of the cells, including their size, shape, complexity, surface markers etc. Flow cytometry operates on the principle of suspending particles (usually cells) in a fluid stream and passing them through a laser beam. These suspended particles or cells passed through a small nozzle in sheath fluid hydro dynamically. The resulting tiny stream of fluid takes cells one at a time and allow to pass through a laser light. Based on the scattering of light helps to reveals the complexity of any cells, through which the size of the cells as well as their granularity will identified. Further, the detectors measure the intensity of the scattered and emitted light, providing information about various cell properties. In terms of the scatter parameters, forward scatter (FSC) provides information about the size of the cells or particles, while side scatter (SSC) gives insights into cell granularity and complexity. Other than this the flow cytometry can also help to identified/characterized the specific cells by the help of fluorochrome tagged dyes/antibodies against the specific cell surface epitopes/receptor or molecules whether it will be intracellular or intercellular.

3 Journey of Flow Cytometry from Cell to Neurotoxicity Analysis

The development of flow cytometry represents a significant advancement in biological sciences, having its origins in World War II-era innovations. Initially motivated by the U.S. Army's need to detect bacterial infections or bio warfare agents in aerosols, early developments included Gucker et al.'s 1947 photo electronic counter for airborne particle detection (Gucker Jr. et al., 1947). By 1950, Coons and Kaplan had pioneered the use of fluorescein isothiocyanate (FITC) in antibody labelling (Coons & Kaplan, 1950). In 1949, Coulter's breakthrough in measuring cell volume through electrical impedance variations in saline suspensions laid foundational principles, considering the ionic nature of sheath fluid and cell suspensions and the insulating properties of cell lipid layers. They got success to measure cell volume in a saline suspension based on their electrical resistance variation proportionality due to the fact that sheath fluid and cell suspension are ionic liquid and the cell surrounded by the lipid layer. Further, the year 1965 marked the debut of the prototype flow cytometer by Mack Fulwyler, integrating principles of cell segregation and Trypan blue uptake for live/dead cell differentiation (Fulwyler, 1965). This period also saw the development of fluorescence-based cytometers, including the air-cooled argon laser-driven Cyto-fluorograph. The first commercial cytometer, known as the Cyto-graph, employed a He-Ne laser system and was introduced in 1970. Subsequently, in 1973, Hullet et al. developed a rapid cell sorter (Hulett et al., 1973). Their enhanced cell analyser and sorter can now pinpoint and segregate viable cells based on their fluorescence intensity, light-scattering properties, or both. This advanced system examines each cell individually as it flows through two laser beams within a narrow liquid jet. The jet transforms into uniform droplets, charging and deflecting droplets with target cells via an electric field. Capable of handling thousands of cells per second, it achieves enrichment factors up to 500, with over 90% purity and viability (Hulett et al., 1973). In 1974, first commercial flow cytometer came into existence which is known as Hemalog D (Technicon, Tarrytown, New York, United States of America) which was invented for the analysis of white blood and for their classification (Mansberg et al., 1974; Ornstein, 1987). In the same year, Becton-Dickinson commenced the commercial production of the FACS-II cytometer, capable of measuring forward scatter and 530 nm emission fluorescence. The term "flow cytometry," formalized in 1976 at a conference in Pensacola, Florida, quickly gained acceptance in scientific discourse. In neuroscience, flow cytometry has been instrumental in analysing and characterizing neuronal cells, using various neuronal markers to assess neurotoxicity. Early neurotoxicity assessments included distinguishing between benign and malignant brain tumors, as elucidated by Frederiksen et al.'s DNA content distribution studies (Frederiksen et al., 1978). Further research by Kawamoto et al. and Hoshino et al. used flow cytometry for comparative DNA content analysis in normal versus benign/malignant human brain tissues, contributing significantly to the understanding of neurotoxicity (Kawamoto et al., 1979). Flow cytometer's role extends to evaluating cell cytotoxicity and apoptosis, crucial in understanding chemotherapeutic agents' efficacy. The technology has elucidated the actions of various antiglioma agents and has been pivotal in assessing the repurposing potential of antipsychotic drugs for CNSM treatment. Recent advancements include Mass Cytometry applications, like the targeted treatment of glioma stem cells with boron neutron capture therapy agents, showcasing flow cytometer's ongoing evolution and impact in biological research.

The journey for the use of flow cytometry in analysing the neurotoxicity and their associated research has been a remarkable one, spanning from its early applications in the late 1980s to the present day in 2025. Over this period, flow cytometry has evolved as a powerful tool for elucidating the intricate mechanisms underlying neurotoxicity and has contributed significantly to our understanding of the effects of various agents on the nervous system. Originally developed by Leonard Herzenberg and colleagues during the 1960s, this technology was intended for haematological analyses and primarily catered to immunological research (Herzenberg et al., 2002). However, by the 1980s, as the field of neurotoxicology matured and began emphasizing the significance of cellular-level responses to toxic agents, flow cytometry found a new role. Researchers utilized this tool to gain a deeper understanding of neural cell populations, as well as the onset of apoptosis and other cell cycle disturbances induced by toxicants (Das et al., 2021). The next decade, the 1990s, ushered in a wave of technological enhancements for flow cytometry, including the capability for multicolour analysis. This allowed researchers to simultaneously probe multiple cellular markers, enabling detailed insights into specific neural cell subpopulations and their reactions to neuro-toxicants (Perfetto et al., 2004). The 2000s were marked by the convergence of flow cytometry with molecular biology techniques. This integration paved the way for comprehensive studies that combined cellular analysis with molecular insights, providing a holistic view of the neurotoxic effects of various agents (Vo, 1988). Badie et al., in 2000 examined the role of dexamethasone as a mechanism to reduce the damaged caused by the inflammation reaction (Badie et al., 2000). In a similar line, Mc Larena and her colleagues evaluated the Neural Stem cells (NSCs) through flow cytometry. They separates the single cells treated with cytokines and further the cells were examined for nestin, beta tubulin and GFAP expression, where GFAP is specific for Astrocytes, nestin for neurons (Bhattacharya et al., 2003; Kawaguchi et al., 2001; McLaren et al., 2001). Bilsland and group in 2006 identified a method for cyclic cells in hippocampus of mice by using the dye BrDU on fixed tissue of hippocampus (Bilsland et al., 2006). More recently, the tool has been complemented by the rise of multi-parametric analysis and sophisticated bioinformatics applications. This has transformed flow cytometry into a potent instrument, capable of deciphering intricate cellular dynamics within the framework of neuro-toxicology, and promises to further elevate the depth and breadth of research in this domain (Maecker et al., 2012).

In 2005, Behbahani et al. utilized flow cytometry, enhanced by Annexin V and Propidium Iodide staining, to detect the neurotoxic effects of Tumor Necrosis Factors (TNFs) on neuronal populations, enabling the quantification of apoptotic and necrotic events within these cells. These findings have implications for the identification of compounds with potential therapeutic efficacy in neurodegenerative diseases (Behbahani et al., 2005). Subsequently, Kipnis et al. (2012) and Schwartz and Kipnis (2011) investigated the cognitive interactions between the central nervous system and immune system, identifying specific T lymphocytes with prospective cognitive function (Kipnis et al., 2012; Schwartz & Kipnis, 2011). Through the use of an immunodeficient mouse model, their research revealed that peripheral T lymphocytes, along with innate immune cells such as microglia and infiltrating monocytes, contribute to variations in brain activity and behavioural responses, as compared to immunocompetent controls.

Over recent decades, numerous studies employing flow cytometry have examined the neurotoxic effects of various chemicals, focusing on both the characterization of neuronal properties and the neurotoxic impacts of these substances. Van Breedam et al. (2022) proposed the use of iPSC-derived luminescent neurospheroids for high-throughput screening in the context of ischemic stroke treatment (Van Breedam et al., 2022). Cheng et al. (2019) demonstrated the neuroprotective effects of tanshinone IIA against oxaliplatin-induced neurotoxicity through mechanisms involving mitochondrial protection and autophagy. Flow cytometry was employed to assess mitochondrial membrane potential using the JC-1 dye and to examine the neurotoxic impact on Neuro 2a mouse neuroblastoma cell lines (Cheng et al., 2019). Liu et al. (2021) elucidated the role of Tetrachlorobenzoquinone (TCBQ) in inducing ferroptosis in PC12 cells exposed to TCBQ (Liu et al., 2021). Moreover, Xu et al. (2022) investigated cadmium-induced neurotoxicity in microglial activation within zebrafish larvae using flow cytometry to delineate the toxic effects (Xu et al., 2022).

In recent years, the neurotoxicity of various chemical nanoparticles has been characterized through flow cytometry. Guo et al. (2023) identified the neurotoxic effects of ethyl acetate extract of Zanthoxylum armatum DC, which induces reactive oxygen species (ROS)-mediated mitochondrial apoptosis in the SH-SY5Y cell line (Guo et al., 2023). Concurrently, Sun et al. (2023) investigated the neurotoxic effects of Melittin, assessing its impact on mitochondrial oxidative phosphorylation; apoptosis and ROS/mitochondrial membrane potential were analyzed via flow cytometer (Sun et al., 2023). Furthermore, Zhang et al. (2023) reported on the efficacy of Lipoxin A4 methyl ester in mitigating ketamine-induced neurotoxicity in SH-SY5Y cells, utilizing both flow cytometry and additional methodologies (Zhang et al., 2023).

The technique allowed for the quantitative assessment of various cellular parameters, including cell viability, apoptosis, and oxidative stress. These early studies laid the foundation for subsequent research by providing insights into the effects of toxic substances on neural cell populations. As we progress into 2025, flow cytometry has become an indispensable tool in neurotoxicity research. The technology has advanced significantly, with the introduction of high-dimensional flow cytometry, spectral flow cytometry, and mass cytometry (CyTOF). These advancements have enabled researchers to simultaneously analyze multiple cellular parameters at a single-cell level, providing a more comprehensive understanding of neurotoxic mechanisms. Flow cytometry has also been integrated with other molecular and imaging techniques, allowing for a holistic approach to neurotoxicity studies. Moreover, its application in neuroinflammation research has been instrumental in deciphering the immune responses in the nervous system following exposure to toxic agents. Other than these study there are several others which elaborate the role of flow cytometer in the assessment of neurotoxicity among humans, which can further help to investigate the key mechanism as well as therapeutic interventions against the neurodegenerative disorders.

4 Different Methods to Access the Neurotoxicity Through Flow Cytometer

Flow cytometer offers a versatile platform for assessing neurotoxicity by measuring various cellular and molecular parameters. The basic protocol for preparing the samples till analyse/assess the neurotoxicity among the human samples/cells during an *in vitro* condition. The use of flow cytometry to assess neurotoxicity provides a way to quantify and characterize cellular changes in neural cell populations following exposure to potential toxicants. Initially, research starts their procedure with the culturing neuronal cells or glial cells depending on the target of interest. They ensure that the cells are in the exponential growth phase. Expose the cells to varying concentrations of the neurotoxicant or compound of interest for a specific duration. It's good to have a control group that is not exposed to the toxicant. After the exposure duration, gently trypsinize (if adherent cells) and collect the cells. Centrifuge to pellet the cells and wash them with phosphate-buffered saline (PBS). Depending on the parameter of interest, choose the appropriate fluorescent dyes or antibodies or even through different available dyes. The acquire results were further analyse for the identification of the effect of neurotoxicant on different neuronal cellular model. Load the stained cells into the flow cytometer. The parameter or appropriate detector and filter settings to capture the fluorescence of the dyes/antibodies used. The data will be collected from a minimum of 10,000 cells for each sample to ensure statistical validity (Flow Cytometry Protocol, 2023; Flow Cytometry Rare Cell Analysis, 2021). The appropriate controls such as unstained, single-stained, to set the compensation and gating parameters. The data from the samples were analysed by using their appropriate flow cytometer software or any other kind of third party providing analysing software such as Flowjo (Flowjo Tutorials) and FCS Express (FCS Express) etc., which is using the FCS file generated from the instruments . In the analysis part, a research will correlate the flow cytometric data with other assays or results if available. A schematic representation outlining the fundamental preparation procedures and sequential steps for flow cytometry analysis is presented in Figure 2. This diagram illustrates the critical phases, from sample preparation to data acquisition, highlighting key methodological steps to ensure accurate and reproducible flow cytometric analysis.

In scientific research, it is imperative that findings from any study are amenable to independent verification or contradiction. The Minimum Information about a Flow Cytometry Experiment (MIFlowCyt) guideline delineates essential criteria for the documentation and dissemination of information pertaining to flow cytometry experiments. This encompasses a comprehensive description of the experiment's scope, sample characteristics, instrumentation details, and data analysis procedures. MIFlowCyt aims to standardize the reporting of clinical, biological, and technical aspects of flow cytometry experiments. It achieves this by defining requisite data elements and providing a structured format for capturing pertinent information (ISAC | MIFlowCyt, 2024; Lee et al., 2008; Spidlen et al., 2012). A critical aspect of conducting flow cytometry experiments involves adhering to fundamental procedural steps. Each stage of the experiment requires careful consideration of specific

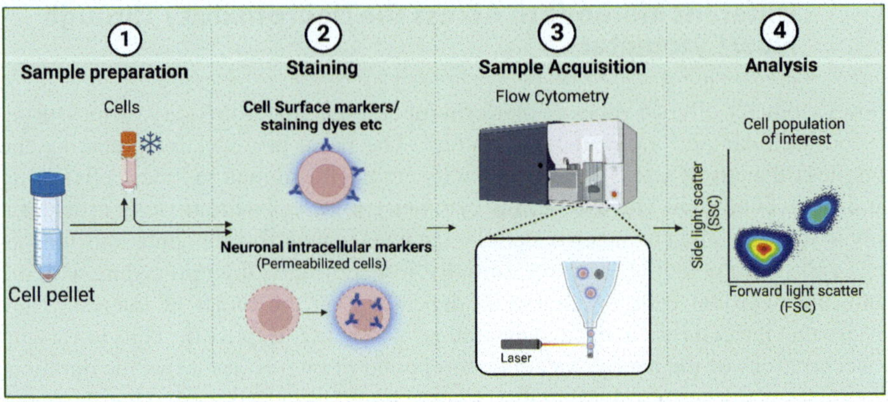

Fig. 2 Schematic representation for the workflow of flow cytometry analysis. (Created through BioRender)

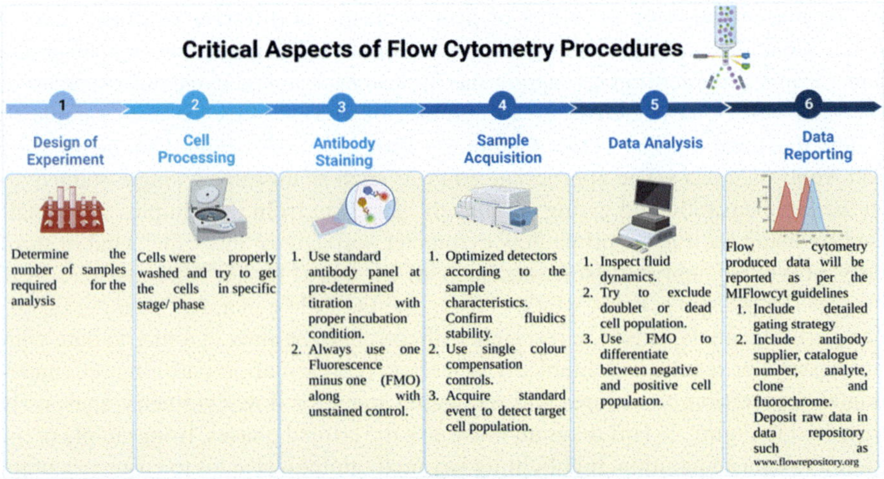

Fig. 3 Essential factors for researchers employing flow cytometry techniques

factors. Figure 3 enumerates these key factors in relation to the respective procedural stages of flow cytometry experiments.

Various methods and assays can be utilized to assess neurotoxicity using flow cytometry, as illustrated in Figure 4. These approaches enable the quantitative analysis of cellular responses to toxicants by evaluating parameters such as cell viability, apoptosis, oxidative stress, and mitochondrial membrane potential etc. By applying specific dyes, fluorescent markers or probes that target distinct cellular events, flow cytometry facilitates a comprehensive evaluation of neurotoxic effects on cellular health and function.

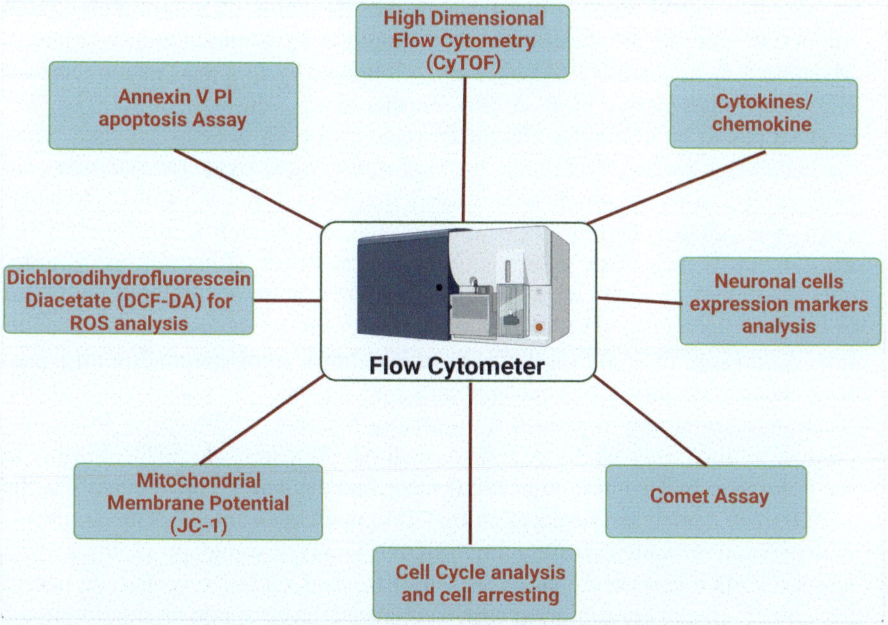

Fig. 4 Different approaches for the analysis of neurotoxicity through flow cytometer. (Created through BioRender)

There are several methods for assessing the neurotoxicity which are described below:

A. **Apoptosis assay:** The Annexin V/Propidium Iodide (PI) apoptosis assay is a cornerstone technique in neurotoxicity research, serving as a reliable method to delineate viable, apoptotic, and necrotic cell populations. This assay leverages the affinity of Annexin V for phosphatidylserine (PS), which translocate from the inner to the outer leaflet of the plasma membrane early in apoptosis. Annexin V binding, therefore, becomes a hallmark of apoptosis, indicating the loss of membrane asymmetry characteristic of apoptotic cells. In contrast, necrotic cells, with compromised membrane integrity, allow the nucleic acid-binding dye propidium iodide (PI) to penetrate, which is normally excluded from viable and early apoptotic cells. By employing these two stains in tandem, researchers can accurately assess the extent of apoptosis and necrosis, providing critical insights into the neurotoxic effects of various substances (Crowley et al., 2016; Rieger et al., 2011; van Engeland et al., 1998). First, Ni et al. in 1996 suggested the preventive effect of *Ginkgo biloba* extract on apoptosis in rat cerebellar neuronal cells induced by hydroxyl radicals and were used flow cytometry to analyse the apoptosis in cerebral neuronal cells with/without exposure of the *Ginkgo biloba* extract (Ni et al., 1996). In 1999, Mirua et al., shows the occurrence of apoptosis was preceded by the accumulation of cells in G2/M after MeHg treatment. Exposure to colchicine, a well-characterized mitotic inhibitor, also caused G2/M-phase arrest followed by the appearance of apoptotic cells. These results suggest that G2/M-phase arrest through the disruption of microtubules is an important event in the development of apoptosis by Methylmercury (MeHg). MeHg treatment led to G2/M-phase arrest followed by apoptosis in nonneuronal HeLa cells also (Miura et al., 1999). Further, during the year of 2000 by Wei et al., it has to be found that the nitric oxide apoptosis in neuronal cells. They studied the apoptotic effect due to the exposure of nitric oxide (NO) in neuronal cells subjected to treatment were quantitatively analyzed via flow cytometry. (Wei et al., 2000). During 2000 to 2023, lots of research were done for the assessment of neurotoxic effect/neuroprotective effect of certain drugs/molecules/chemicals. A study of Scheid et al., in 2023 shows the effect of argon gas in the neuronal cells (SH-SY5Y) pre exposed with the rotenone and preconditioned exposure of argon on the cells which shows argon preconditioning protects against apoptosis of neuronal cells and mediates its action via Toll-like receptors. Argon may represent a promising therapeutic alternative in various clinical settings, such as the treatment of stroke (Scheid et al., 2023). Further in the same year by Li and group (2023) identified the neural cell apoptosis in the MPTP treated mice and reveals that PDCD4 knockdown enhances the apoptosis in neural cell through activating the PI3K/AKT/mTOR signal, providing a novel target for PD treatment (Li et al., 2023b). Lim et al. (2023) revealed the neurotoxicity of ethanol consumption which enhanced the ceramide production in the cytosol of the neural cells, which further induces the mitochondrial mediated apoptosis in the neural cells which were evaluated by the flow cytometry (Lim et al., 2023). This assay is most frequent and versatile option to assess the neurotoxic studies, allowing for the detailed analysis of cell death mechanisms and the potential cytotoxic effects of various compounds on neural cell.

B. **Oxidative Stress and Reactive Oxygen Species (ROS) Assessment through Dihydroethidium (DHE) or 2′, 7′-Dichlorofluorescin diacetate (DCF-DA):** These dyes can be used to measure intracellular ROS levels, which are often elevated in response to neurotoxic agents. Dihydroethidium (DHE) and 2′,7′-Dichlorofluorescin diacetate (DCF-DA) are vital fluorescent probes employed in neurotoxicity studies to measure reactive oxygen species (ROS) within cells, which is a key indicator of oxidative stress—a critical factor in neurodegeneration. DHE is specifically sensitive to superoxide anions, allowing for the detection of these species within neurons, while DCF-DA is a general ROS indicator, which upon cellular esterase action and subsequent oxidation, emits fluorescence. The employment of these probes in fluorescence-based assays such as flow cytometry enables the quantification of intracellular ROS levels, providing insights into the oxidative state of neurons under various toxic conditions. These assessments are essential for delineating the mechanisms of neurotoxicity and the consequent neuronal damage that might contribute to neurodegenerative processes (Lee & Jang, 2021). In this method of neurotoxicity assessment, we follow the standardized procedure as per manufacturer's protocol. In this, the cells will be plated (seeding density 2.5×10^4 cells/cm^2) in FBS-supplemented medium w/o phenol red onto a cell culture vessel, incubate it for 30 min at room temperature. After incubation, cells were washed one time with

1× buffer/PBS, then the cells were incubated with DCF-DA 10 μM (final concentration should be 5 mM) for 30 min at 37 ° C protected from light. Then the cells data will be acquire through flow cytometer by using specified voltage settings

Tamariello et al. in 2000 reported first time for the detection of ROS Reactive oxygen species (ROS) are necessary for programmed cell death (PCD) in neurons, but the underlying ROS-producing enzymes have not been identified. NADPH oxidase produces ROS, although the expression of its five subunits are thought to be restricted largely to non-neuronal cells (Tammariello et al., 2000). They show the NADPH oxidase subunits are present in neurons. Moreover, both an NADPH oxidase inhibitor, diphenyleneiodonium, and NAPDH oxidase genetic deficiency inhibit apoptosis in a classic model of PCD, i.e., NGF-deprived sympathetic neurons. Overall, results indicate that NADPH oxidase is unexpectedly present in neurons and can contribute to neuronal apoptosis. Mousavi et al. (2010) shows the protective effect of saffron extract and crocin on ROS mediated neurotoxicity by using PC12 cells. They have shown that the glucose (13.5 and 27 mg/mL) reduced the cell viability of PC12 cells after 4 days. Saffron extract (5 and 25 mg/mL), crocin (10 and 50 μM) and GSH (10 μM) could decrease this toxicity. Glucose toxicity was consistent with increased ROS production which reduced by saffron, crocin and GSH pre-treatment. Their results suggest saffron and its carotenoid crocin could be potentially useful in diabetic neuropathy treatment *via* flow cytometry (Mousavi et al., 2010). Cheng et al. during 2019 demonstrated the Tanshinone IIA ameliorates oxaliplatin-induced neurotoxicity via mitochondrial protection and autophagy promotion in Mouse brain neuroma cell line N2a cells by using various techniques including the flow cytometry. Their results showed that oxaliplatin promoted N2a cell apoptosis in a concentration-dependent manner. Tanshinone IIA and oxaliplatin co-treated cells exhibited a lower percentage of apoptosis compared with the oxaliplatin alone group. Further, by using the DCFDA, they revealed that oxaliplatin significantly increased the intracellular ROS levels compared with the control group. However, tanshinone IIA alone had no obvious effect on ROS, but co-treatment with oxaliplatin and transhinone IIA reduced the ROS levels through flow cytometry (Cheng et al., 2019). Further, in 2019, it has to be reported by the Coronel et al. that (E)-Nicotinaldehyde O-Cinnamyloxime, which is a well-known Nicotine Analog, Attenuates Neuronal Cells Death Against Rotenone-Induced Neurotoxicity. They evaluated the effect of nicotine, nicotine analogs and rotenone on cell viability and reactive oxygen species (ROS) production in the SH-SY5Y neuronal cell line to validate possible protective effects. We observed that the research work done by Jurado-Coronel et. al., in 2019, pre-treatment with nicotine or (E)-nicotinaldehyde O-cinnamyloxime (10 μM) improved cell viability and diminished ROS production in SH-SY5Y cells insulted with rotenone (Jurado-Coronel et al., 2019). Further, Firdaus et al. (2019) suggested that the thymoquinone meditional uses against the arsenic induced neurotoxicity and they evaluated the ROS activity through DCFDA in flow cytometry by using the SH-SY5Y cell line (Firdaus et al., 2019).

C. **Mitochondrial Membrane Potential (ΔΨm):** Flow cytometer can assess changes in ΔΨm using fluorescent probes like JC-1, TMRE (tetramethylrhodamine, ethyl ester), or Rhodamine 123. Altered mitochondrial function can be an early sign of neurotoxicity. The assessment of mitochondrial membrane potential (ΔΨm) is a critical aspect of evaluating neurotoxicity, and flow cytometry offers a precise and efficient means to achieve this. Changes in ΔΨm are indicative of mitochondrial dysfunction, a hallmark of neurotoxic damage. By using specific fluorescent dyes, such as JC-1 or TMRE, that accumulate in mitochondria in response to their membrane potential, flow cytometry can quantitatively measure these changes in individual cells. A decrease in ΔΨm, often detected as a shift in fluorescence intensity, suggests mitochondrial depolarization, which is a precursor to apoptotic pathways and indicative of cellular stress or damage. This method is particularly relevant in neurotoxicology due to the high energy demand and mitochondrial density in neuronal cells. Qiang et al. reveals the neuroprotective effect of gelsolin against the amyloid-β peptide which can reduced the mitochondrial membrane potential in PC12 cell line (Qiao et al., 2005). During the year of 2010 Xu et al. (2010), shows that melatonin can be able to protect the neuronal cells against neurotoxic insult that can caused by the nickel through the reduction in oxidative stress as well as it can helps to maintaining the mitochondrial function effectively. They used JC1 dye to assess the neurotoxicity of nickel as well as protective effect of melatonin in the Neuro 2a cell line (Xu et al., 2010). Recent studies by Zhao et al. (2023) have demonstrated the utility of this technique in assessing the neurotoxic effects of toxins known as Benzo[b] fluranthene (B[b]F), on the bEnd.3 cells. This study results showed that B[b]F promoted ROS production and trigger mitochondria disruption (Zhao et al., 2023).

D. **DNA Damage and Genotoxicity Assessment:** Assessing neurotoxicity through the evaluation of DNA damage and genotoxicity is a crucial aspect of neurotoxicology research. DNA damage and genotoxicity are indicative of the potential harm toxicants can inflict on neuronal cells, leading to detrimental effects on the nervous system. Various molecular and cellular assays, such as the comet assay and the micronucleus test, have been employed to quantify DNA damage and genotoxicity, providing valuable insights into the mechanistic underpinnings of neurotoxicity (Nakamura et al., 2014). These assessments aid in understanding the complex interplay between toxicants and neuronal DNA, ultimately contributing to the identification of potential neuroprotective strategies and the development of therapeutic interventions in the context of neurodegenerative disorders (Siddique et al., 2014). These contains the different mechanism by which we can assess the neurotoxic effect of different toxicants. They are as follows:

- **Comet Assay:** Flow cytometry can be used to analyze DNA fragmentation via the comet assay. This method quantifies the extent of DNA damage in response to neurotoxic agents.
- **pH2AX (γ-H2AX) Staining**: Detection of phosphorylated H2AX, a marker of DNA double-strand breaks, can be performed using flow cytometry to assess DNA damage.

E. **Cell Cycle Analysis:**

Neurotoxic assessment through cell cycle analysis via flow cytometry is a sophisticated approach that provides valuable insights into the impact of toxicants on neuronal cell proliferation and viability. Flow cytometry, a high-throughput technique, allows for the quantitative assessment of different cell cycle phases, including G0, G1, S, and G2/M, as well as the detection of cells undergoing apoptosis or senescence (Kim & Sederstrom, 2015; Pozarowski & Darzynkiewicz, 2004). Changes in the distribution of neuronal cells across these phases can indicate disruptions in cellular homeostasis and potential neurotoxic effects. Moreover, flow cytometry can be coupled with specific markers to evaluate the expression of key cell cycle regulatory proteins, shedding light on the molecular mechanisms underlying neurotoxicity (Chung et al., 2019; Kim & Sederstrom, 2015). This approach not only contributes to a comprehensive understanding of neurotoxic behaviour but also assists in the development of targeted interventions to mitigate neurodegenerative disorders.

In this assay, we follow the basic protocol as described here. Neuronal cell cultures will be subjected to neurotoxin insults, harvested through trypsinization or suitable methods, and then centrifuged at low speed (e.g., 300–500× *g*) for 5 min. After aspirating the culture medium, the cell pellet will be suspended in phosphate-buffered saline (PBS). For cell fixation, ice-cold 70% ethanol will used to prevent clumping, followed by incubation at 4 °C for a minimum of 1 h or overnight. The fixed cells will centrifuged to remove residual ethanol, and the cell pellet was resuspended in PBS containing propidium iodide (PI) and RNase A (typically at 50 μg/mL PI and 100 μg/mL RNase A concentrations) and incubated in darkness at room temperature for 15–30 min. Stained cells will be analyzed using a flow cytometer with appropriate PI filters (e.g., 488 nm excitation and 610 nm emission) after setting compensation controls and gating. DNA content data were collected, allowing for the identification of cell cycle phases (G0/G1, S, and G2/M). Histograms will be generated to display cell cycle distribution profiles, and the results will be interpreted in the context of the experiment, providing insights into the impact of neurotoxicants or other treatments on cell proliferation and DNA synthesis for subsequent scientific reporting. Flow cytometry can determine alterations in the cell cycle distribution following exposure to neurotoxic agents. It provides insights into cell cycle arrest and DNA damage responses.

F. **Neuroinflammation and Immune Response Analysis:** Neuroinflammation and immune response analysis through flow cytometry represent vital approaches in the field of neurosciences and neurotoxicology. Flow cytometry allows for the precise characterization of immune cell populations within the central nervous system, offering insights into the mechanisms underlying neuroinflammatory processes and their impact on neurodegenerative disorders. Researchers have employed this technique to investigate various aspects of neuroinflammation, such as microglial activation, T cell infiltration, and cytokine expression, providing valuable data for understanding the immune responses in neurotoxicological contexts (Garcia-Gomes et al., 2020). This powerful analytical tool not only aids in unraveling the complexities of neuroinflammatory pathways but also contributes to the development of potential therapeutic interventions targeting neuro-immuno interactions in neurological diseases.

G. **Microglial Activation:** Microglial activation plays a critical role in assessing neurotoxicity, with flow cytometry emerging as a key technique in this domain. Flow cytometry offers a high-throughput, quantitative method to analyze microglial cells, the primary immune cells in the central nervous system. By employing specific markers such as Iba1, CD11b, and TMEM119, researchers can accurately determine the activation state of microglia in response to neurotoxicants (Butovsky et al., 2014; Kettenmann et al., 2011). This method is particularly effective in identifying early stages of microglial activation, which is crucial for understanding the onset of neuroinflammatory and neurodegenerative processes (Colonna & Butovsky, 2017; Ransohoff, 2016). The precision of flow cytometry in measuring these cellular changes provides invaluable insights into the mechanisms of neurotoxicity and aids in the development of potential therapeutic strategies for neurodegenerative diseases (Hickman et al., 2018).

H. **Cytokine and Chemokine Production:** Flow cytometry can quantify the production of proinflammatory and anti-inflammatory cytokines and chemokines by immune and glial cells in response to neurotoxic agents. The assessment of neurotoxicity through the measurement of cytokine and chemokine production is an essential aspect of neurotoxicological studies, and flow cytometry serves as a powerful tool in this context. This technique allows for the precise quantification and characterization of cytokines and chemokines released by neural cells in response to toxicants, which are key indicators of neuroinflammation and cellular stress. By using flow cytometry, researchers can detect and analyze a broad range of inflammatory mediators, such as TNF-α, IL-6, and MCP-1, in a highly sensitive and specific manner (Block et al., 2007; Dutta et al., 2006). These mediators are crucial for understanding the mechanisms of neurotoxicity, as they play significant roles in the initiation and progression of neuroinflammatory and neurodegenerative processes (Glass et al., 2010; Ransohoff, 2016). The ability of flow cytometry to simultaneously measure multiple cytokines and chemokines in individual cells provides a comprehensive overview of the inflammatory response, making it an invaluable tool in the field of neurotoxicity research (Hickman et al., 2013).

I. **High-Dimensional Flow Cytometry:** High-dimensional flow cytometry symbolises a significant advancement in the field of cytometry analysis, enabling the simultaneous measurement of multiple parameters at the single-cell level. This technique expands upon traditional flow cytometry by incorporating a larger number of fluorescent markers, thereby allowing for a more comprehensive profiling of cell populations. High-dimensional flow cytometry employs advanced technologies such as spectral flow cytometry and mass cytometry (CyTOF), which can analyze over 40 parameters simultaneously, providing a more detailed cellular analysis compared to conventional methods (Bendall et al., 2011; Nolan & Condello, 2013). This approach is particularly beneficial in complex biological systems like the immune system, where it can delineate intricate cell subpopulations and functional states (Maecker & Trotter, 2006; Newell & Cheng, 2016). In the context of neurosciences, this technology offers unprecedented insights into the heterogeneity of neural and glial cell populations, aiding in the understanding of various neurological conditions and the effects of neurotoxic substances (Spitzer & Nolan, 2016). High-dimensional flow cytometry's ability to provide a holistic view of cellular dynamics makes it a powerful tool in both basic research and clinical diagnostics.

Employing multiple fluorescent markers and advanced techniques like mass cytometry (CyTOF) allows for the simultaneous measurement of numerous cellular parameters, enabling a comprehensive analysis of complex neurotoxicity mechanisms.

J. **Neuronal Morphology Analysis:** Beyond traditional flow cytometry, flow cytometry coupled with imaging systems can be used to analyse neuronal morphology, dendritic arborisation, and other structural changes caused by neurotoxic agents.

These methods can be tailored to the specific research question and the neurotoxic agent under investigation. Flow cytometer's flexibility in measuring multiple parameters at the single-cell level makes it an invaluable tool in the field of neurotoxicity research, providing a comprehensive understanding of the effects of neurotoxic agents on neuronal and glial cells.

K. **Neurological Biomarker:**

Utilizing Neurological Biomarkers in Flow Cytometry represents a powerful methodology at the forefront of neuroscientific research (Reddy & Abeygunaratne, 2022). This innovative approach enables the precise quantification and characterization of specific biomarkers associated with neurotoxicology and neurodegenerative disorders, shedding light on complex cellular and molecular processes (Koníčková et al., 2022). By leveraging the high-throughput capabilities of flow cytometry, researchers can examine the expression profiles of key biomolecules, such as surface receptors and intracellular signaling molecules, in various neural cell populations (Black et al., 2011; Robinson et al., 2023). The integration of flow cytometry with neurological biomarker analysis not only advances our understanding of neurotoxicological phenomena but also facilitates the development of targeted therapeutic interventions.

Flow cytometry can be used to assess the expression of specific biomarkers associated with neurotoxicity, including neuronal markers, inflammation-related markers, and stress response proteins.

The various neurological markers utilized for flow cytometry analysis are illustrated in the flow chart presented in **Figure** 5.

Based on their effective working during the neuronal differentiation and neuronal cell function, these biomarkers can be broadly subdivided into 3 different types:

1. **Diagnostic Neuronal Markers:**
   A. **Beta-Amyloid and Tau Proteins:** Amyloid-β peptides are proteolytic fragments of the transmembrane amyloid precursor protein, whereas tau is a brain-specific, axon-enriched microtubule-associated protein. Biomarkers for Alzheimer's disease, their presence in cerebrospinal fluid (CSF) or buildup in the brain can be diagnostic. Beta-amyloid and tau proteins play

pivotal roles in assessing neurotoxicity through flow cytometry, contributing to a deeper understanding of neurodegenerative disorders. Beta-amyloid, known for its aggregation in Alzheimer's disease, can be assessed in neural cell populations, shedding light on its role in neurotoxicity (Numaguchi et al., 2023). Additionally, tau protein, implicated in tauopathies, can be examined using flow cytometry to unravel its contribution to neuronal dysfunction and toxicity (Samelson et al., 2023; Tackett, 2023). Combining these biomarkers with flow cytometric analysis enables researchers to explore their interactions and their cumulative effects on neurotoxicity, furthering our knowledge of these devastating disorders.

B. **Alpha-Synuclein:** Alpha-synuclein, a protein closely associated with neurodegenerative disorders like Parkinson's disease, used in assessing neurotoxicity. Alpha-synuclein (α-synuclein) plays a central role in the pathogenesis of Parkinson's Disease (PD), a neurodegenerative disorder characterized by the progressive loss of dopaminergic neurons in the substantia nigra region of the brain. The exact function of alpha-synuclein in normal brain physiology is not completely understood, but its pathological role in Parkinson's Disease is well-documented and multifaceted. In Parkinson's Disease, alpha-synuclein abnormally accumulates and forms aggregates known as Lewy bodies, which are a hallmark of the disease. These aggregates are believed to contribute to neuronal death. Flow cytometry offers a valuable tool to investigate alpha-synuclein's involvement in neurotoxicity-related processes, contributing to a better understanding of its implications (Li et al., 2023a). By quantifying alpha-synuclein levels and studying its aggregation in neural cell populations, flow cytometry enables researchers to discern its role in neurotoxicity mechanisms (Schonhoff et al., 2023).

C. **S100B Protein:** The S100B protein has emerged as a crucial neuronal marker in the field of neurotoxicology, facilitating the assessment of neurotoxicity with remarkable precision. This calcium-binding protein is predominantly expressed in astrocytes and Schwann cells, making it an ideal candidate for evaluating neurotoxic effects. Its utility lies in its extracellular release upon neuronal damage or injury, allowing for non-invasive monitoring in various experimental settings. Researchers in neurosciences and neurotoxicology frequently employ S100B as a reliable biomarker to gauge the extent of neural damage and assess the effectiveness of interventions designed to mitigate neurotoxic effects (Wang et al., 2018). Its use not only enhances our understanding of neurodegenerative disorders but also contributes significantly to the advancement of therapeutic strategies in this critical research domain. Often elevated in CSF after brain injury; it can serve as a marker of neuronal damage (Wang et al., 2018). Assessing the expression of S100B protein in conjunction with flow cytometry represents a powerful approach in the assessment of neurotoxicity. S100B serves as a valuable marker to detect changes in neural cells exposed to potential neurotoxicants. By employing specific antibodies against S100B, researchers can effectively label and isolate neurons or glial cells expressing this protein. Flow cytometry then allows for the quantitative analysis of these labelled cells, enabling the assessment of neurotoxicity-induced alterations in the identified cell populations. This integrated approach not only enhances the precision of neurotoxicity evaluation but also provides valuable insights into the mechanisms underlying neural damage at a cellular level, thereby contributing significantly to our understanding of neurotoxicology.

D. **Neurofilament markers:** Neurofilaments (NFs) are intermediary filaments of neurons that provides rigidity, tensile strength and possibly intracellular transport guidance to axons and dendrites. Neurofilaments (NF) are comprised of neuron-specific intermediate filaments consisting of one light subunit (NF-L) plus either a medium-chain subunit (NF-M) or heavy-chain subunit (NF-H) (Liu et al., 2004). Neurofilament Light Chain (NF-L) has emerged as a pivotal protein neuronal marker in the field of neurotoxicity assessment. This cytoskeletal protein, predominantly found in neurons, plays a crucial role in maintaining neuronal structure and function. In the context of neurotoxicology research, NF-L serves as a valuable biomarker for evaluating the impact of toxicants on the nervous system. Its utility lies in its release into the cerebrospinal fluid and bloodstream following neuronal damage or injury, making it accessible for detection and quantification. Researchers in the neurosciences and neurotoxicology field frequently rely on different kind of markers such as NF-L, NF-M and NF-H measurements to assess the severity of neural damage, monitor disease progression in neurodegenerative disorders, and evaluate the efficacy of therapeutic interventions. This versatile marker not only enhances our ability to detect neurotoxic effects but also contributes significantly to our understanding of the underlying mechanisms involved in neurodegeneration and repair.

E. **14-3-3 Proteins:** Within neurons, 14-3-3 proteins are found distributed across the cytoplasmic compartment, various intracellular organelles, and the plasma membrane. Specific isoforms of 14-3-3 proteins are notably concentrated in synaptic regions, where they play a critical role in governing synaptic transmission and synaptic plasticity (Akefe et al., 2023; Baxter et al., 2002; Broadie et al., 1997; Martin et al., 1994). The 14-3-3 proteins, widely distributed throughout the body, are associated with numerous diseases, ranging from cancer to neurodegenerative disorders. Despite lacking direct enzymatic activity, they play a central role in various signalling pathways by serving as pivotal hubs for protein-protein interactions. Their presence in CSF is associated with Creutzfeldt-Jakob disease, a prion disorder.

F. **SOX:** SOX is a gene that encodes a transcription factor featuring an HMG-box (high mobility group) DNA-binding domain and plays a pivotal role primarily in neurogenesis. It belongs to the SOX gene family, specifically the SOXB1 group, along with SOX2 and SOX3, all of which share transcription factors related to SRY, the determinant of testis development. SOX1 holds significance in the development of the central nervous system, especially in neurogenesis, and also plays a role in eye development, where it functions redundantly with SOX3 and, to a lesser extent, SOX2. Additionally, SOX1 contributes to maintaining the identity of neural progenitor cells. Its expression is confined to the neuroectoderm in proliferating progenitor cells during tetrapod embryo development. The initiation of neuroectoderm formation coincides with the activation of the SOX1 gene. In ectodermal cells committed to specific cell fates, SOX1 is among the earliest transcription factors to be expressed, with its presence typically detected during the late head fold stage (Nitta et al., 2006; Pevny et al., 1998). SOX1, a transcription factor crucial in neural development, can be employed as a valuable marker in assessing neurotoxicity through flow cytometry. Its expression serves as a biomarker for neural stem cells, making it possible to monitor their presence and differentiation. Flow cytometry analysis of SOX1 levels can indicate alterations in neural cell populations caused by neurotoxicants. Furthermore, SOX1 can be integrated into panels of markers to assess cell viability, proliferation, and gene expression changes. Overall, SOX1's presence and expression levels in neural cell populations, as determined through flow cytometry, provide essential insights into the impact of neurotoxic substances on neural development and function. Further, Renu et al. used this marker for the characterization of the human iPSCs through immunocytochemistry which further can be used for the characterized through flow cytometry (Negi et al., 2023).

G. **MAP2:** The proteins within this family are believed to play a critical role in the assembly of microtubules, a fundamental process in the formation of neurons. Similar genes in rats and mice produce neuron-specific cytoskeletal proteins that are highly concentrated in dendrites, indicating their involvement in shaping and maintaining dendritic morphology during the development of neurons. Numerous alternatively spliced variants, each encoding distinct isoforms, have been documented. All variants of MAP2 share a common core sequence that includes microtubule (MT) binding domains, consisting of 18 amino acid sequences also present in other MT-associated proteins like MAP Tau and MAP1B. It is believed that MAP2 isoforms play a crucial role in microtubule assembly, a pivotal process in neurogenesis. MAP2 functions by enhancing the stability of microtubule growth through crosslinking microtubules with intermediate filaments and other microtubules. These MAP2 isoforms are specific to neurons and are prominently located in dendrites and perikarya, indicating their involvement in shaping and stabilizing neuronal morphology during the course of neuron development.

2. **Therapeutic Neuronal Markers:** The therapeutic neuronal markers can serve as targets for drug delivery, help monitor the efficacy of treatments, or even act as biomarkers for the progression of neurological diseases.

A. **BDNF (Brain-Derived Neurotrophic Factor):** Brain-Derived Neurotrophic Factor (BDNF) plays a significant role in assessing neurotoxicity via flow cytometry. BDNF is a neurotrophins that promotes the growth, survival, and differentiation of neurons, making it a crucial biomarker for evaluating neural health. In neurotoxicity studies, researchers can utilize flow cytometry to measure BDNF levels in neuronal cultures or brain tissues. By employing specific antibodies against BDNF and flow cytometry techniques, they can quantitatively assess changes in BDNF expression, secretion, or uptake by neurons, providing insights into the impact of neurotoxicants on neurotrophic support. This approach not only allows for the direct measurement of BDNF in neural cells but also enables the investigation of its downstream effects on neural survival and function, contributing to a comprehensive evaluation of neurotoxicity. A target for therapeutic agents aiming to enhance neurogenesis and neuroplasticity in conditions such as depression and neurodegenerative disorders.

B. **NMDA Receptors:** NMDA receptors (N-methyl-D-aspartate receptors) play a critical role in assessing neurotoxicity through flow cytometry. These receptors are key components of excitatory synapses and are highly expressed in neurons. NMDA receptors are involved in synaptic plasticity and are sensitive to the effects of neurotoxicants. By combining flow cytometry with NMDA receptor-related assays, researchers can gain a better understanding of how neurotoxicants affect neuronal function and viability, contributing to the assessment of neurotoxicity and potential therapeutic interventions. Targeted by drugs for the treatment of Alzheimer's disease and other neurodegenerative disorders to modulate glutamatergic signalling.

C. **Dopamine Markers:** Dopaminergic neurons produce dopamine, a monoamine neurotransmitter with roles in neurological functions such as mood and reward. The progressive loss of dopaminergic neurons is the cause of many of the motor symptoms associated with Parkinson's disease. A widely used treatment for Parkinson's disease is Levodopa, which is a pure form of the precursor to dopamine, L-DOPA. Targets for the treatment of Parkinson's disease, such as the dopamine transporter (DAT) or enzymes involved in dopamine metabolism.

3. **Dual Role Markers:**
   A. **GDNF (Glial cell line-Derived Neurotropic Factor):** While Glial cell line–derived neurotropic factor (GDNF) demonstrates the capacity to safeguard and restore dopamine neurons in vitro and in animal models of Parkinson's disease (PD), the clinical application of GDNF is hindered by challenges associated with its pharmacokinetic profile. While serving as a potential therapeutic agent to support neuronal survival, its expression levels can also indicate the health of certain neuron populations.

   B. **Inflammatory Cytokines (TNF-alpha, IL-6):** While indicative of neuroinflammation and used in diagnosis, they are also targets for therapeutic intervention to reduce inflammation-related neuronal damage.

   C. **Endogenous miRNAs:** Certain microRNAs can serve as biomarkers for disease states and are also being explored as therapeutic agents due to their regulatory roles in gene expression.

Other than these biomarkers several studies shows that more neuronal biomarkers is using now a days to access the neurotoxicity generated through any of the mean.

The utility of these markers in diagnostics and therapeutics is an active area of research, with many studies aiming to better understand their roles in the pathology of neurodegenerative diseases and their potential as therapeutic targets.

5 Limitations of Flow Cytometry in Assessing the Neurotoxicity

Though the flow cytometry is a powerful technique to assess the cell populations and their characteristics in various fields, including neurotoxicology. However, there are notable limitations to its application in assessing neurotoxicity. In general, flow cytometry primarily analyses individual cells in suspension, which means it cannot effectively measure interactions or connectivity between neurons and other cells in the nervous system, critical aspects in neurotoxicology. Secondly, the technique requires a homogenous cell suspension, which can be challenging to obtain from complex neural tissues without altering the cells' physiological state. This disruption may affect the accuracy of the results, especially in the context of delicate neuronal interactions and functions. Furthermore, the sensitivity of flow cytometry to detect subtle changes in neuronal function is another limitation. While it can effectively identify cell death and major phenotypic changes, subtle changes in neuronal function, which are often crucial in the early stages of neurotoxicity, might not be detected. This limitation can lead to an underestimation of the early and potentially reversible effects of neurotoxicants.

Other than this some of the common limitation can be observe for the assessment of cells in flow cytometry.

5.1. **Challenges in Detecting Rare Cell Populations** While flow cytometry excels in identifying rare cell populations, detecting cells at extremely low frequencies remains a significant challenge (Hedley & Keeney, 2013). This limitation stems from the inherent difficulty in distinguishing rare cells amidst a vast majority of other cells.

5.2. **Complexities in Analysing Heterogeneous Cell Populations** Flow cytometry enables the analysis of multiple parameters on individual cells. However, interpreting results becomes particularly challenging when dealing with heterogeneous cell populations. This complexity arises from the presence of multiple cell types exhibiting similar phenotypes, which can blur the distinctions between different cell populations.

5.3. **Limitations in Assessing Intracellular Components** A notable limitation of flow cytometry is its constrained capacity in analysing intracellular elements, such as proteins and cytokines within cells. To access these intracellular components, cells must undergo permeabilization, a process that can compromise cell integrity and affect the accuracy of the results.

5.4. **Data Analysis and Reporting Challenges** Flow cytometry generates a substantial volume of data, which can present difficulties in both analysis and reporting. Particularly challenging is the interpretation of data involving diverse cell types with overlapping characteristics. Additionally, the methods required for data analysis are often intricate and demand significant time investment, complicating the process further (Betters, 2015).

The complexity and cost of flow cytometry equipment and the need for specialized technical expertise can be a barrier in some research settings, limiting its accessibility and widespread use in neurotoxicology research. This factor is crucial, especially in longitudinal studies where consistent and repeated measurements are required.

6 Summary

The application of flow cytometry in elucidating the intricate mechanisms of neurotoxicity has proven to be exceptionally valuable in the field of neurosciences and neurotoxicology. This advanced technique offers a high degree of sensitivity and specificity in detecting and quantifying cellular responses at the individual cell level, allowing for a more detailed understanding of the neurotoxic effects induced by various agents. Through the identification and analysis of specific cellular markers, flow cytometry has facilitated insights into the dynamic changes in neural cell populations, including microglia and neurons, under toxic stress. Furthermore, its ability to analyze the expression of cytokines and chemokine has provided critical information on the inflammatory responses associated with neurotoxicity. This technology has not only enhanced our comprehension of the cellular and molecular pathways involved in neurotoxic damage but also opened avenues for the development of targeted therapeutic strategies to mitigate or prevent neurotoxic effects. As research continues to evolve, flow cytometry is expected to remain a cornerstone in the investigation of neurotoxic mechanisms, contributing significantly to our understanding and management of neurodegenerative diseases and neurotoxic exposures.

7 Resources to Be Followed

There are several reputable websites and online resources associated with flow cytometry that can provide valuable information, protocols, and resources for researchers and professionals in the field. Here are some of them:

7.1. **International Society for Advancement of Cytometry (ISAC):** The ISAC website (https://isac-net.org/) is a hub for cytometry-related information, including conferences, educational materials, and resources for flow cytometry researchers.

7.2. **Flow Cytometry Basics (Thermo Fisher Scientific):** Thermo Fisher Scientific offers a comprehensive guide to flow cytometry basics on their website (https://www.thermofisher.com/us/en/home/life-science/cell-analysis/flow-cytometry/flow-cytometry-learning-center/flow-cytometry-basics.html). It's a valuable resource for beginners.

7.3. **BD Biosciences Flow Cytometry Resource Center:** BD Biosciences (https://www.bdbiosciences.com/us/resources/flow-cytometry) provides a wide range of flow cytometry resources, including protocols, webinars, and educational materials.

7.4. **CytoForum (Beckman Coulter):** Beckman Coulter's CytoForum (https://www.beckman.com/resources/learning-centers/cytoforum) is an online community for flow cytometry enthusiasts, offering discussions, protocols, and expert insights.

7.5. **Antibody Suppliers:** Many antibody suppliers, such as BioLegend, Thermo Fisher Scientific), Abcam, R & D Biosciences Becton Dickenson (BD Biosciences), Beckman Coulter (BC) and others, provide online resources to help researchers select antibodies for their flow cytometry panels. These resources often include detailed product information and suggested panel configurations.

7.6. **FlowJo:** FlowJo (https://www.flowjo.com/) is a popular software used for flow cytometry data analysis. Their website provides information about the software, training resources, and support.

7.7. **FCS Express:** This powerful tool was developed by Denovo Software (https://denovosoftware.com). This software provides detailed analysis suite related to all kind of flow cytometry experiments such as Apoptosis, Cell Cycle, Multicolour immunophenotyping, Autophagy, ROS, MMP etc.

7.8. **Flowcytometry UK:** Flowcytometry UK (http://www.flowcytometryuk.org/) is a community website for flow cytometer experts in the United Kingdom. It provides information about meetings, courses, and resources related to flow cytometry.

7.9. **Cytometry Development Workshop (CDW):** CDW (https://cdwflow.org/) offers educational workshops and online resources for flow cytometry, focusing on data analysis and interpretation.

7.10. **Flow Cytometry Core Facilities:** Many universities and research institutions have their own flow cytometry core facilities with dedicated websites. These facilities often provide information on services, equipment, and training.

7.11. **PubMed:** While not a specific website, PubMed (https://pubmed.ncbi.nlm.nih.gov/) is a valuable resource for accessing research articles and publications related to flow cytometry. You can search for specific topics or find the latest research in the field.

8 Website for Flowcytometry Panel Designing

Designing flow cytometry panels can be a crucial aspect of planning experiments in this field. While there may not be a single website dedicated solely to panel design, there are various tools and resources available that can help you in this process. Here are some resources and software commonly used for designing flow cytometry panels

8.1. **BD Multicolor Panel Selector:** BD Biosciences offers a Multicolor Panel Selector tool (https://www.bdbiosciences.com/en-us/applications/research/multicolor-panel-selector) that assists in designing multicolor flow cytometry

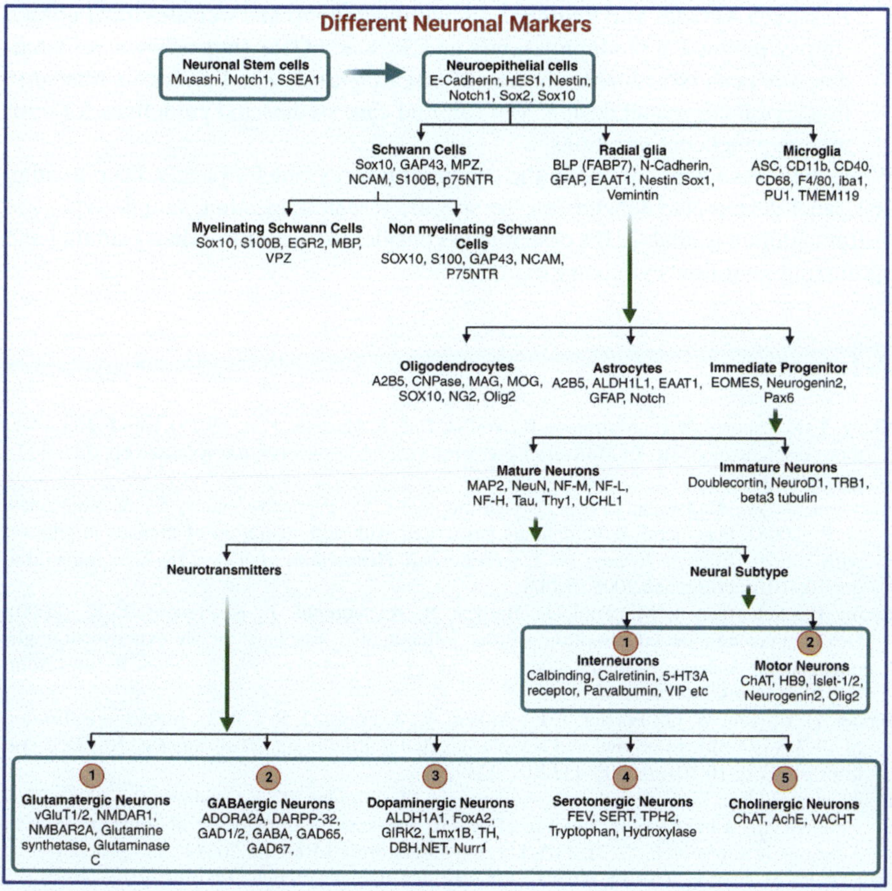

Fig. 5 Illustrative representation of the different neuronal biomarkers in different stages of their maturations. (Created through Biorender)

panels. It helps you select fluorochrome-conjugated antibodies for your specific experimental needs.

8.2. **FluoroFinder:** FluoroFinder (https://www.fluorofinder.com/) is a web-based platform that helps researchers design flow cytometry panels. It offers a database of antibodies and fluorochromes, allowing you to create customized panels.

8.3. **Cytometry by Time of Flight (CyTOF):** If you're working with CyTOF technology, you can use the CyTOF Panel Designer tool provided by Fluidigm (https://www.fluidigm.com/products/hyperion-imaging-systems/cytof-panel-designer) for panel design.

8.4. **Antibody Suppliers:** Many antibody suppliers, such as BioLegend, Thermo Fisher Scientific), Abcam, R & D Biosciences Becton Dikenson (BD Biosciences), Beckman Coulter (BC) and others, provide online resources to help researchers select antibodies for their flow cytometry panels. These resources often include detailed product information and suggested panel configurations.

8.5. **Research Articles and Protocols:** Scientific journals and websites like PubMed (https://pubmed.ncbi.nlm.nih.gov/) and ResearchGate (https://www.researchgate.net/) can be valuable sources of research articles and protocols related to flow cytometry panel design. You can find case studies and guidelines for optimizing panel configurations.

While these resources can aid in the design of flow cytometry panels, keep in mind that panel design also depends on the specific research question, sample type, and instrumentation available. It's essential to consider these factors when planning and optimizing your flow cytometry experiments.

References

Akefe, I. O., Osborne, S. L., Matthews, B., Wallis, T. P., & Meunier, F. A. (2023). Lipids and secretory vesicle exocytosis. In *Molecular mechanisms of neurotransmitter release* (pp. 357–397). Springer.

Argyriou, A. A., Koltzenburg, M., Polychronopoulos, P., Papapetropoulos, S., & Kalofonos, H. P. (2008). Peripheral nerve damage associated with administration of taxanes in patients with cancer. *Critical Review on Oncology and Hematology, 66*(3), 218–228. https://doi.org/10.1016/j.critrevonc.2008.01.008

Badie, B., Schartner, J. M., Paul, J., Bartley, B. A., Vorpahl, J., & Preston, J. K. (2000). Dexamethasone-induced abolition of the inflammatory response in an experimental glioma model: A flow cytometry study. *Journal of Neurosurgery, 93*(4), 634–639. https://doi.org/10.3171/jns.2000.93.4.0634

Baxter, H. C., Liu, W. G., Forster, J. L., Aitken, A., & Fraser, J. R. (2002). Immunolocalisation of 14-3-3 isoforms in normal and scrapie-infected murine brain. *Neuroscience, 109*(1), 5–14. https://doi.org/10.1016/s0306-4522(01)00492-4

Behbahani, H., Rickle, A., Concha, H., Ankarcrona, M., Winblad, B., & Cowburn, R. (2005). Flow cytometry as a method for studying effects of stressors on primary rat neurons. *Journal of Neuroscience Research, 82*(3), 432–441. https://doi.org/10.1002/jnr.20634

Betters, D. M. (2015). Use of Flow Cytometry in Clinical Practice. *Journal of the Advanced Practitioner in Oncology 6* (5) 10.6004/jadpro https://doi.org/10.6004/jadpro.2015.6.5 10.6004/jadpro.2015.6.5.4

Bendall, S. C., Simonds, E. F., Qiu, P., Amir el, A. D., Krutzik, P. O., Finck, R., Bruggner, R. V., Melamed, R., Trejo, A., Ornatsky, O. I., Balderas, R. S., Plevritis, S. K., Sachs, K., Pe'er, D., Tanner, S. D., & Nolan, G. P. (2011). Single-cell mass cytometry of differential immune and drug responses across a human hematopoietic continuum. *Science, 332*(6030), 687–696. https://doi.org/10.1126/science.1198704

Bhat, A. H., Dar, K. B., Anees, S., Zargar, M. A., Masood, A., Sofi, M. A., & Ganie, S. A. (2015). Oxidative stress, mitochondrial dysfunction and neurodegenerative diseases; a mechanistic insight. *Biomedicine & Pharmacotherapy, 74*, 101–110. https://doi.org/10.1016/j.biopha.2015.07.025

Bhattacharya, S., Jackson, J. D., Das, A. V., Thoreson, W. B., Kuszynski, C., James, J., Joshi, S., & Ahmad, I. (2003). Direct identification and enrichment of retinal stem cells/progenitors by Hoechst dye efflux assay. *Investigative Ophthalmology & Visual Science, 44*(6), 2764–2773. https://doi.org/10.1167/iovs.02-0899

Bilge, S. (2022). Neurotoxicity, types, clinical manifestations, diagnosis and treatment. In S. Sabuncuoglu (Ed.), *Neurotoxicity-new advances*. IntechOpen.

Bilsland, J. G., Haldon, C., Goddard, J., Oliver, K., Murray, F., Wheeldon, A., Cumberbatch, J., McAllister, G., & Munoz-Sanjuan, I. (2006). A rapid method for the quantification of mouse hippocampal neurogenesis in vivo by flow cytometry. Validation with conventional and enhanced immunohistochemical methods. *Journal of Neuroscience Methods, 157*(1), 54–63. https://doi.org/10.1016/j.jneumeth.2006.03.026

Black, C. B., Duensing, T. D., Trinkle, L. S., & Dunlay, R. T. (2011). Cell-based screening using high-throughput flow cytometry. *Assay and Drug Development Technologies, 9*(1), 13–20. https://doi.org/10.1089/adt.2010.0308

Block, M. L., Zecca, L., & Hong, J. S. (2007). Microglia-mediated neurotoxicity: Uncovering the molecular mechanisms. *Nature Review Neuroscience, 8*(1), 57–69. https://doi.org/10.1038/nrn2038

Broadie, K., Rushton, E., Skoulakis, E. M., & Davis, R. L. (1997). Leonardo, a Drosophila 14-3-3 protein involved in learning, regulates presynaptic function. *Neuron, 19*(2), 391–402. https://doi.org/10.1016/s0896-6273(00)80948-4

Brown, G. C., & Bal-Price, A. (2003). Inflammatory neurodegeneration mediated by nitric oxide, glutamate, and mitochondria. *Molecular Neurobiology, 27*(3), 325–355. https://doi.org/10.1385/MN:27:3:325

Bushnell, P. J., Boyes, W. K., Shafer, T. J., Bale, A. S., & Benignus, V. A. (2007). Approaches to extrapolating animal toxicity data on organic solvents to public health. *Neurotoxicology, 28*(2), 221–226. https://doi.org/10.1016/j.neuro.2006.03.013

Butovsky, O., Jedrychowski, M. P., Moore, C. S., Cialic, R., Lanser, A. J., Gabriely, G., Koeglsperger, T., Dake, B., Wu, P. M., Doykan, C. E., Fanek, Z., Liu, L., Chen, Z., Rothstein, J. D., Ransohoff, R. M., Gygi, S. P., Antel, J. P., & Weiner, H. L. (2014). Identification of a unique TGF-beta-dependent molecular and functional signature in microglia. *Nature Neuroscience, 17*(1), 131–143. https://doi.org/10.1038/nn.3599

Carbon, M., Hsieh, C. H., Kane, J. M., & Correll, C. U. (2017). Tardive dyskinesia prevalence in the period of second-generation antipsychotic use: A meta-analysis. *Journal of Clinical Psychiatry, 78*(3), e264–e278. https://doi.org/10.4088/JCP.16r10832

Cecil, K. M., Brubaker, C. J., Adler, C. M., Dietrich, K. N., Altaye, M., Egelhoff, J. C., Wessel, S., Elangovan, I., Hornung, R., Jarvis, K., & Lanphear, B. P. (2008). Decreased brain volume in adults with childhood lead exposure. *PLoS Medicine, 5*(5), e112. https://doi.org/10.1371/journal.pmed.0050112

Cheng, W., Xiang, W., Wang, S., & Xu, K. (2019). Tanshinone IIA ameliorates oxaliplatin-induced neurotoxicity via mitochondrial protection and autophagy promotion. *American Journal of Translational Research, 11*(5), 3140–3149.

Chung, Y. P., Yen, C. C., Tang, F. C., Lee, K. I., Liu, S. H., Wu, C. C., Hsieh, S. S., Su, C. C., Kuo, C. Y., & Chen, Y. W. (2019). Methylmercury exposure induces ROS/Akt inactivation-triggered endoplasmic reticulum stress-regulated neuronal cell apoptosis. *Toxicology, 425*, 152245. https://doi.org/10.1016/j.tox.2019.152245

Clifford, D. B., & Ances, B. M. (2013). HIV-associated neurocognitive disorder. *Lancet Infectious Diseases, 13*(11), 976–986. https://doi.org/10.1016/s1473-3099(13)70269-x

Colonna, M., & Butovsky, O. (2017). Microglia function in the central nervous system during health and neurodegeneration. *Annual Review of Immunology, 35*, 441–468. https://doi.org/10.1146/annurev-immunol-051116-052358

Coons, A. H., & Kaplan, M. H. (1950). Localization of antigen in tissue cells: II. Improvements in a method for the detection of antigen by means of fluorescent antibody. *The Journal of Experimental Medicine, 91*(1), 1–13. https://doi.org/10.1084/jem.91.1.1

Costa, L. G. (2006). Current issues in organophosphate toxicology. *Clinica Chimica Acta, 366*(1–2), 1–13. https://doi.org/10.1016/j.cca.2005.10.008

Costa, L. G., Giordano, G., Guizzetti, M., & Vitalone, A. (2008). Neurotoxicity of pesticides: A brief review. *Frontiers in Bioscience, 13*, 1240–1249. https://doi.org/10.2741/2758

Crowley, L. C., Marfell, B. J., Scott, A. P., & Waterhouse, N. J. (2016). Quantitation of apoptosis and necrosis by annexin V binding, propidium iodide uptake, and flow cytometry. *Cold Spring Harbour Protocol, 2016*(11). https://doi.org/10.1101/pdb.prot087288

Cunha-Oliveira, T., Rego, A. C., & Oliveira, C. R. (2008). Cellular and molecular mechanisms involved in the neurotoxicity of opioid and psychostimulant drugs. *Brain Research Reviews, 58*(1), 192–208. https://doi.org/10.1016/j.brainresrev.2008.03.002

Dai, C., Xiao, X., Li, J., Ciccotosto, G. D., Cappai, R., Tang, S., Schneider-Futschik, E. K., Hoyer, D., Velkov, T., & Shen, J. (2019). Molecular mechanisms of neurotoxicity induced by polymyxins and chemoprevention. *ACS Chemical Neuroscience, 10*(1), 120–131. https://doi.org/10.1021/acschemneuro.8b00300

Das, N., Raymick, J., & Sarkar, S. (2021). Role of metals in Alzheimer's disease. *Metabolic Brain Disease, 36*(7), 1627–1639. https://doi.org/10.1007/s11011-021-00765-w

de Graaf, M. T., de Jongste, A. H., Kraan, J., Boonstra, J. G., Sillevis Smitt, P. A., & Gratama, J. W. (2011). Flow cytometric characterization of cerebrospinal fluid cells. *Cytometry Part B: Clinical Cytometry, 80*(5), 271–281. https://doi.org/10.1002/cyto.b.20603

Dutta, R., McDonough, J., Yin, X., Peterson, J., Chang, A., Torres, T., Gudz, T., Macklin, W. B., Lewis, D. A., Fox, R. J., Rudick, R., Mirnics, K., & Trapp, B. D. (2006). Mitochondrial dysfunction as a cause of axonal degeneration in multiple sclerosis patients. *Annals of Neurology, 59*(3), 478–489. https://doi.org/10.1002/ana.20736

El-Hajjar, L., Ali Ahmad, F., & Nasr, R. (2023). A guide to flow cytometry: Components, basic principles, experimental design, and cancer research applications. *Current Protocols, 3*(3), e721. https://doi.org/10.1002/cpz1.721

FCS Express. Retrieved from: https://denovosoftware.com/. Accessed on Jan 08 2024.

Feder, J., & Goverman, M. (2022). *Do antipsychotics have neurotoxic effects in youth?* Retrieved from: https://www.thecarlatreport.com/blogs/2-the-carlat-psychiatry-podcast/post/3651-do-antipsychotics-have-neurotoxic-effects-in-youth-. Accessed on 08 Jan 2024.

Firdaus, F., Zafeer, M. F., Anis, E., Ahmad, F., Hossain, M. M., Ali, A., & Afzal, M. (2019). Evaluation of phyto-medicinal efficacy of thymoquinone against Arsenic induced mitochondrial dysfunction and cytotoxicity in SH-SY5Y cells. *Phytomedicine, 54*, 224–230. https://doi.org/10.1016/j.phymed.2018.09.197

Flow Cytometry Protocol. (2023). Accessed on 08 Jan 2024.

Flow Cytometry Rare Cell Analysis. (2021). Accessed on 08 Jan 2024.

Flowjo Tutorials. Retrieved from: https://www.flowjo.com/solutions/flowjo/downloads/tutorials. Accessed on 08 Jan 2024.

Frederiksen, P., Reske-Nielsen, E., & Bichel, P. (1978). Flow cytometry in tumours of the brain. *Acta Neuropathologica, 41*(3), 179–183. https://doi.org/10.1007/BF00690432

Fu, N., Yu, J., Zhu, L., Yang, L., Ma, L., He, J., Yu, H., Liu, J., Tian, Y., & Xu, J. (2023). Role of miR-219a-5p in regulating NMDAR in nonylphenol-induced synaptic plasticity damage. *Ecotoxicology Environmental Safety, 252*, 114576. https://doi.org/10.1016/j.ecoenv.2023.114576

Fulwyler, M. J. (1965). Electronic separation of biological cells by volume. *Science, 150*(3698), 910–911. https://doi.org/10.1126/science.150.3698.910

Gallagher, R. (2007). Opioid-induced neurotoxicity. *Canadian Family Physician, 53*(3), 426–427.

Garcia-Gomes, M. S. A., Zanatto, D. A., Yamamoto, P. K., Wadt, D., Antiorio, A., Aleman-Laporte, J., Alexandre-Ribeiro, S. R., Marson, G. A., Guizzo, C., Massironi, S. M. G., Bernardi, M. M., & Mori, C. M. C. (2020). A simple and fast battery test for phenotypic characterization of mice. *Bio-Protocol, 10*(7), e3568. https://doi.org/10.21769/BioProtoc.3568

Glass, C. K., Saijo, K., Winner, B., Marchetto, M. C., & Gage, F. H. (2010). Mechanisms underlying inflammation in neurodegeneration. *Cell, 140*(6), 918–934. https://doi.org/10.1016/j.cell.2010.02.016

Godwin, B., Frank, C., Molnar, F., Dyks, D., & Akter, R. (2022). Identification and management of opioid-induced neurotoxicity in older adults. *Canadian Family Physician, 68*(4), 269–270. https://doi.org/10.46747/cfp.6804269

Grandjean, P., & Landrigan, P. J. (2006). Developmental neurotoxicity of industrial chemicals. *Lancet, 368*(9553), 2167–2178. https://doi.org/10.1016/s0140-6736(06)69665-7

Grill, M. F., & Maganti, R. K. (2011). Neurotoxic effects associated with antibiotic use: Management considerations. *British Journal of Clinical Pharmacology, 72*(3), 381–393. https://doi.org/10.1111/j.1365-2125.2011.03991.x

Gucker, F. T., Jr., O'Konski, C. T., Pickard, H. B., & Pitts, J. N., Jr. (1947). A photoelectronic counter for colloidal particles1. *Journal of the American Chemical Society, 69*(10), 2422–2431. https://doi.org/10.1021/ja01202a053

Guilarte, T. R. (2013). Manganese neurotoxicity: New perspectives from behavioral, neuroimaging, and neuropathological studies in humans and non-human primates. *Frontiers in Aging Neuroscience, 5*, 23. https://doi.org/10.3389/fnagi.2013.00023

Guo, J., Yang, N., Zhang, J., Huang, Y., Xiang, Q., Wen, J., Chen, Y., Hu, T., Qiuyan, L., & Rao, C. (2023). Neurotoxicity study of ethyl acetate extract of Zanthoxylum armatum DC. on SH-SY5Y based on ROS mediated mitochondrial apoptosis pathway. *Journal of Ethnopharmacology, 319*, 117321. https://doi.org/10.1016/j.jep.2023.117321

Harry, G. J., Hooth, M. J., Vallant, M., Behl, M., Travlos, G. S., Howard, J. L., Price, C. J., McBride, S., Mervis, R., & Mouton, P. R. (2014). Developmental neurotoxicity of 3, 3′, 4, 4′-tetrachloroazobenzene with thyroxine deficit: Sensitivity of glia and dentate granule neurons in the absence of behavioral changes. *Toxics, 2*(3), 496–532. https://doi.org/10.3390/toxics2030496

Hedley B. D., Keeney M. (2013). Technical issues: flow cytometry and rare event analysis. *Summary International Journal of Laboratory Hematology 35*(3) 344–350. https://doi.org/10.1111/ijlh.2013.35.issue-3 10.1111/ijlh.12068

Herzenberg, L. A., Parks, D., Sahaf, B., Perez, O., Roederer, M., & Herzenberg, L. A. (2002). The history and future of the fluorescence activated cell sorter and flow cytometry: A view from Stanford. *Clinical Chemistry, 48*(10), 1819–1827. https://doi.org/10.1093/clinchem/48.10.1819

Hickman, S. E., Kingery, N. D., Ohsumi, T. K., Borowsky, M. L., Wang, L. C., Means, T. K., & El Khoury, J. (2013). The microglial sensome revealed by direct RNA sequencing. *Nature Neurosciences, 16*(12), 1896–1905. https://doi.org/10.1038/nn.3554

Hickman, S., Izzy, S., Sen, P., Morsett, L., & El Khoury, J. (2018). Microglia in neurodegeneration. *Nature Neurosciences, 21*(10), 1359–1369. https://doi.org/10.1038/s41593-018-0242-x

Hulett, H. R., Bonner, W. A., Sweet, R. G., & Herzenberg, L. A. (1973). Development and application of a rapid cell sorter. *Clinical Chemistry, 19*(8), 813–816. https://doi.org/10.1093/clinchem/19.8.813

ISAC | MIFlowCyt. (2024). Retrieved from: https://isac-net.org/general/custom.asp?page=MIFlowCyt. Accessed on 08 Jan 2024.

Jomova, K., & Valko, M. (2011). Advances in metal-induced oxidative stress and human disease. *Toxicology, 283*(2–3), 65–87. https://doi.org/10.1016/j.tox.2011.03.001

Jurado-Coronel, J. C., Loaiza, A. E., Diaz, J. E., Cabezas, R., Ashraf, G. M., Sahebkar, A., Echeverria, V., Gonzalez, J., & Barreto, G. E. (2019). (E)-nicotinaldehyde O-cinnamyloxime, a nicotine analog, attenuates neuronal cells death against rotenone-induced neurotoxicity. *Molecular Neurobiology, 56*(2), 1221–1232. https://doi.org/10.1007/s12035-018-1163-0

Kawaguchi, A., Miyata, T., Sawamoto, K., Takashita, N., Murayama, A., Akamatsu, W., Ogawa, M., Okabe, M., Tano, Y., Goldman, S. A., & Okano, H. (2001). Nestin-EGFP transgenic mice: Visualization of the self-renewal and multipotency of CNS stem cells. *Molecular and Cellular Neuroscience, 17*(2), 259–273. https://doi.org/10.1006/mcne.2000.0925

Kawamoto, K., Herz, F., Wolley, R. C., Hirano, A., Kajikawa, H., & Koss, L. G. (1979). Flow cytometric analysis of the DNA distribution in human brain tumors. *Acta Neuropathology, 46*(1–2), 39–44. https://doi.org/10.1007/BF00684802

Keifer, M. C., & Firestone, J. (2007). Neurotoxicity of pesticides. *Agromedicine, 12*(1), 17–25. https://doi.org/10.1300/J096v12n01_03

Kettenmann, H., Hanisch, U. K., Noda, M., & Verkhratsky, A. (2011). Physiology of microglia. *Physiological Reviews, 91*(2), 461–553. https://doi.org/10.1152/physrev.00011.2010

Kim, K. H., & Sederstrom, J. M. (2015). Assaying cell cycle status using flow cytometry. *Current Protocols in Molecular Biology, 111*, 28.26.21–28.26.11. https://doi.org/10.1002/0471142727.mb2806s111

Kipnis, J., Gadani, S., & Derecki, N. C. (2012). Pro-cognitive properties of T cells. *Nature Reviews Immunology, 12*(9), 663–669. https://doi.org/10.1038/nri3280

Koníčková, D., Menšíková, K., Tučková, L., Hényková, E., Strnad, M., Friedecký, D., Stejskal, D., Matěj, R., & Kaňovský, P. (2022). Biomarkers of neurodegenerative diseases: Biology, taxonomy, clinical relevance, and current research status. *Biomedicines, 10*(7). https://doi.org/10.3390/biomedicines10071760

Le Govic, Y., Demey, B., Cassereau, J., Bahn, Y. S., & Papon, N. (2022). Pathogens infecting the central nervous system. *PLoS Pathogens, 18*(2), e1010234. https://doi.org/10.1371/journal.ppat.1010234

Lee, D. K., & Jang, H. D. (2021). Carnosic acid attenuates an early increase in ROS levels during adipocyte differentiation by suppressing translation of Nox4 and inducing translation of antioxidant enzymes. *International Journal of Molecular Sciences, 22*(11). https://doi.org/10.3390/ijms22116096

Lee, J. A., Spidlen, J., Boyce, K., Cai, J., Crosbie, N., Dalphin, M., Furlong, J., Gasparetto, M., Goldberg, M., Goralczyk, E. M., Hyun, B., Jansen, K., Kollmann, T., Kong, M., Leif, R., McWeeney, S., Moloshok, T. D., Moore, W., Nolan, G., Nolan, J., Nikolich-Zugich, J., Parrish, D., Purcell, B., Qian, Y., Selvaraj, B., Smith, C., Tchuvatkina, O., Wertheimer, A., Wilkinson, P., Wilson, C., Wood, J., Zigon, R., Scheuermann, R. H., & Brinkman, R. R. (2008). MIFlowCyt: The minimum information about a flow cytometry experiment. *Cytometry Part A, 73*(10), 926–930. https://doi.org/10.1002/cyto.a.20623

Li, M. R., Men, S. H., Wang, Z. Y., Liu, C., Zhou, G. R., & Yan, Z. G. (2023a). The application of human-derived cell lines in neurotoxicity studies of environmental pollutants. *Science of the Total Environment, 912*, 168839. https://doi.org/10.1016/j.scitotenv.2023.168839

Li, Y., Pang, J., Wang, J., Dai, G., Bo, Q., Wang, X., & Wang, W. (2023b). Knockdown of PDCD4 ameliorates neural cell apoptosis and mitochondrial injury through activating the PI3K/AKT/mTOR signal in Parkinson's disease. *Journal of Chemical Neuroanatomy, 129*, 102239. https://doi.org/10.1016/j.jchemneu.2023.102239

Lim, J. R., Chae, C. W., Park, J. Y., Jung, Y. H., Yoon, J. H., Kim, M. J., Lee, H. J., Choi, G. E., & Han, H. J. (2023). Ethanol-induced ceramide production causes neuronal apoptosis by increasing MCL-1S-mediated ER-mitochondria contacts. *Neurobiology of Disease, 177*, 106009. https://doi.org/10.1016/j.nbd.2023.106009

Lin, M. T., & Beal, M. F. (2006). Mitochondrial dysfunction and oxidative stress in neurodegenerative diseases. *Nature, 443*(7113), 787–795. https://doi.org/10.1038/nature05292

Lipton, S. A., & Rosenberg, P. A. (1994). Excitatory amino acids as a final common pathway for neurologic disorders. *New England Journal of Medicine, 330*(9), 613–622. https://doi.org/10.1056/NEJM199403033300907

Liu, Q., Xie, F., Siedlak, S. L., Nunomura, A., Honda, K., Moreira, P. I., Zhua, X., Smith, M. A., & Perry, G. (2004). Neurofilament proteins in neurodegenerative diseases. *Cellular and Molecular Life Sciences, 61*(24), 3057–3075. https://doi.org/10.1007/s00018-004-4268-8

Liu, Z., Lv, X., Yang, B., Qin, Q., Song, E., & Song, Y. (2021). Tetrachlorobenzoquinone exposure triggers ferroptosis contributing to its neurotoxicity. *Chemosphere, 264*(Pt 1), 128413. https://doi.org/10.1016/j.chemosphere.2020.128413

Maecker, H. T., & Trotter, J. (2006). Flow cytometry controls, instrument setup, and the determination of positivity. *Cytometry Part A, 69*(9), 1037–1042. https://doi.org/10.1002/cyto.a.20333

Maecker, H. T., McCoy, J. P., & Nussenblatt, R. (2012). Standardizing immunophenotyping for the human immunology project. *Nature Review of Immunology, 12*(3), 191–200. https://doi.org/10.1038/nri3158

Mansberg, H. P., Saunders, A. M., & Groner, W. (1974). The Hemalog D white cell differential system. *Journal of Histochemistry and Cytochemistry, 22*(7), 711–724. https://doi.org/10.1177/22.7.711

Martin, H., Rostas, J., Patel, Y., & Aitken, A. (1994). Subcellular localisation of 14-3-3 isoforms in rat brain using specific antibodies. *Journal of Neurochemistry, 63*(6), 2259–2265. https://doi.org/10.1046/j.1471-4159.1994.63062259.x

Maurice, T., Grégoire, C., & Espallergues, J. (2006). Neuro (active) steroids actions at the neuromodulatory sigma1 (σ1) receptor: Biochemical and physiological evidences, consequences in neuroprotection. *Pharmacology Biochemistry Behavior, 84*(4), 581–597. https://doi.org/10.1016/j.pbb.2006.07.009

McEwen, B. S., Bowles, N. P., Gray, J. D., Hill, M. N., Hunter, R. G., Karatsoreos, I. N., & Nasca, C. (2015). Mechanisms of stress in the brain. *Nature Neurosciences, 18*(10), 1353–1363. https://doi.org/10.1038/nn.4086

McLaren, F. H., Svendsen, C. N., Van der Meide, P., & Joly, E. (2001). Analysis of neural stem cells by flow cytometry: Cellular differentiation modifies patterns of MHC expression. *Neuroimmunololgy, 112*(1–2), 35–46. https://doi.org/10.1016/s0165-5728(00)00410-0

Miura, K., Koide, N., Himeno, S., Nakagawa, I., & Imura, N. (1999). The involvement of microtubular disruption in methylmercury-induced apoptosis in neuronal and nonneuronal cell lines. *Toxicology and Applied Pharmacology, 160*(3), 279–288. https://doi.org/10.1006/taap.1999.8781

Mousavi, S. H., Tayarani, N. Z., & Parsaee, H. (2010). Protective effect of saffron extract and crocin on reactive oxygen species-mediated high glucose-induced toxicity in PC12 cells. *Cellular and Molecular Neurobiology, 30*(2), 185–191. https://doi.org/10.1007/s10571-009-9441-z

Nakamura, J., Mutlu, E., Sharma, V., Collins, L., Bodnar, W., Yu, R., Lai, Y., Moeller, B., Lu, K., & Swenberg, J. (2014). The endogenous exposome. *DNA Repair (Amst), 19*, 3–13. https://doi.org/10.1016/j.dnarep.2014.03.031

Neafsey, P., Ginsberg, G., Hattis, D., & Sonawane, B. (2009). Genetic polymorphism in cytochrome P450 2D6 (CYP2D6): Population distribution of CYP2D6 activity. *Journal of Toxicology and Environmental Health, Part B, 12*(5–6), 334–361. https://doi.org/10.1080/10937400903158342

Needleman, H. (2004). Lead poisoning. *Annual Review of Medicine, 55*, 209–222. https://doi.org/10.1146/annurev.med.55.091902.103653

Negi, R., Srivastava, A., Srivastava, A., Pandeya, A., Vatsa, P., Ansari, U., & Pant, A. (2023). Proteome architecture of human-induced pluripotent stem cell-derived three-dimensional organoids as a tool for early diagnosis of neuronal disorders. *Indian Journal Of Pharmacology, 55*(2), 108–118. https://doi.org/10.4103/ijp.ijp_56_23

Newell, E. W., & Cheng, Y. (2016). Mass cytometry: Blessed with the curse of dimensionality. *Nature Immunology, 17*(8), 890–895. https://doi.org/10.1038/ni.3485

Ni, Y., Zhao, B., Hou, J., & Xin, W. (1996). Preventive effect of Ginkgo biloba extract on apoptosis in rat cerebellar neuronal cells induced by hydroxyl radicals. *Neuroscience Letters, 214*(2–3), 115–118. https://doi.org/10.1016/0304-3940(96)12897-4

Nitta, K. R., Takahashi, S., Haramoto, Y., Fukuda, M., Onuma, Y., & Asashima, M. (2006). Expression of Sox1 during Xenopus early embryogenesis. *Biochemical Biophysical Research Communication, 351*(1), 287–293. https://doi.org/10.1016/j.bbrc.2006.10.040

Nolan, J. P., & Condello, D. (2013). Spectral flow cytometry. *Current Protocol in Cytometry, Chapter 1*, 1.27.1–1.27.13. https://doi.org/10.1002/0471142956.cy0127s63

Numaguchi, Y., Tsukakoshi, K., Takeuchi, N., Suzuki, Y., Ikebukuro, K., & Kawano, R. (2023). Real-time monitoring of the Aβ42 monomer-to-oligomer channel transition using a lipid bilayer system. *PNAS Nexus*, pgad437. https://doi.org/10.1093/pnasnexus/pgad437

Olney, J., Labruyere, J., Wang, G., Wozniak, D., Price, M., & Sesma, M. (1991). NMDA antagonist neurotoxicity: Mechanism and prevention. *Science, 254*(5037), 1515–1518. https://doi.org/10.1126/science.1835799

Ornstein, L. (1987). Tenuous but contingent connections. *Electrophoresis, 8*(1), 3–13. https://doi.org/10.1002/elps.1150080103

Perfetto, S. P., Chattopadhyay, P. K., & Roederer, M. (2004). Seventeen-colour flow cytometry: Unravelling the immune system. *Nature Review of Immunology, 4*(8), 648–655. https://doi.org/10.1038/nri1416

Pevny, L. H., Sockanathan, S., Placzek, M., & Lovell-Badge, R. (1998). A role for SOX1 in neural determination. *Development, 125*(10), 1967–1978. https://doi.org/10.1242/dev.125.10.1967

Pozarowski, P., & Darzynkiewicz, Z. (2004). Analysis of cell cycle by flow cytometry. *Methods in Molecular Biology, 281*, 301–311. https://doi.org/10.1385/1-59259-811-0:301

Qiao, H., Koya, R. C., Nakagawa, K., Tanaka, H., Fujita, H., Takimoto, M., & Kuzumaki, N. (2005). Inhibition of Alzheimer's amyloid-beta peptide-induced reduction of mitochondrial membrane potential and neurotoxicity by gelsolin. *Neurobiology of Aging, 26*(6), 849–855. https://doi.org/10.1016/j.neurobiolaging.2004.08.003

Radio, N. M., & Mundy, W. R. (2008). Developmental neurotoxicity testing in vitro: Models for assessing chemical effects on neurite outgrowth. *Neurotoxicology, 29*(3), 361–376. https://doi.org/10.1016/j.neuro.2008.02.011

Ransohoff, R. M. (2016). A polarizing question: Do M1 and M2 microglia exist? *Nature Neurosciences, 19*(8), 987–991. https://doi.org/10.1038/nn.4338

Reddy, D. S., & Abeygunaratne, H. N. (2022). Experimental and clinical biomarkers for progressive evaluation of neuropathology and therapeutic interventions for acute and chronic neurological disorders. *International Journal of Molecular Sciences, 23*(19). https://doi.org/10.3390/ijms231911734

Richardson, J. R., Fitsanakis, V., Westerink, R. H. S., & Kanthasamy, A. G. (2019). Neurotoxicity of pesticides. *Acta Neuropathology, 138*(3), 343–362. https://doi.org/10.1007/s00401-019-02033-9

Rieger, A. M., Nelson, K. L., Konowalchuk, J. D., & Barreda, D. R. (2011). Modified annexin V/propidium iodide apoptosis assay for accurate assessment of cell death. *Journal of Visualized Experiments*, (50). https://doi.org/10.3791/2597

Robinson, J. P., Ostafe, R., Iyengar, S. N., Rajwa, B., & Fischer, R. (2023). Flow cytometry: The next revolution. *Cells, 12*(14). https://doi.org/10.3390/cells12141875

Sainio, M. A., Sr. (2015). Neurotoxicity of solvents. *Handbook on Clinical Neurology, 131*, 93–110. https://doi.org/10.1016/b978-0-444-62627-1.00007-x

Samelson, A. J., Ariqat, N., McKetney, J., Rohanitazangi, G., Bravo, C. P., Goodness, D., Tian, R., Grosjean, P., Abskharon, R., Eisenberg, D., Kanaan, N. M., Gan, L., Condello, C., Swaney, D. L., & Kampmann, M. (2023). CRISPR screens in iPSC-derived neurons reveal principles of tau proteostasis. *bioRxiv*. https://doi.org/10.1101/2023.06.16.545386

Sapolsky, R. M. (2000). Glucocorticoids and hippocampal atrophy in neuropsychiatric disorders. *Archives of General Psychiatry, 57*(10), 925–935. https://doi.org/10.1001/archpsyc.57.10.925

Scheid, S., Lejarre, A., Wollborn, J., Buerkle, H., Goebel, U., & Ulbrich, F. (2023). Argon preconditioning protects neuronal cells with a Toll-like receptor-mediated effect. *Neural Regeneration Research, 18*(6), 1371–1377. https://doi.org/10.4103/1673-5374.355978

Schonhoff, A. M., Figge, D. A., Williams, G. P., Jurkuvenaite, A., Gallups, N. J., Childers, G. M., Webster, J. M., Standaert, D. G., Goldman, J. E., & Harms, A. S. (2023). Border-associated macrophages mediate the neuroinflammatory response in an alpha-synuclein model of Parkinson disease. *Nature Communications, 14*(1), 3754. https://doi.org/10.1038/s41467-023-39060-w

Schwartz, M. T., & Calvert, J. F. (1990). Potential neurologic toxicity related to ciprofloxacin. *Dicp, 24*(2), 138–140. https://doi.org/10.1177/106002809002400204

Schwartz, M., & Kipnis, J. (2011). A conceptual revolution in the relationships between the brain and immunity. *Brain, Behavior, Immunity, 25*(5), 817–819. https://doi.org/10.1016/j.bbi.2010.12.015

Shapiro, H. (2003). In A. R. Liss (Ed.), *Practical flow cytometry* (4th ed.). Wiley.

Shlosberg, D., Benifla, M., Kaufer, D., & Friedman, A. (2010). Blood–brain barrier breakdown as a therapeutic target in traumatic brain injury. *Nature Reviews of Neurology, 6*(7), 393–403. https://doi.org/10.1038/nrneurol.2010.74

Siddique, Y. H., Khan, W., Khanam, S., Jyoti, S., Naz, F., Singh, B. R., & Naqvi, A. H. (2014). Toxic potential of synthesized graphene zinc oxide nanocomposite in the third instar larvae of transgenic Drosophila melanogaster (hsp70-lacZ) Bg 9. *BioMed Research International, 2014*, 1–10. https://doi.org/10.1155/2014/382124

Singh, T., & Yadav, S. (2020). Role of microRNAs in neurodegeneration induced by environmental neurotoxicants and aging. *Ageing Research Reviews, 60*, 101068. https://doi.org/10.1016/j.arr.2020.101068

Slanzi, A., Iannoto, G., Rossi, B., Zenaro, E., & Constantin, G. (2020). In vitro models of neurodegenerative diseases. *Frontiers in Cell and Development Biology, 8*, 328. https://doi.org/10.3389/fcell.2020.00328

Spencer, P. S., & Lein, P. J. (2014). Neurotoxicity. In P. Wexler (Ed.), *Encyclopedia of toxicology*. Academic Press.

Spidlen, J., Breuer, K., & Brinkman, R. (2012). Preparing a minimum information about a flow cytometry experiment (MIFlowCyt) compliant manuscript using the International Society for Advancement of Cytometry (ISAC) FCS file repository (FlowRepository.org). *Current Protocols in Cytometry, 61*(1), 10.18.11–10.18.26. https://doi.org/10.1002/0471142956.cy1018s61

Spitzer, M. H., & Nolan, G. P. (2016). Mass cytometry: Single cells, many features. *Cell, 165*(4), 780–791. https://doi.org/10.1016/j.cell.2016.04.019

Srivastava, A., Srivastava, A. K., Pandeya, A., & Pant, A. B. (2023). Pesticide mediated silent neurotoxicity and its unmasking: An update on recent progress. *Toxicology, 500*, 153665. https://doi.org/10.1016/j.tox.2023.153665

Sun, D., Li, S., Huang, H., & Xu, L. (2023). Neurotoxicity of melittin: Role of mitochondrial oxidative phosphorylation system in synaptic plasticity dysfunction. *Toxicology, 497-498*, 153628. https://doi.org/10.1016/j.tox.2023.153628

Tackett, A. (2023). *Clearance of beta-amyloid 42 (Aβ-42) and tau proteins in Alzheimers Disease (AD)* (Masters of Science). University of Alaska Fairbanks. Retrieved from https://scholarworks.alaska.edu/handle/11122/14742. Accessed on 08 Jan 2024.

Tammariello, S. P., Quinn, M. T., & Estus, S. (2000). NADPH oxidase contributes directly to oxidative stress and apoptosis in nerve growth factor-deprived sympathetic neurons. *Neurosciences, 20*(1), RC53. https://doi.org/10.1523/JNEUROSCI.20-01-j0006.2000

Van Breedam, E., Nijak, A., Buyle-Huybrecht, T., Di Stefano, J., Boeren, M., Govaerts, J., Quarta, A., Swartenbroekx, T., Jacobs, E. Z., & Menten, B. (2022). Luminescent human iPSC-derived neurospheroids enable modeling of neurotoxicity after oxygen–glucose deprivation. *Neurotherapeutics, 19*(2), 550–569. https://doi.org/10.1007/s13311-022-01212-z

van Engeland, M., Nieland, L. J., Ramaekers, F. C., Schutte, B., & Reutelingsperger, C. P. (1998). Annexin V-affinity assay: A review on an apoptosis detection system based on phosphatidylserine exposure. *Cytometry, 31*(1), 1–9. https://doi.org/10.1002/(sici)1097-0320(19980101)31:1<1::aid-cyto1>3.0.co;2-r

Verstappen, C. C., Heimans, J. J., Hoekman, K., & Postma, T. J. (2003). Neurotoxic complications of chemotherapy in patients with cancer: Clinical signs and optimal management. *Drugs, 63*(15), 1549–1563. https://doi.org/10.2165/00003495-200363150-00003

Vo, S. (1988). Flow cytometric analysis of human bone marrow. IV. Differential quantitative expression of T-200 common leukocyte antigen during normal hemopoiesis. *Journal of Immunology, 140*(6), 1861–1867. https://doi.org/10.4049/jimmunol.140.6.1861

Wang, K. K., Yang, Z., Zhu, T., Shi, Y., Rubenstein, R., Tyndall, J. A., & Manley, G. T. (2018). An update on diagnostic and prognostic biomarkers for traumatic brain injury. *Expert Review of Molecular Diagnostics, 18*(2), 165–180. https://doi.org/10.1080/14737159.2018.1428089

Was, H., Borkowska, A., Bagues, A., Tu, L., Liu, J. Y. H., Lu, Z., Rudd, J. A., Nurgali, K., & Abalo, R. (2022). Mechanisms of chemotherapy-induced neurotoxicity. *Frontiers in Pharmacology, 13*, 750507. https://doi.org/10.3389/fphar.2022.750507

Wei, T., Chen, C., Hou, J., Xin, W., & Mori, A. (2000). Nitric oxide induces oxidative stress and apoptosis in neuronal cells. *Biochimica et Biophysica Acta, 1498*(1), 72–79. https://doi.org/10.1016/s0167-4889(00)00078-1

Witcher, K. G., Eiferman, D. S., & Godbout, J. P. (2015). Priming the inflammatory pump of the CNS after traumatic brain injury. *Trends in Neurosciences, 38*(10), 609–620. https://doi.org/10.1016/j.tins.2015.08.002

Xu, S. C., He, M. D., Zhong, M., Zhang, Y. W., Wang, Y., Yang, L., Yang, J., Yu, Z. P., & Zhou, Z. (2010). Melatonin protects against nickel-induced neurotoxicity in vitro by reducing oxidative stress and maintaining mitochondrial function. *Journal of Pineal Research, 49*(1), 86–94. https://doi.org/10.1111/j.1600-079X.2010.00770.x

Xu, Y., Zhao, H., Wang, Z., Gao, H., Liu, J., Li, K., Song, Z., Yuan, C., Lan, X., Pan, C., & Zhang, S. (2022). Developmental exposure to environmental levels of cadmium induces neurotoxicity and activates microglia in zebrafish larvae: From the perspectives of neurobehavior and neuroimaging. *Chemosphere, 291*(Pt 1), 132802. https://doi.org/10.1016/j.chemosphere.2021.132802

Zareifopoulos, N., Skaltsa, M., Dimitriou, A., Karveli, M., Efthimiou, P., Lagadinou, M., Tsigkou, A., & Velissaris, D. (2021). Converging dopaminergic neurotoxicity mechanisms of antipsychotics, methamphetamine and levodopa. *European Review for Medical Pharmacological Sciences, 25*(13).

Zhang, R., Wang, X., Xie, Z., Cao, T., Jiang, S., & Huang, L. (2023). Lipoxin A4 methyl ester attenuated ketamine-induced neurotoxicity in SH-SY5Y cells via regulating leptin pathway. *Toxicology In Vitro, 89*, 105581. https://doi.org/10.1016/j.tiv.2023.105581

Zhao, S., Bao, Q., Ma, G., Yao, Y., Xie, L., & Xiong, J. (2023). Benzo[b]fluoranthene (B[b]F) affects apoptosis, oxidative stress, mitochondrial membrane potential and expressions of blood-brain barrier markers in microvascular endothelial cells. *Toxicology In Vitro, 86*, 105522. https://doi.org/10.1016/j.tiv.2022.105522

Imagination of Hematotoxicity/Immunotoxicity Without Flow Cytometry

Sunil Babu Gosipatala, Komal Dixit, Alka Singh, Shruti R. Kuril, and Pushpa Reddy

Abstract

In the evolving landscape of toxicity assessment, the surge in demand for heightened precision and environmentally conscious methodologies has become increasingly palpable. This chapter navigates the evolving landscape of hematotoxicity and immunotoxicity assessment within the realm of drug discovery. It explores advanced methodologies surpassing the limitations of traditional methods. The journey begins with a historical overview, highlighting the transformative role of flow cytometry, and unfolds the narrative of its indispensability. Imagine a scenario without flow cytometry and delve into novel approaches-DNA methylation analysis, RNA sequencing, and multiplexed immunofluorescence-each offering unique insights into immune cell profiling and drug effects on tissues. *In silico* predictions, leveraging artificial intelligence, and *in vitro* assays, including high-throughput screening for hematotoxicity, emerge as potent tools in toxicity assessment. The exploration extends to chemico-methods, *ex vivo* assays, and in-depth insights into *in vitro* hematotoxicity assays. The abstract underscores the pivotal role of omics technologies-metabolomics, genomics, transcriptomics, and proteomics-in decoding toxicogenomics. It envisions a future where a holistic approach reshapes the assessment of hematotoxicity and immunotoxicity in drug development, aligning with the transformative

S. B. Gosipatala (✉) · K. Dixit · A. Singh · S. R. Kuril
Department of Biotechnology, Babasaheb Bhimrao Ambedkar University, Lucknow, Uttar Pradesh, India
e-mail: gsbabu@bbau.ac.in

P. Reddy (✉)
Department of Biochemistry, Indian Academy Degree College-Autonomous, Bangalore, Karnataka, India
e-mail: pushpa.lifesciences@iadc.ac.in

AB Pant et al. (eds.), *Flow Cytometry: Applications in Cellular and Molecular Toxicology*, https://doi.org/10.1007/978-981-97-9758-5_8

potential of emerging methodologies. As tradition meets innovation, this abstract provides a glimpse into the synergy propelling drug development towards precision and efficacy. Despite significant advancements, persistent challenges such as high costs and the absence of standardized databases are underscored. The chapter concludes by emphasizing the collective impact of these diverse methodologies, pushing the boundaries of our understanding of hepatotoxic and immunotoxic phenomena in the pursuit of more effective drug development.

Keywords

Hematotoxicity · Immunotoxicity · Flow cytometry · Drug discovery · Paradigm shift

1 Introduction

In the world of toxicology research and regulations, understanding how various substances affect our blood and immune systems is crucial. The changing lifestyles can also be linked to an increased demand on daily requirements and to fulfil these demands, there is a need for a huge rise in food, industry and commercial sectors. These in turn release tremendous amount of harmful chemicals, drugs, or agents. Exposure to and accumulation of different industrial and agricultural agents can have harmful effects, particularly on the immune system (immunotoxic) and blood (hematotoxic). There are several different hematotoxicity mechanisms, and some substances or medications may have direct or indirect effects on blood cells. They could attach to the surface of cells, suppress or stimulate cells, or kill cell precursors (Krishnan & Pelekis, 1995).

Our traditional assessing tool, flow cytometry, considered as one of the powerful and versatile technique for analyzing single cells or particles in suspension, has been a cornerstone of hematotoxicity and immunotoxicity research. It enables researchers to assess various immune and blood cell populations, providing a wealth of data for toxicological evaluation. However, there are circumstances where the use of flow cytometry is impractical or unavailable. In some cases, researchers may be limited by resource constraints, equipment availability, or the nature of the sample, such as archival samples from historical studies. Acknowledging the importance of exploring hematotoxicity and immunotoxicity in the absence of flow cytometry is crucial for widening the scope of research engagement. Such an approach fosters inclusivity in research methodologies, offering insights into viable alternative techniques. It must be recognized, though, that these methods may lack the sensitivity and specificity inherent in flow cytometry and might not encompass all facets of hematotoxicity and immunotoxicity. Nevertheless, these alternatives hold significance as viable options in contexts where flow cytometry is either unavailable or impractical.

Immunotoxicity, the study of the adverse effects on the immune system, can be due to occupational, inadvertent, or therapeutic exposure to a variety of environmental chemicals or biologic materials. Depending on the chemical or agent involved, immunotoxic interactions in humans can have different mechanisms. Immunotoxicants can produce immunologic suppression, which modifies host resistance, or a marked increase in a particular immune response, which can result in unfavourable health outcomes like autoimmune diseases, allergic contact dermatitis, and respiratory tract allergies (National Research Council (US) Subcommittee on Immunotoxicology, 1992). Thus, the systems such as hematotoxicity and immunotoxicity, play a vital role in understanding our knowledge on health and disease. One can do even better if the knowledge on the methods assessing different aspects of immunotoxicity and hematotoxicity is pooled.

This chapter gives a brief outline on the various strategies available for the measurement of immunotoxicity and hematotoxicity, its evolution starting from the very old conventional methods to flow cytometric techniques, with a more detailed understanding of both immunotoxic and hematotoxic effects including in-silico studies, as it enhances the thoroughness and accuracy of evaluating the toxicity effects compared to relying solely on conventional techniques which might lack specificity and depth, potentially missing nuanced immune and blood-related impacts that could otherwise be captured. Certainly, when assessing immunotoxicity and hematotoxicity in the presence versus absence of flow cytometry, a notable difference in comprehensive analysis emerges. This is because currently a plenty of novel assessing methods have emerged. Thus, there is a necessity to broaden our knowledge on the new tools, and thereby saving the time, cost and labour. Thus, this chapter further emphasizes the necessity of a paradigm shift in toxicological evaluation from the traditional dependence on flow cytometry to modernized methods making us imagine about how without a flow cytometry, toxicity assays can be performed in a more advanced way, as shown in Fig. 1.

2 Deciphering the Complexity of Hematotoxicity (Causes and Mechanisms)

Hematotoxicity, encompassing a spectrum of adverse effects on blood and blood-forming tissues, poses significant challenges in both medical and environmental contexts. Understanding its causes and mechanisms is crucial for effective assessment and management. Flow cytometry has historically played a pivotal role in hematotoxicity assessment, offering detailed insights into cellular characteristics and functions. However, despite its utility, flow cytometry presents certain limitations and challenges that warrant exploration of alternative or complementary methods.

3 Causes of Hematotoxicity

In the realm of toxicology, hematotoxicity stems from a myriad of sources, including pharmaceuticals, chemicals, and radiation, all capable of impairing blood cell survival, formation, and function. This impairment can arise from direct

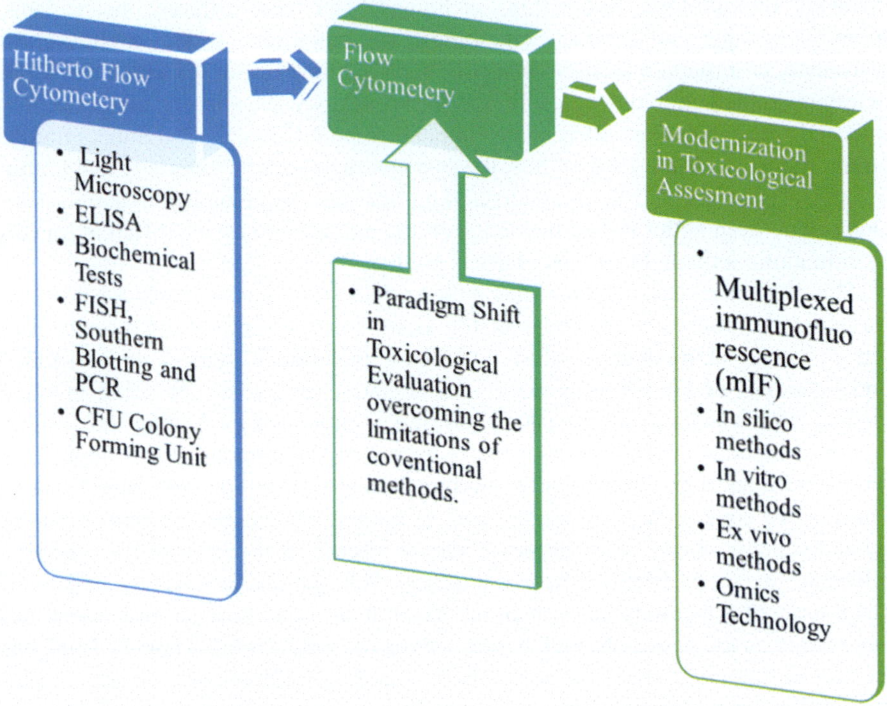

Fig. 1 Paradigm shift in toxicological evaluation from conventional to modernized methods

cytotoxicity to blood cells or the targeting of bone marrow progenitor/stem cells (Bloom et al., 2013). Notably, the hematotoxic effects of these agents often necessitate dosage restrictions, particularly in the context of administering anti-neoplastic agents to cancer patients. Such limitations can significantly challenge therapeutic efficacy, as achieving optimal dosage becomes precarious due to the compromised blood system (Wang et al., 2022).

Moreover, the concomitant use of radiation therapy, which also targets bone marrow, further exacerbates hematotoxicity, amplifying its impact (Bloom, 1993). Adding to the complexity, medications such as non-steroidal anti-inflammatory drugs (NSAIDs) and anticoagulants have been identified as culprits in causing bleeding disorders or interfering with platelet function. Furthermore, infections such as HIV or hepatitis C directly impede blood cell production and function, compounding the hematotoxic burden (Budinsky, 2003). In the pursuit of understanding and mitigating hematotoxicity, toxicologists confront a multifaceted challenge, navigating the intricate interplay of various agents and their deleterious effects on the blood system. Such insights are pivotal not only in clinical settings but also in guiding regulatory frameworks and therapeutic interventions aimed at safeguarding human health.

4 Mechanisms

Understanding the mechanisms underlying hematotoxicity unveils a labyrinth of intricacies crucial for effective intervention. Figure 2 illustrates key mechanisms pivotal in comprehending hematotoxicity. From bone marrow myelosuppression to the suppression of marrow activity, disruptions in blood cell production emerge as a central concern. Various agents, spanning chemicals, drugs, and infections, can incite these mechanisms, impeding the production of essential blood cells (Krishnan & Pelekis, 1995; Friberg & Karlsson, 2003).

Hematopoiesis, the finely regulated process of blood cell production within bone marrow, emerges as a vulnerable target, precipitating issues such as erythropoiesis suppression, hemolytic anemia, leukopenia, and thrombocytopenia (Bloom, 1993; Budinsky, 2003). Complications extend to aplastic anemia, erythema, coagulation-related disorders, apoptosis, and necrosis (Haschek-Hock et al., 2022). Contributing to hematotoxicity are additional factors such as alcohol and its metabolite, acetaldehyde. Their hematotoxic effects can lead to an elevation in mean corpuscular volume (MCV) (Dasgupta, 2015). Advanced methodologies, as demonstrated in benzene-induced hematotoxicity studies, offer a glimpse into the intricacies that shed light on these complex processes (Yoon et al., 2003). Such depth of understanding is paramount for toxicologists striving to unravel the multifaceted nature of hematotoxicity and devise targeted interventions for its mitigation.

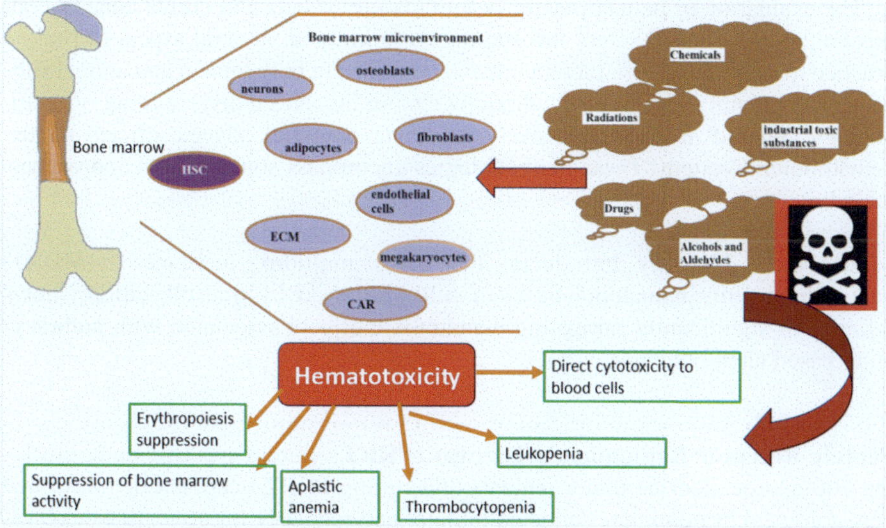

Fig. 2 Mechanisms involved in hematotoxicity

5 Unravelling Hematotoxicity: Assessment Approaches and Methodologies

In the realm of medical investigation, professionals diligently seek indicators of hematotoxicity in patients, ranging from hemorrhage to altered white blood cell counts or anemia (Bloom, 1993). These clinical observations often serve as crucial signals, pointing towards hematotoxic effects induced by medications, substances, or underlying illnesses. A pivotal method in elucidating hematotoxicity involves the microscopic examination of blood smears. Through staining and scrutiny of blood smears, researchers and clinicians can discern aberrations in blood cell morphology and count, including the presence of immature or anomalous cells, offering valuable insights into the etiology of hematotoxicity (Gulati et al., 2013).

Preclinical toxicology studies frequently rely on animal models to assess the hematotoxicity of drugs and chemicals. By administering substances to animals and monitoring their blood parameters, researchers can gauge potential hematotoxic effects prior to human trials. These studies aid in distinguishing between primary and secondary hematotoxicity, with a focus on primary hematotoxicity, which can pose significant challenges in drug development and clinical application (Bloom, 1993). With advancements in cell culture techniques and molecular biology, in vitro assays have emerged as indispensable tools for evaluating hematotoxicity (Pessina et al., 2005). These assays allow researchers to expose isolated blood cells or bone marrow cells to test substances or environmental toxins, assessing their effects on cell proliferation, differentiation, and other hematological parameters.

The evaluation of hematopoietic system toxicity stands as a cornerstone in both preclinical and clinical safety investigations, forming an integral aspect of routine practice to monitor diverse therapeutic interventions in both human and animal subjects. The pursuit of early hematotoxicity testing involves a diverse array of methodologies aimed at comprehensively assessing potential adverse effects on the hematopoietic system. These methodologies encompass sophisticated approaches, including:

MDR1 Gene Therapy Introducing the human multidrug resistance 1 (MDR1) gene into healthy hematopoietic stem cells, followed by transplantation, shows promise in significantly mitigating hematotoxic effects associated with anthracycline-based chemotherapy.

Peptide Receptor Radionuclide Therapy (PRRT) A retrospective study involving 100 neuroendocrine tumor patients undergoing PRRT treatment demonstrated hematological evaluations conducted initially and regularly for up to 10 weeks following the completion of the fourth cycle. The treatment regimen consisted of four cycles administered every 10 weeks, with a targeted dosage.

Intensity Modulated Radiotherapy (IMRT)/Volumetric Modulated Arc Radiotherapy (VMAT) Technique A study conducted from 2013 to 2019 on 40 squamous cell carcinoma patients receiving treatment at University Hospital Motol's Oncology Department utilized advanced radiotherapy techniques (IMRT/VMAT) to effectively target tumors and reduce skin, stomach, urinary, and blood-related side effects. Lower doses to the bone marrow (below 27 Gy) were associated with fewer blood-related issues.

6 Immunotoxicity (Toxicology of the Immune Cells)

Immunotoxicity, the intricate interplay of hazardous agents with the immune system, unfolds as a multifaceted phenomenon influenced by a spectrum of substances, including air pollution, industrial chemicals, heavy metals, pharmaceuticals, UV radiation, agrochemicals, and certain biological materials. These immunotoxic substances, termed xenobiotics, exert their effects on both the innate and adaptive immune systems (Haschek-Hock et al., 2022; Rooney et al., 2012). The resulting manifestations—ranging from immunosuppression and hypersensitivity to autoimmunity—arise from dysregulated immune function, altered signaling pathways, and immune cell destruction (Gulati & Ray, 2009; Bou Zerdan et al., 2021; Wang et al., 2022).

7 Understanding the Disruption: Immunotoxicity (Mechanisms, Causes, Implications)

The intricate decay of the immune system involves shared mechanisms, each contributing to a complex interplay of malfunctioning cells and misguided reactions. From disruption of immune cell function and molecular mimicry to cytokine signaling modification and induction of autoimmunity, these mechanisms intricately shape the landscape of immunotoxicity (Delves & Roitt, 1998). Figure 3 visually guides this exploration, offering detailed representations of shared mechanisms underlying immunotoxicity.

Comprehending the varied agents causing immunotoxicity is crucial for developing effective treatments and preventive measures. By minimizing exposure to these threats and adopting a healthy lifestyle, we can fortify our immune system's ability to safeguard against infections and illnesses (Semwal et al., 2022).

8 Unravelling Immunotoxicity: Testing and Assessment Methodologies

Recent advances have introduced a suite of innovative methodologies revolutionizing immunotoxicity assessment. These techniques employ intricate cellular analyses and functional assays, providing nuanced insights into the impacts of substances

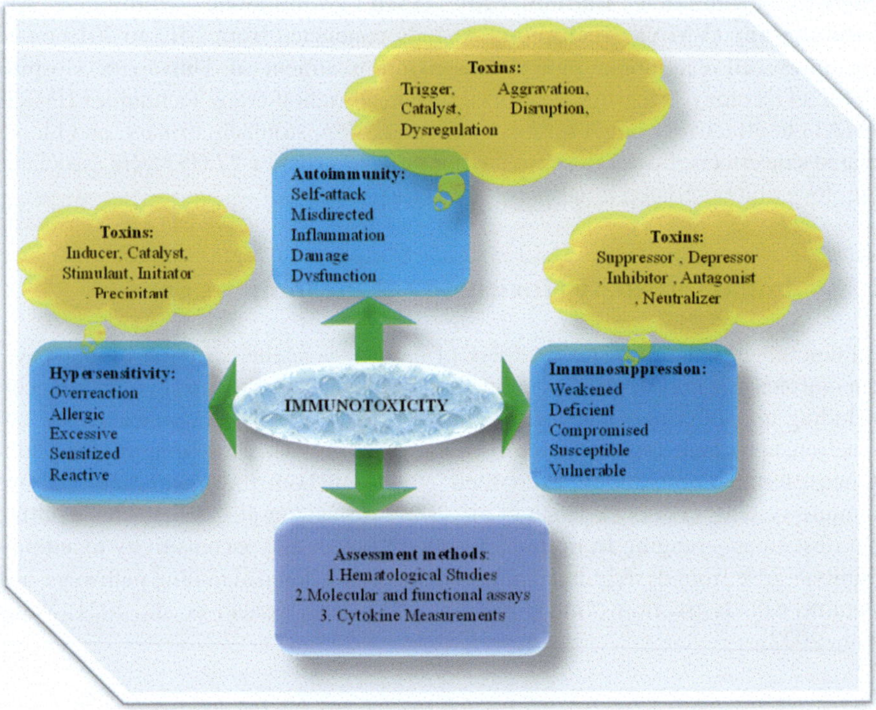

Fig. 3 Unveiling immunotoxicity mechanisms: A visual exploration

on immune function. Such approaches collectively enhance the depth and breadth of our understanding in immunotoxicology evaluation. Further elaboration on select techniques follows below, offering insight into their pivotal role in contemporary immunotoxicity assessment:

- Fluorescent Cell Chip: This in vitro immunotoxicity testing employs a Fluorescent Cell Chip integrating cell lines derived from lymphocytes, mast cells, and monocytes-macrophages. Genetically modified through transfection with diverse EGFP cytokine reporter gene constructs, these cells amplify the array of detectable immune responses for comprehensive assessment. Utilizing immediate-early response genes could further enhance the Fluorescent Cell Chip, providing insights into cellular stress responses (Cochran, 1993). For instance, the c-fos gene swiftly responds to stimuli like heat shock, UVB, growth factors, cAMP, and heavy metals (Chen et al., 1998).
- Developmental Immunotoxicity Testing (DIT): With the escalating incidence of immune-based childhood ailments, scientists prioritize the adoption of DIT testing to unearth potential environmental risk factors linked to these conditions. This specialized methodology probes the timing, extent, and impact of external agents on the intricate development of the immune system.

- Natural Killer (NK) Assay: Assessing NK cell function is pivotal in immunotoxicity testing. Evaluating NK cell activity involves gauging cytolytic function against target tumor cells through assays like the 51Cr-release test. Additionally, quantifying NK cell counts in human peripheral blood or animal spleen using flow cytometry provides valuable insights into immune responses. Conversely, the pursuit of immunostimulation hinges on the detection of heightened responses in these facets (Exon et al., 1986).
- Cytokine Gene Expression: Alterations in cytokine gene expression within immune and non-immune cellular domains coincide with specific manifestations of immunotoxicity.
- For instance, the elevation of interleukin-4 (IL-4) expression in lymphocytes and mast cells induced by heavy metal exposure (Oliveira et al., 1995) or the augmented IL-5 expression in lymphocytes triggered by amino acid derivatives (Yamaoka et al., 1994) underscores their immunotoxic effects observed in vivo.
- Host Resistance Assay: Offering a holistic view of innate, adaptive, and regulatory immune functions, host resistance assays serve as the benchmark in immunotoxicity testing. Comprehensive and targeted host resistance models provide invaluable perspectives on how varying doses induce protective immune responses.

Other techniques for measuring immunotoxicity include Local Lymph Node Assay (LLNA), Organotypic Cultures, and Immunotoxicogenomics (Luebke et al., 2006; Baken et al., 2008). These testing methods illuminate the intricate impacts of substances on the immune system, aiding in comprehensive evaluations for immunotoxic effects and targeted investigations into immune mechanisms.

9 Regulatory Guidelines for the Screening of Hematotoxicity and Immunotoxicity

Due to their potential to impact the immune system and blood systems, respectively, hematotoxicity and immunotoxicity are two crucial components of evaluating the safety of pharmaceuticals. When compared to other toxicities, hematotoxicity happens at comparatively lower exposure levels to chemical substances (Yamada et al., 2022). If an effect leads to unwanted harm to group of cells (body's selfimmunity, allergic reactions, or long-lasting irritation) or compromises immunity that the host needs to protect body from contagious or malignant disease (immunosuppression), it is considered to be immunotoxic or adverse (Immunotoxicocity Testing Guidance, 1999). Several regulatory guidelines are available for calculating the danger of hematotoxicity and immunotoxicity association with medications and biologics in non-clinical research. Some of the main guidelines are:

FDA Guidance for Immunotoxicity Testing Guidance (Immunotoxicocity Testing Guidance, 1999)

WHO Guidance for immunotoxicity risk assessment for chemicals (Guidance for immunotoxicity risk assessment for chemicals, 2012)

FDA Guidance for Nonclinical Safety Evaluation of the Immunotoxic Potential of Drugs and Biologics (Nonclinical Evaluation of the Immunotoxic Potential of Pharmaceuticals, 2023)

EHA/EBMT Consensus Recommendations on Immune Effector Cell—Associated Hematotoxicity (Rejeski et al., 2023)

EMA guideline on the immunogenicity assessment of therapeutic proteins (Committee for Medicinal Products for Human Use, 2017)

10 Chronicles in the Assessment of Toxicity

Researchers and medical professionals have evaluated toxicity using a variety of techniques throughout history, which has laid the groundwork for our current understanding. These earlier research works have greatly advanced our understanding and have influenced the development of contemporary evaluation techniques, which have changed over time. The pioneering initiatives in early toxicity testing encompass the strategic formulation and execution of sophisticated screening modalities geared towards the prompt discernment of putative toxicity during the nascent phases of drug or therapy development. Such concerted efforts are poised to expeditiously pinpoint and redress adverse effects on the hematopoietic system at the incipient stages of preclinical and clinical assessments, thereby augmenting the safety paradigm of emergent interventions prior to extensive scrutiny or implementation. There have been many approaches before and after the arrival of cutting-edge technologies like flow cytometry for the assessment of Toxicity.

11 Flow Cytometry Integration Ensuring Robust Analysis

Flow cytometry serves as a cornerstone in evaluating hematotoxicity and immunotoxicity, offering a comprehensive analysis of immune responses and cell phenotyping (Espenschied et al., 2018; Rawat et al., 2019).

Its ability to measure cellular parameters such as proliferation, activation, differentiation, cytokine production, apoptosis, phagocytosis, oxidative burst, DNA damage, mitochondrial membrane potential, and intracellular signaling has been instrumental in advancing our understanding (Phadnis-Moghe & Kaminski, 2017).

- Cellular proliferation: Flow cytometry can be used to measure the proliferation of immune cells in response to a stimulus, which is an important indicator of immunotoxicity
- Cellular activation: Flow cytometry can detect changes in the expression of cell surface markers, such as CD25 and CD69, which are indicative of immune cell activation

- Cellular differentiation: Flow cytometry can identify and quantify different cell populations based on their surface markers, allowing for the detection of changes in cell differentiation
- Cytokine production: Flow cytometry can measure intracellular cytokine levels in immune cells, providing information about the functional status of the immune system
- Apoptosis and cell death: Flow cytometry can detect changes in the expression of apoptotic markers, such as Annexin V and propidium iodide, allowing for the measurement of cell death and apoptosis
- Phagocytosis: Flow cytometry can be used to measure the phagocytic activity of immune cells, such as macrophages and neutrophils, which is an important indicator of their functional status
- Oxidative burst: Flow cytometry can measure the production of reactive oxygen species (ROS) by immune cells, which is an important indicator of their antimicrobial activity.
- DNA damage: Flow cytometry can detect changes in DNA content, such as the formation of DNA adducts or DNA strand breaks, which are indicative of genotoxicity.
- Mitochondrial membrane potential: Flow cytometry can measure changes in mitochondrial membrane potential, which is an important indicator of cellular health and function.
- Intracellular signaling: Flow cytometry can measure changes in the phosphorylation status of intracellular signaling molecules, providing insights into the mechanisms of action of immunotoxic agents.

Flow cytometry allows for the examination of effects of immunotoxic agents on target cells of the immune system and offers a more detailed understanding of cellular dynamics, activation markers, and intricate immune responses (Burchiel et al., 1997). The discovery and implementation of flow cytometry revolutionized the field of toxicity assessment. Flow cytometry enabled researchers to analyze multiple parameters of individual cells simultaneously, offering a high-throughput approach to studying cellular characteristics, activation markers, and immune responses. The technology's ability to distinguish between different cell types based on surface markers and functional parameters brought unprecedented depth to toxicity studies.

Flow cytometry is a technology that provides rapid multi-parametric analysis of single cells in solution. It utilizes lasers as light sources to produce both scattered and fluorescent light, which are then converted into electronic signals and analyzed by a computer. This allows for the simultaneous characterization of mixed populations of cells from various sources, such as blood, bone marrow, spleen, mucosal tissues, and solid tumors. Additionally, flow cytometry is a powerful tool with applications in immunology, virology, molecular biology, cancer biology, and infectious disease monitoring (McKinnon, 2018).

12 Assessment of Toxicity Before Flow Cytometry

Before the advent of flow cytometry, the assessment of hematotoxicity and immunotoxicity relied on conventional methods. These methods, while informative, had limitations in providing detailed insights into cellular dynamics, activation markers, and intricate immune responses. They often lacked the precision and multiparametric capabilities of modern technologies.

Some of these methods included:

1. **Cell counting**: This technique was used to determine the number of cells in each sample, but it did not provide detailed insights into cellular dynamics or activation markers.
2. **Microscopy**: Microscopic examination of cells allowed for the visualization of cellular structures and the identification of specific cell types, but it did not offer a quantitative assessment of cellular responses.
3. **Biochemical assays**: These assays were used to measure specific biochemical parameters in cells, but they did not provide a comprehensive understanding of cellular dynamics or immune responses.

13 Interplay Between Hematotoxicity and Immunotoxicity

Hematotoxicity and Immunotoxicity are the major facet in the domain of immunological responses. People might think of these two aspects as a lone part but little do they know, both of them go hand in hand. The whole orchestra of immune response does not utterly depend upon the lymphatic system but it is amalgamated with the circulatory system also. So, any effect that could affect the immune system also goes with circulatory system too. The inception of studies in the domain of Hematotoxicity and Immunotoxicity has paved a way into many micro-physiological systems. HSCs, central to the physiological cascade, generate blood and immune cells in the bone marrow, guided by complex regulatory networks governed by transcription factors. These cells give rise to myeloid and lymphoid lineages, generating various progenitor cells in different anatomical locations (Rothenberg, 2000). Xenobiotics entering the body not only suppress immune responses but also alter blood component thresholds, exemplifying the amalgamation of the lymphatic and circulatory systems (Picker & Siegelman, 1999).

Research in hematotoxicity and immunotoxicity has evolved from initial investigations of anaphylaxis to a dynamic and extensive discipline of immunotoxicology. The assessment of these phenomena provides critical insights into the effects of substances on the blood and immune system, influencing individual health. Environmental contaminants, like tetrachloroethylene, induce changes in cell counts, emphasizing their impact on immune and blood systems (Emara et al., 2010). Evaluation of nucleic acid-based therapeutics poses challenges in immunotoxicity, hematotoxicity, and formulation, as observed in chimeric antigen receptor T-cell therapy (Dobrovolskaia & McNeil, 2015; Rejeski et al., 2021).

Understanding hematotoxicity and immunotoxicity without relying on flow cytometry sheds light on the intricate interplay between these realms. The journey underscores the need for alternative testing methods in immunotoxicity assessment (Boverhof et al., 2014), providing profound insights into the delicate harmony of the body's defense mechanisms.

Having dissected the intricate interplay between hematotoxicity and immunotoxicity, we now pivot towards the critical juncture of assessing these intertwined phenomena. It's like we've discovered a connection where what happens in one area affects the other. This brings us to the part where we explore different ways to understand these toxicities.

14 Case Studies Highlighting Flow Cytometry's Influence

The evaluation of immunotoxicity and hematotoxicity has evolved significantly, demanding more precise methods. This chapter explores the transition from traditional to modern flow cytometry techniques and their influence on immunotoxicity assessment.

- **Pre-Flow Cytometry Era**

Flow cytometry, starting with Fulwyler's cell sorter in 1965 (Fulwyler, 1965), has undergone substantial evolution from its early pioneers like Ehrlich and Lazarus (Ehrlich & Lazarus, 1900). Their work not only studied ocular fluid dynamics but also introduced fluorescein, laying foundational concepts for subsequent advancements in flow cytometry techniques. Early methods relied on polychromatic flow cytometry (PFC), limiting detection to one fluorochrome per detector and imposing constraints on signal collection.

Early developments by Moldavan, Gucker, O'Konski, and Crosland-Taylor contributed essential concepts to the technology, including cell counting, photomultiplier-based particle counting, and hydrodynamic focusing (Moldavan, 1934; Gucker Jr. & O'Konski, 1949; Crosland-Taylor, 1953).

Flow cytometry stands as an indispensable tool for biologists, immunologists, and physicians, tracing its roots back to pivotal milestones in scientific history. The critical juncture emerged with Coons conjugation of an antibody to fluorescein, marking the initiation of immunologically based fluorescence conjugation to antibodies (Coons et al., 1941). However, the transformative leap was exemplified by Hulett, who merged flow and fluorescence (Hulett et al., 1969). Their pioneering work laid the groundwork for the fusion of flow technology with fluorescence, culminating in the seamlessly integrated flow cytometry tools used by clinicians and researchers worldwide today.

- **Transition from PFC to Spectral Flow Cytometry (SFC)**

To overcome these limitations, Spectral Flow Cytometry (SFC) emerged, utilizing full-spectrum detection and spectral unmixing principles. Initially proposed as a hypothesis, SFC was developed with a 32-channel full-spectral detector system, allowing for comprehensive spectral analysis for each cell.

15 Modern Approaches: Integration of In-Silico Models Including Enhanced Computational Methods

In today's scientific landscape, the integration of cutting-edge in-silico methods and enhanced computational techniques has revolutionized research and development. These modern approaches harness the power of computer simulations, modeling, and data analytics to drive innovation across various fields, offering faster, more cost-effective, and predictive avenues for analysis and discovery.

Some of the modern techniques are discussed as below:

Cytometric Bead Array (CBA) The multiplex BD™ Cytometric Bead Array (CBA) utilizes fluorescently labeled beads coupled with antibodies to assess multiple proteins within a single sample accurately and quantitatively. So that it can specifically capture a particular type of molecule present within biological fluids including sera, plasma, tears, tissue culture supernatants. CBA Kits are engineered to accurately and quantitatively assess multiple proteins concurrently within a single sample. These kits are meticulously designed for optimal performance when analysing specific analytes across various sample types, including tissue culture supernatants, EDTA-treated plasma and serum samples, and cell lysates.

Ultraviolet 320 nm Laser Excitation Despite initial cost constraints, recent advancements in UV laser technology have facilitated its integration into flow cytometers, offering enhanced connectivity and compatibility with smaller analytical platforms, typically operating within the range of 325–365 nm, are not commonly used in cytometry. The rarity of their utilization primarily stems from their cost and the limited number of applications necessitating this wavelength. However, the advent of Brilliant Ultraviolet (BUV) fluorochromes has notably amplified the significance of this previously niche excitation wavelength. While DPSS UV lasers persist as large and expensive relative to modern solid-state laser standards, recent developments have introduced this laser technology in a more compact form factor than the current 355 nm units. This compact size not only addresses space constraints but also enhances connectivity, providing compatibility with smaller cytometers and fostering easier integration within various analytical platforms (Telford et al., 2017).

Quantitative Structure Activity Relationship (QSAR) A statistical model correlating chemical structures with biological activities, enabling predictive analysis of potential cytokines or pharmaceutical molecules. As a rigorously validated statistical model, QSAR embodies a mathematical framework that reliably correlates the intricate relationship between chemical structures and their associated activity profiles. QSAR is the most frequently used modeling approach to explore the dependency of biological, toxicological, or other types of activities/properties of chemicals on their molecular features. In the past two decades, QSAR modeling has been used extensively in drug discovery process (Tenorio-Borroto et al., 2012).

Quantitative Structure In vitro-In vivo Relationship (QSIIR) A novel modeling approach focusing on various animal toxicity endpoints, employing cheminformatics methodologies to predict target properties based on environmental agents' characteristics.

The primary focus of QSIIR remains centered on biological activities, encompassing various animal toxicity endpoints. However, what sets QSIIR apart is its adaptable approach to "descriptors" and model interpretation, which vary based on the specific characteristics—chemical, biological, and genomic—of environmental agents being assessed. By aiming to predict identical target properties through diverse traits of environmental agents, QSIIR presents a promising avenue to delve deeply into understanding the interplay between these agents and their resulting impacts. This approach, leveraging cheminformatics, holds the potential to unravel the intricate connections between chemical compositions, biological responses, and genetic factors, advancing our comprehension of toxicological pathways from the source to the ultimate outcomes.

16 Paradigm Shift: Embracing Modern Approaches for Immunotoxicity and Hematotoxicity Evaluation

The rise in environmental chemicals has necessitated a shift from traditional animal-based toxicity assays to combined in vitro assays and computational toxicology. Initiatives like the Tox21 project leverage in vitro assay techniques and High-Throughput Screening (HTS) technology to identify compounds with detrimental health impacts efficiently.

In 2007, the U.S. National Research Council's report, "Toxicity Testing in the 21st Century: A Vision and a Strategy," advocated for a shift from animal experiments to combined in vitro assays and computational toxicology. This approach aimed to explore interactions between pollutants and biomolecules, unravelling key toxicity pathways. This marked a beginning of paradigm shift in toxicity testing. The Tox21 project, initiated in 2008, hinged significantly on established and dependable in vitro assay techniques to efficiently evaluate the potential harmful effects of specific chemicals on health. These assays actually provide rich experiences and clear development ideas for HTS (High-Throughput Screening) of environmental toxic chemicals, and the first step is to establish an effective and stable in vitro assay. Currently, Tox21 has adopted 232 different bioassays.

According to the experience of HTS for drugs, these assays can be divided into two categories:

1. Target based
2. Cell Phenotype-based methods.

Target-based methods aim at specific molecules like receptors (e.g., estrogen receptors (ERs), androgen receptors (ARs), and aryl hydrocarbon receptors (AhRs)), enzymes (e.g., acetylcholinesterase (AChE), aromatase, and cytochrome P450), and DNA. When dealing with pollutants whose toxicity mechanisms remain unclear, cell phenotype-based methods emerge as a more suitable option. These methods leverage diverse cells as screening tools, primarily detecting cytotoxicity, functional impairment, specific functional proteins, secretions, and alterations in signaling pathways.

Modern computational methods offer faster, cost-effective, and predictive avenues for assessing immunotoxicity and hematotoxicity, marking a paradigm shift in toxicity testing towards more efficient and reliable approaches.

17 Investigating Hematotoxicity and Immunotoxicity: Uncovering the Intricacies Without Flow Cytometry

Flow cytometry is an effective tool for many applications, but it is especially useful in clinical diagnostics and biomedical research. Depending on the particular situation, its absence may have serious repercussions and call for the adoption of alternative strategies with different considerations.

1. **Limited Cell Characterization:** Individual cell characteristics such as size, surface markers, and internal components are all well-documented by flow cytometry. Researcher and clinician struggles could be as follows in its absence:
   - Distinguish between cell populations: Research on immunological response, cancer diagnosis, and cell development are all impacted by the challenge of recognizing and characterizing various cell types within a mixture sample.
   - Assessing cell activation markers and functional parameters such as cytokine production can be achieved through the use of flow cytometry. Understanding the reactions of cells in different biological processes becomes difficult in the absence of it.
   - Selective cell population isolation is made possible by flow cytometry, which is essential for subsequent investigations such as single-cell analysis and functional assays. Research focusing on particular cell types may be hampered by the lack of this ability.
2. **Decreased Diagnostic Accuracy:** Leukemia, lymphoma, and autoimmune disorders are just a few of the diseases that flow cytometry is vital for diagnosing. Lack of it may result in:

- Misdiagnosis or delayed diagnosis: It can be challenging to identify and categorize disease-specific cell populations, which increases the risk of misdiagnosis or delayed diagnosis.
- Restricted options for treatment: Providing precise descriptions of immune and tumor cells is essential for developing customized treatment plans. Treatment choices might be less accurate in the absence of flow cytometry.
- Reduced ability to track the progression of the disease: Flow cytometry can track changes in cell populations over time to track how the disease responds to treatment. Lack of it can make it more difficult to evaluate the effectiveness of treatment and modify therapy as necessary.

3. **Impaired Advancement of Research:** Flow cytometry is a flexible instrument utilized in many different fields of study.

   Lack of it can affect:

- Research on cell signaling, cell-cell interactions, and cellular responses to stimuli may be severely hindered, so research progress should be slowed down.
- Restrict the range of the study questions: Some research questions might become unanswerable, especially if they call for a thorough examination of individual cells.
- Decrease the effectiveness of research: Large cell populations can be subjected to high-throughput analysis using flow cytometry. Alternative approaches might take longer and be less effective.

Hence, flow cytometry is an effective tool for many applications, but it is especially useful in clinical diagnostics and biomedical research. Depending on the particular situation, its absence may have serious repercussions and call for the adoption of alternative strategies with different considerations.

18 Limitations in Imagination of Flow Cytometry in Hematotoxicity and Immunotoxicity Assessment

Despite its transformative impact, flow cytometry came with its set of limitations.

Technical Limitations

Flow cytometry requires viable cells in suspension, which can limit the types of samples that can be analyzed. Additionally, the laser can only analyze one cell at a time, which can be time-consuming and limit the throughput of the analysis. The need for single-cell suspensions can pose challenges when working with tissues or heterogeneous samples (Betters, 2015).

Limited availability of reagents

The availability of commercially or otherwise available antibodies to cell surface markers is limited for many animal species, which can limit the analysis of animal samples using flow cytometry (Hunka et al., 2020).

Quasi-quantitative nature
Flow cytometry is a quasi-quantitative technique, which can limit the accuracy of the results (Newton & Dobrovolskaia, 2022). This is because the fluorescence intensity of the cells is affected by factors such as the instrument settings, the quality of the antibodies, and the autofluorescence of the cells.

Difficulty in detecting rare cell populations
Although flow cytometry is useful in detecting rare cell populations, it can be challenging to detect cells that are present at very low frequencies (Hedley & Keeney, 2013).

Limited ability to analyze complex cell populations
Although flow cytometry allows for the simultaneous analysis of multiple parameters on individual cells, it can be challenging to analyze complex cell populations. This is because the interpretation of the results can be complicated by the presence of multiple cell types with overlapping phenotypes.

Difficulty in analyzing intracellular components
Flow cytometry is limited in its ability to analyze intracellular components, such as intracellular proteins and cytokines. This is because the cells need to be permeabilized to allow the antibodies to access the intracellular components, which can affect the integrity of the cells. Challenges in data analysis and reporting: The increased amount of data generated in flow cytometry can pose challenges in data analysis and reporting. The interpretation of the results can be complicated by the presence of multiple cell types with overlapping phenotypes, and the data analysis methods can be complex and time-consuming (Betters, 2015).

19 Flow Cytometry Presence Versus Absence: Comparative Impact Analysis

The ability to precisely characterize individual cells based on their size, surface markers, and internal components made possible by flow cytometry has transformed research and clinical diagnostics. The insights this analysis provides into cell populations, their roles in different biological processes, and their functions are unparalleled. When flow cytometry is compared with and without it, its extensive influence on a wide range of fields is evident.

Presence:

- **Improved comprehension of cell heterogeneity:** Flow cytometry makes it possible for researchers to recognize and distinguish between various cell populations within a sample, which is essential for researching the growth, differentiation, and functionality of cells (Abbas et al., 2014).
- **Cellular parameter analysis:** Cellular parameters can be analyzed at high throughput. Thousands of cells can be measured per second for surface markers,

DNA content, and protein expression, among other internal components. This allows for the effective analysis of large datasets (Shapiro et al., 1998).

- **Characterization of a cell's function:** Flow cytometry can evaluate cytokine production, cell activation markers, and other functional parameters, giving information about cellular responses to stimuli and signaling pathways in cells (Reinherz et al., 1979).
- **Isolation of particular cell populations for later analysis**: Single-cell analysis, functional tests, and research on uncommon cell populations are made easier by classifying cells according to their properties (Herzenberg et al., 1976).
- **Precision medicine advancement:** By identifying disease-specific cell populations and their biomarkers, flow cytometry opens the door to customized therapeutic strategies (Betters, 2015).
- **Prognosis and disease subtype identification**: Extensive cell population characterization aids in distinguishing between disease subtypes and forecasts patient outcomes, which guides treatment choices (Abbas et al., 2014).
- **Treatment response monitoring:** Flow cytometry can monitor alterations in cell populations over time to evaluate the efficacy of therapy and direct modifications to treatment (Levine et al., 2021).
- **Better patient care:** Higher patient outcomes and survival rates are eventually the result of accurate diagnosis, disease classification, and treatment monitoring (Bonilla et al., 2021).

Absence:

- **Inadequate knowledge of cell diversity:** Research on cell-cell interactions, tissue development, and disease processes is hampered by the difficulty of characterizing complex cell mixtures (Abbas et al., 2014).
- **Dependency on methods that aren't as effective:** Compared to flow cytometry, other techniques such as immunohistochemistry and cell culture assays are frequently more time-consuming and have lower sensitivity and specificity (Reinherz et al., 1979).
- **Restricted capacity to investigate uncommon cell populations:** Research on rare cell types and their role in development and disease is hampered by the difficulty of identifying and isolating them (Herzenberg et al., 1976).
- **Impeded comprehension of cellular processes:** It becomes less accurate and efficient to analyze signaling pathways, cell activation, and other functional parameters (Shapiro et al., 1998).
- Precision medicine will advance more slowly if cells are not thoroughly characterized. This will make the development of biomarkers and tailored treatment plans less possible.

The use of flow cytometry greatly improves research capacities by allowing in-depth analyses of cell populations, their activities, and their roles in both health and disease. It is essential for precise diagnosis, individualized care, and better patient outcomes in clinical settings. Even though its absence can be compensated for with other strategies, it can limit our ability to comprehend, identify, and treat a wide range of illnesses. There is no denying the importance of flow cytometry, and expanding scientific understanding and enhancing healthcare delivery depend on its continuous development and accessibility.

20 Absence of Flow Cytometry: Challenges

The absence of this highly sophisticated technique may be a little troublesome as a consequence of which following challenges would be encountered:

1. **Lack of sensitivity and specificity**: Flow cytometry's ability to detect rare cell populations and differentiate between subsets based on specific markers is unparalleled. Alternative techniques such as microscopy or biochemical assays (which might not be able to discern between various types of leukocytes or platelets, or might not be able to capture the dynamic interactions between molecules and cells) may struggle to capture subtle changes indicative of hematotoxicity or immunotoxicity. For instance, while hematology analyzers may quantify cell counts, they may overlook cellular activation or apoptosis (Dobrovolskaia & McNeil, 2015). In contrast, flow cytometry provides precise lineage differentiation and subset quantification, enhancing sensitivity and specificity in toxicity assessment. For instance, based on their surface markers, flow cytometry can differentiate between various lymphocyte types (B and T cells, non-killer cells) and their subsets (CD4+, CD8+, regulatory T cells), whereas microscopy or histology can only identify lymphocytes based on their morphology (Takahashi et al., 2012).
2. **Lack of multiplexing and automation**: Flow cytometry's capacity for high-throughput analysis of complex samples sets a benchmark in cellular analysis. However, alternative methods may require more time and resources to achieve comparable data output. Manual processing inherent in some assays can introduce bias and human error (Dobrovolskaia & McNeil, 2015). Mass cytometry (CyTOF) and single-cell RNA sequencing (scRNA-seq) offer promising alternatives, enabling multiparametric analysis and overcoming limitations of traditional flow cytometry.
3. **Lack of standardization, validation and reproducibility**: Standardized procedures in flow cytometry ensure data accuracy and consistency. Without such standards, variability in sample preparation and analysis may compromise results. Alternative methods must undergo rigorous validation and standardization to ensure reliability and comparability across studies. Moreover, some of the techniques for detecting hematotoxicity and immunotoxicity that do not use flow cytometry may not be well-validated or standardized (Hua et al., 2021). Mass cytometry and scRNA-seq, while promising, require standardized protocols to maximize their utility in hematotoxicity assessment.

4. **Lack of reliable and standardized method**: The absence of a standardized approach to hematotoxicity assessment poses challenges in evaluating chemical effects on blood-forming organs and cells. In vitro methods offer sensitivity and applicability but may not fully capture in vivo complexities (Hua et al., 2021). Use alternative methods that can provide similar information as flow cytometry, such as mass cytometry, also referred to as cytometry by time-of-flight (CyTOF), is a novel approach that combines the quantitative strength and the selective nature of plasma-mass spectrometry which shows inductive-coupleness with the cellular analysis principles of conventional fluorescence-based flow cytometry (Atkuri et al., 2015; Zhang et al., 2020). High-dimensional time-of-flight mass cytometry (CyTOF) has become a popular method for discovering biomarkers and conducting extensive immune profiling by simultaneously identifying more than 40 parameters (Gadalla et al., 2019). With the help of this potent high-throughput technology, individual cells can be quantified and classified by analyzing up to 50 protein markers per cell (Cheng et al., 2022). Novel biomarkers and indicators, coupled with advanced technologies like mass cytometry and RNA sequencing, hold promise in enhancing the accuracy and reliability of toxicity assessment. For example, researchers may use gene expression profiling, proteomics, metabolomics, or epigenetics to reveal the molecular changes induced by chemicals in blood cells or immune cells (Takahashi et al., 2012).

By expanding our imagination and exploring new possibilities, we can gain a deeper understanding of these toxicities and develop more effective strategies for their assessment and management. This journey invites us to envision a pathway forward that transcends the traditional reliance on flow cytometry, prompting us to explore alternative strategies with a keen awareness of the challenges and implications that lie ahead.

21 Revolutionizing the Assessment: Viable Ersatzes for Flow Cytometry in the Modern Era

Flow cytometry has some limitations, which includes the need for the intact and fresh cells, the dependence on surface markers, and the lack of official guidance documents for method validation in the regulatory context (Du et al., 2015).

In light of flow cytometry's limitations, researchers are exploring alternative techniques to assess hematotoxicity and immunotoxicity that might yield data that is comparable or complementary. These include:

Epigenetic Analysis Novel approaches based on DNA methylation analysis offer insights into immune cell subtypes not detected by traditional methods. This technique accommodates samples under various conditions, making it suitable for clinical and epidemiological research (Salas et al., 2022).

Transcriptomic Profiling RNA sequencing of peripheral blood mononuclear cells (PBMCs) provides valuable information on immune cell proportions and functions. Despite challenges in sample handling and storage, transcriptomic analysis identifies novel biomarkers and pathways associated with toxicity (Oliveira et al., 1995).

Multiplexed Immunofluorescence (mIF) Fluorescence microscopy-based techniques offer spatial data on protein expression in tissues, circumventing limitations of flow cytometry. mIF provides valuable insights into tissue-level interactions and drug effects. When it comes to immune profiling in a regulatory context, the two aforementioned techniques provide a potent substitute for traditional flow cytometry. They can facilitate the assessment of novel drug entities' effects on the immune system and offer more thorough information about an individual's immune status in both health and disease. They do, however, have certain drawbacks, including the requirement for bioinformatics analysis, the lack of standardization, and the uncertainty surrounding the clinical relevance (Salas et al., 2022; med space).

In Silico Methods Computational approaches using machine learning algorithms predict hematotoxic and immunotoxic effects of chemicals based on molecular properties. While cost-effective and ethical, these methods require validation on robust datasets for widespread adoption (Hua et al., 2021).

***In Vitro* Methods** These assays quantify the impact of chemicals on hematological or immunological parameters by utilizing isolated cells, tissues, or organs. The most popular methods for assessing in vitro immunotoxicity include repertoire sequencing, multiplexing, quantitative reverse transcription-polymerase chain reaction (qRT-PCR), transcriptomics, enzyme-linked immunosorbent assay (ELISA), and multiplexing to find out about things like antibody titers, cytokine/chemokine expression and secretion levels, upregulation of surface markers, and T cell (Corsini & House, 2018).

In order to evaluate the blood toxicity of lineage-specific (blood forming progenitor cells) at various stages of their transformation, an in vitro hematotoxicity assay based on liquid culture and microplates can be utilized. Several benefits of this assay include its suitability for high-throughput screening and its flexibility in allowing the addition of extra endpoints for mechanistic studies (Mahalingaiah et al., 2018).

Some of the techniques which helps us in the assessment of hematotoxicity and immunotoxicity including flow cytometry are shown in Fig. 4.

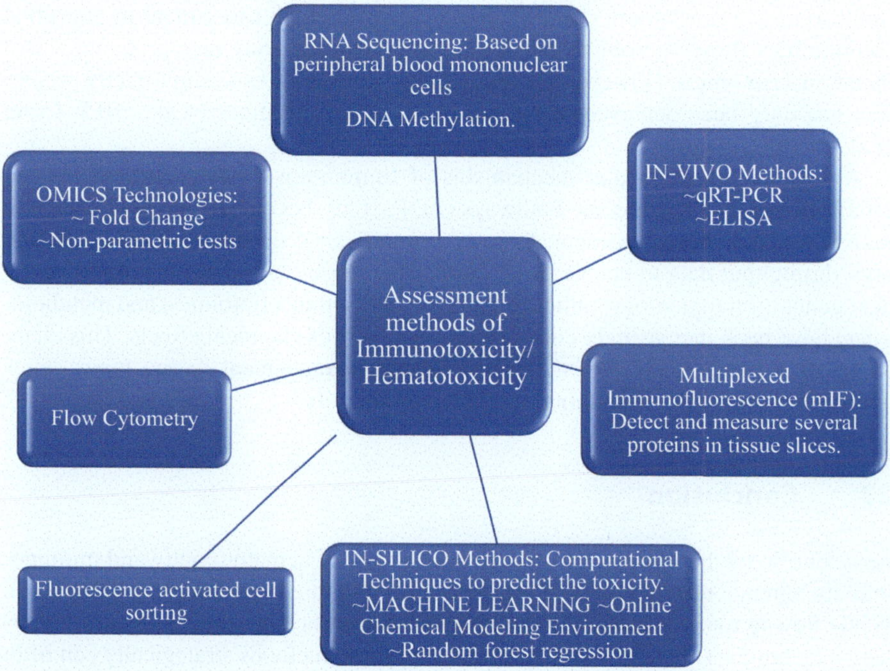

Fig. 4 Methods for the evaluation of immunotoxicity and hematotoxicity

22 The Role of Omics Technologies: A Sumptuous Technology for Toxicogenomics

Omics technologies, including genomics, transcriptomics, proteomics, and metabolomics (Joseph, 2017; Heijne et al., 2005), offer comprehensive insights into biological responses to chemical exposure. Integrating multiple omics approaches enhances our understanding of complex toxicity mechanisms and facilitates the development of adverse outcome pathways (AOPs). Challenges such as cost, data standardization, and analysis pipelines persist but are outweighed by the potential for transformative advancements in toxicology.

Whilst the area of systems biology is over 50 years old (Bose, 2013), only recently have advancements in high-throughput technologies led to the generation of absolutely large data sets to assess the functioning of a biological system in a meaningful way (Feng et al., 2009). These data types and the methods used to analyse them are generally grouped under the 'omics' label. In an ideal world, toxicogenomics would be amalgamated with these four types of readouts, with the future omics layers of biological information, thereby capturing a closer look to the 'complete' biological response of a system to a compound treatment (Sturla et al., 2014). Future directions of omics in toxicology are envisioned to include the application of novel omics methods such as integration of multiple omics approaches (multi-omics), translatomics, and including chemical risk assessment to microbial communities (host-microbiota metabolite interactions) using metaproteomics and metatranscriptomics. However, the inculcation of multiple omics in toxicity prediction has only been achieved in a handful of studies (Wilmes et al., 2015; Craig et al., 2006).

As explicating molecular mechanisms of Immunotoxicity/Hematotoxicity and establishing Adverse Outcome Pathways (AOPs) have become the pivot of chemical toxicology and risk assessment, there is a significance need for high content and high throughput data to enable these goals. To comply with these processes of the next generation toxicology, omics (e.g., proteomics, transcriptomics, and metabolomics) have been increasingly employed for analysis of chemical effects. This chapter gives a brief of the recent methodological advancement in the main omics approaches applied in Immunotoxicity/Hematotoxicity.

23 Conclusion

As scientific methodologies evolve, the assessment of hematotoxicity and immunotoxicity enters a new era marked by innovation and interdisciplinary collaboration. While flow cytometry remains a cornerstone, researchers have a spectrum of alternative techniques at their disposal. Combining these methods strategically can offer a more comprehensive and nuanced understanding of cellular responses to environmental factors. In envisioning a future without the exclusive reliance on flow cytometry, researchers are actively exploring and integrating diverse technologies. This multi-faceted approach holds promise for addressing the challenges posed by traditional methods and advancing our ability to assess complex toxicological endpoints. As technology progresses, the synergy of various techniques will likely redefine the landscape of hematotoxicity and immunotoxicity evaluation, pushing the boundaries of knowledge and contributing to improved safety assessments in diverse settings.

References

Abbas, A. K., Lichtman, A. H., & Pillai, S. (2014). *Cellular and molecular immunology E-Book: Cellular and molecular immunology E-Book*. Elsevier Health Sciences.

Atkuri, K. R., Stevens, J. C., & Neubert, H. (2015). Mass cytometry: A highly multiplexed single-cell technology for advancing drug development. *Drug Metabolism and Disposition: The Biological Fate of Chemicals, 43*(2), 227–233. https://doi.org/10.1124/dmd.114.060798

Baken, K. A., Pennings, J. L., Jonker, M. J., Schaap, M. M., de Vries, A., van Steeg, H., Breit, T. M., & van Loveren, H. (2008). Overlapping gene expression profiles of model compounds provide opportunities for immunotoxicity screening. *Toxicology and Applied Pharmacology, 226*(1), 46–59. https://doi.org/10.1016/j.taap.2007.08.026

Betters, D. M. (2015). Use of flow cytometry in clinical practice. *Journal of the Advanced Practitioner in Oncology, 6*(5), 435–440. https://doi.org/10.6004/jadpro.2015.6.5.4

Bloom, J. C. (1993). Principles of hematotoxicology: Laboratory assessment and interpretation of data. *Toxicologic Pathology, 21*(2), 130–134. https://doi.org/10.1177/019262339302100203

Bloom, J. C., Schade, A. E., & Brandt, J. T. (2013). Toxic responses of the blood. In C. D. Klaassen (Ed.), *Casarett and Doull's toxicology: The basic science of poisons* (8th ed.). McGraw Hill. https://accesspharmacy.mhmedical.com/content.aspx?bookid=958&sectionid=53483733

Bonilla, D. L., Reinin, G., & Chua, E. (2021). Full spectrum flow cytometry as a powerful technology for cancer immunotherapy research. *Frontiers in Molecular Biosciences, 7*, 612801. https://doi.org/10.3389/fmolb.2020.612801

Bose, B. (2013). Systems biology: A biologist's viewpoint. *Progress in Biophysics and Molecular Biology, 113*(3), 358–368. https://doi.org/10.1016/j.pbiomolbio.2013.07.001

Bou Zerdan, M., Moussa, S., Atoui, A., & Assi, H. I. (2021). Mechanisms of immunotoxicity: Stressors and evaluators. *International Journal of Molecular Sciences, 22*(15), 8242. https://doi.org/10.3390/ijms22158242

Boverhof, D. R., Ladics, G., Luebke, B., Botham, J., Corsini, E., Evans, E., Germolec, D., Holsapple, M., Loveless, S. E., Lu, H., van der Laan, J. W., White, K. L., Jr., & Yang, Y. (2014). Approaches and considerations for the assessment of immunotoxicity for environmental chemicals: A workshop summary. *Regulatory Toxicology and Pharmacology: RTP, 68*(1), 96–107. https://doi.org/10.1016/j.yrtph.2013.11.012

Budinsky, R. A. (2003, September). Hematotoxicity: Chemically induced toxicity of the blood. In *Principles of toxicology: Environmental and industrial applications* (2nd ed., pp. 87–109). https://doi.org/10.1002/0471231800.ch4

Burchiel, S. W., Kerkvliet, N. L., Gerberick, G. F., Lawrence, D. A., & Ladics, G. S. (1997). Assessment of immunotoxicity by multiparameter flow cytometry. *Toxicological Sciences, 38*(1), 38–54. https://doi.org/10.1093/toxsci/38.1.38

Chen, W., Borchers, A. H., Dong, Z., Powell, M. B., & Bowden, G. T. (1998). UVB irradiation-induced activator protein-1 activation correlates with increased c-fos gene expression in a human keratinocyte cell line. *The Journal of Biological Chemistry, 273*(48), 32176–32181. https://doi.org/10.1074/jbc.273.48.32176

Cheng, L., Karkhanis, P., Gokbag, B., Liu, Y., & Li, L. (2022). DGCyTOF: Deep learning with graphic cluster visualization to predict cell types of single cell mass cytometry data. *PLoS Computational Biology, 18*(4), e1008885. https://doi.org/10.1371/journal.pcbi.1008885

Cochran, B. H. (1993). Regulation of immediate early gene expression. *NIDA Research Monograph, 125*, 3–24.

Committee for Medicinal Products for Human Use. (2017). *Guideline on immunogenicity assessment of therapeutic proteins*. European Medicines Agency.

Coons, A. H., Creech, H. J., & Jones, R. (1941). Immunological Properties of an antibody containing a fluorescent group. *Proceedings of the Society for Experimental Biology and Medicine*. https://doi.org/10.3181/00379727-47-13084P

Corsini, E., & House, R. V. (2018). Evaluating cytokines in immunotoxicity testing. In *Immunotoxicity testing: Methods and protocols* (pp. 297–314). Springer.

Craig, A., Sidaway, J., Holmes, E., Orton, T., Jackson, D., Rowlinson, R., Nickson, J., Tonge, R., Wilson, I., & Nicholson, J. (2006). Systems toxicology: Integrated genomic, proteomic and metabonomic analysis of methapyrilene induced hepatotoxicity in the rat. *Journal of Proteome Research, 5*(7), 1586–1601. https://doi.org/10.1021/pr0503376

Crosland-Taylor, P. J. (1953). A device for counting small particles suspended in a fluid through a tube. *Nature, 171*(4340), 37–38. https://doi.org/10.1038/171037b0

Dasgupta, A. (2015, December). Chapter 6-Mean corpuscular volume and carbohydrate-deficient transferrin as alcohol biomarkers. In *Alcohol and its biomarkers-clinical aspects and laboratory determination* (pp. 139–162). Elsevier. https://doi.org/10.1016/B978-0-12-800339-8.00006-7

Delves, P. J., & Roitt, I. M. (1998). *Encyclopedia of immunology* (pp. 145–145). Academic Press.

Dobrovolskaia, M. A., & McNeil, S. E. (2015). Immunological and hematological toxicities challenging clinical translation of nucleic acid-based therapeutics. *Expert Opinion on Biological Therapy, 15*(7), 1023–1048. https://doi.org/10.1517/14712598.2015.1014794

Du, L., Grover, A., Ramanan, S., & Litwin, V. (2015). The evolution of guidelines for the validation of flow cytometric methods. *International Journal of Laboratory Hematology, 37*(Suppl 1), 3–10. https://doi.org/10.1111/ijlh.12344

Ehrlich, P., & Lazarus, A. (1900). *Histology of the blood: Normal and pathological*. The University Press.

Emara, A. M., Abo El-Noor, M. M., Hassan, N. A., & Wagih, A. A. (2010). Immunotoxicity and hematotoxicity induced by tetrachloroethylene in Egyptian dry cleaning workers. *Inhalation Toxicology, 22*(2), 117–124. https://doi.org/10.3109/08958370902934894

Espenschied, S. T., Tighe, R. M., & Gowdy, K. M. (2018). Flow cytometry for the immunotoxicologist. *Methods in Molecular Biology (Clifton, N.J.), 1803*, 183–197. https://doi.org/10.1007/978-1-4939-8549-4_12

Exon, J. H., Koller, L. D., Talcott, P. A., O'Reilly, C. A., & Henningsen, G. M. (1986). Immunotoxicity testing: An economical multiple-assay approach. *Fundamental and Applied Toxicology, 7*(3), 387–397. https://doi.org/10.1016/0272-0590(86)90088-6

Feng, Y., Mitchison, T. J., Bender, A., Young, D. W., & Tallarico, J. A. (2009). Multi-parameter phenotypic profiling: Using cellular effects to characterize small-molecule compounds. Nature reviews. *Drug Discovery, 8*(7), 567–578. https://doi.org/10.1038/nrd2876

Friberg, L. E., & Karlsson, M. O. (2003). Mechanistic models for myelosuppression. *Investigational New Drugs, 21*(2), 183–194. https://doi.org/10.1023/a:1023573429626

Fulwyler, M. J. (1965). Electronic separation of biological cells by volume. *Science (New York, N.Y.), 150*(3698), 910–911. https://doi.org/10.1126/science.150.3698.910

Gadalla, R., Noamani, B., MacLeod, B. L., Dickson, R. J., Guo, M., Xu, W., Lukhele, S., Elsaesser, H. J., Razak, A. R. A., Hirano, N., McGaha, T. L., Wang, B., Butler, M., Guidos, C. J., Ohashi, P. S., Siu, L. L., & Brooks, D. G. (2019). Validation of CyTOF against flow cytometry for immunological studies and monitoring of human cancer clinical trials. *Frontiers in Oncology, 9*, 415. https://doi.org/10.3389/fonc.2019.00415

Gucker, F. T., Jr., & O'Konski, C. T. (1949). Electronic methods of counting aerosol particles. *Chemical Reviews, 44*(2), 373–388. https://doi.org/10.1021/cr60138a009

Guidance for immunotoxicity risk assessment for chemicals. (2012). *IPCS harmonization project document no. 10*.

Gulati, K., & Ray, A. (2009, December). Immunotoxicity. In *Handbook of toxicology of chemical warfare agents* (pp. 595–609). https://doi.org/10.1016/B978-012374484-5.00040-7

Gulati, G., Song, J., Florea, A. D., & Gong, J. (2013). Purpose and criteria for blood smear scan, blood smear examination, and blood smear review. *Annals of Laboratory Medicine, 33*(1), 1–7. https://doi.org/10.3343/alm.2013.33.1.1

Haschek-Hock, W. M., Rousseaux, C. G., Wallig, M. A., & Bolon, B. (2022). New: Haschek and Rousseaux's handbook of toxicologic pathology, Volume 1: Principles and practice of toxicologic pathology, 4th edition – 2021. *International Journal of Toxicology, 41*(3), 253–254. https://doi.org/10.1177/10915818221099126

Hedley, B. D., & Keeney, M. (2013). Technical issues: Flow cytometry and rare event analysis. *International Journal of Laboratory Hematology, 35*(3), 344–350. https://doi.org/10.1111/ijlh.12068

Heijne, W. H., Kienhuis, A. S., van Ommen, B., Stierum, R. H., & Groten, J. P. (2005). Systems toxicology: Applications of toxicogenomics, transcriptomics, proteomics and metabolomics in toxicology. *Expert Review of Proteomics, 2*(5), 767–780. https://doi.org/10.1586/14789450.2.5.767

Herzenberg, L. A., Sweet, R. G., & Herzenberg, L. A. (1976). Fluorescence-activated cell sorting. *Scientific American, 234*(3), 108–117. https://doi.org/10.1038/scientificamerican0376-108

Hua, Y., Shi, Y., Cui, X., & Li, X. (2021). In silico prediction of chemical-induced hematotoxicity with machine learning and deep learning methods. *Molecular Diversity, 25*(3), 1585–1596. https://doi.org/10.1007/s11030-021-10255-x

Hulett, H. R., Bonner, W. A., Barrett, J., & Herzenberg, L. A. (1969). Cell sorting: Automated separation of mammalian cells as a function of intracellular fluorescence. *Science (New York, N.Y.), 166*(3906), 747–749. https://doi.org/10.1126/science.166.3906.747

Hunka, J., Riley, J. T., & Debes, G. F. (2020). Approaches to overcome flow cytometry limitations in the analysis of cells from veterinary relevant species. *BMC Veterinary Research, 16*(1), 83. https://doi.org/10.1186/s12917-020-02299-2

Immunotoxicocity Testing Guidance. (1999, May). Docket Number: FDA-2020-D-0957 Issued by: Center for Devices and Radiological Health.

Joseph, P. (2017). Transcriptomics in toxicology. *Food and Chemical Toxicology, 109*, 650–662.

Krishnan, K., & Pelekis, M. (1995). Hematotoxic interactions: Occurrence, mechanisms and predictability. *Toxicology, 105*(2–3), 355–364. https://doi.org/10.1016/0300-483x(96)83476-7

Levine, L. S., Hiam-Galvez, K. J., Marquez, D. M., Tenvooren, I., Madden, M. Z., Contreras, D. C., et al. (2021). Single-cell analysis by mass cytometry reveals metabolic states of early-activated CD8+ T cells during the primary immune response. *Immunity, 54*(4), 829–844.

Luebke, R. W., Holsapple, M. P., Ladics, G. S., Luster, M. I., Selgrade, M., Smialowicz, R. J., Woolhiser, M. R., & Germolec, D. R. (2006). Immunotoxicogenomics: The potential of genomics technology in the immunotoxicity risk assessment process. *Toxicological Sciences, 94*(1), 22–27. https://doi.org/10.1093/toxsci/kfl074

Mahalingaiah, P. K., Palenski, T., & Van Vleet, T. R. (2018). An in vitro model of hematotoxicity: Differentiation of bone marrow-derived stem/progenitor cells into hematopoietic lineages and evaluation of lineage-specific hematotoxicity. *Current Protocols in Toxicology, 76*(1), e45. https://doi.org/10.1002/cptx.45

McKinnon, K. M. (2018). Flow cytometry: An overview. *Current Protocols in Immunology, 120*(1), 5–1.

Moldavan, A. (1934). Photo-electric technique for the counting of microscopical cells. *Science (New York, N.Y.), 80*(2069), 188–189. https://doi.org/10.1126/science.80.2069.188

National Research Council (US) Subcommittee on Immunotoxicology. (1992). *Biologic markers in immunotoxicology*. National Academies Press (US). https://doi.org/10.17226/1591

Newton, H. S., & Dobrovolskaia, M. A. (2022). Immunophenotyping: Analytical approaches and role in preclinical development of nanomedicines. *Advanced Drug Delivery Reviews, 185*, 114281. https://doi.org/10.1016/j.addr.2022.114281

Nonclinical Evaluation of the Immunotoxic Potential of Pharmaceuticals. (2023, June). Docket Number: FDA-2019-D-5607 Issued by: Center for Drug Evaluation and Research.

Oliveira, D. B., Gillespie, K., Wolfreys, K., Mathieson, P. W., Qasim, F., & Coleman, J. W. (1995). Compounds that induce autoimmunity in the brown Norway rat sensitize mast cells for mediator release and interleukin-4 expression. *European Journal of Immunology, 25*(8), 2259–2264. https://doi.org/10.1002/eji.1830250822

Pessina, A., Malerba, I., & Gribaldo, L. (2005). Hematotoxicity testing by cell clonogenic assay in drug development and preclinical trials. *Current Pharmaceutical Design, 11*(8), 1055–1065. https://doi.org/10.2174/1381612053381648

Phadnis-Moghe, A. S., & Kaminski, N. E. (2017). Immunotoxicity testing using human primary leukocytes: An adjunct approach for the evaluation of human risk. *Current Opinion in Toxicology, 3*, 25–29. https://doi.org/10.1016/j.cotox.2017.04.005

Picker, L. J., & Siegelman, M. H. (1999). Lymphoid tissues and organs. In W. E. Paul (Ed.), *Fundamental immunology* (4th ed., p. 145). Lippincott-Raven.

Rawat, A., Arora, K., Shandilya, J., Vignesh, P., Suri, D., Kaur, G., et al. (2019). Flow cytometry for diagnosis of primary immune deficiencies—A tertiary center experience from North India. *Frontiers in Immunology, 10*, 2111.

Reinherz, E. L., Nadler, L. M., Rosenthal, D. S., Moloney, W. C., & Schlossman, S. F. (1979). T-cell-subset characterization of human T-CLL. *Blood, 53*(6), 1066–1075.

Rejeski, K., Perez, A., Sesques, P., Hoster, E., Berger, C., Jentzsch, L., Mougiakakos, D., Frölich, L., Ackermann, J., Bücklein, V., Blumenberg, V., Schmidt, C., Jallades, L., Fehse, B., Faul, C., Karschnia, P., Weigert, O., Dreyling, M., Locke, F. L., von Bergwelt-Baildon, M., et al. (2021). CAR-HEMATOTOX: A model for CAR T-cell-related hematologic toxicity in relapsed/refractory large B-cell lymphoma. *Blood, 138*(24), 2499–2513. https://doi.org/10.1182/blood.2020010543

Rejeski, K., Subklewe, M., Aljurf, M., Bachy, E., Balduzzi, A., Barba, P., et al. (2023). Immune effector cell–associated hematotoxicity: EHA/EBMT consensus grading and best practice recommendations. *Blood, 142*(10), 865–877.

Rooney, A. A., Luebke, R. W., Selgrade, M. K., & Germolec, D. R. (2012). Immunotoxicology and its application in risk assessment. *Experientia Supplementum, 101*, 251–287. https://doi.org/10.1007/978-3-7643-8340-4_9

Rothenberg, E. V. (2000). Stepwise specification of lymphocyte developmental lineages. *Current Opinion in Genetics & Development, 10*(4), 370–379. https://doi.org/10.1016/s0959-437x(00)00098-8

Salas, L. A., Zhang, Z., Koestler, D. C., Butler, R. A., Hansen, H. M., Molinaro, A. M., Wiencke, J. K., Kelsey, K. T., & Christensen, B. C. (2022). Enhanced cell deconvolution of peripheral blood using DNA methylation for high-resolution immune profiling. *Nature Communications, 13*(1), 761. https://doi.org/10.1038/s41467-021-27864-7

Semwal, R., Semwal, R. B., Lehmann, J., & Semwal, D. K. (2022). Recent advances in immunotoxicity and its impact on human health: Causative agents, effects and existing treatments. *International Immunopharmacology, 108*, 108859. https://doi.org/10.1016/j.intimp.2022.108859

Shapiro, H. M., Perlmutter, N. G., & Stein, P. G. (1998). A flow cytometer designed for fluorescence calibration. *Cytometry, 33*(2), 280–287.

Sturla, S. J., Boobis, A. R., FitzGerald, R. E., Hoeng, J., Kavlock, R. J., Schirmer, K., Whelan, M., Wilks, M. F., & Peitsch, M. C. (2014). Systems toxicology: From basic research to risk assessment. *Chemical Research in Toxicology, 27*(3), 314–329. https://doi.org/10.1021/tx400410s

Takahashi, M., Tsujimura, N., Yoshino, T., Hosokawa, M., Otsuka, K., Matsunaga, T., & Nakasono, S. (2012). Assessment of benzene-induced hematotoxicity using a human-like hematopoietic lineage in NOD/Shi-scid/IL-2Rγnull mice. *PloS One, 7*(12), e50448. https://doi.org/10.1371/journal.pone.0050448

Telford, W., Stickland, L., & Koschorreck, M. (2017). Ultraviolet 320 nm laser excitation for flow cytometry. *Cytometry Part A: The Journal of the International Society for Analytical Cytology, 91*(4), 314–325. https://doi.org/10.1002/cyto.a.23066

Tenorio-Borroto, E., Peñuelas-Rivas, C. G., Vásquez-Chagoyán, J. C., Prado-Pradoa, F. J., García-Mera, X., & González-Díaz, H. (2012). Immunotoxicity, flow cytometry, and chemoinformatics: Review, bibliometric analysis, and new QSAR model of drug effects over macrophages. *Current Topics in Medicinal Chemistry, 12*(16), 1815–1833.

Wang, X., Li, N., Ma, M., Han, Y., & Rao, K. (2022). Immunotoxicity in vitro assays for environmental pollutants under paradigm shift in toxicity tests. *International Journal of Environmental Research and Public Health, 20*(1), 273. https://doi.org/10.3390/ijerph20010273

Wilmes, A., Bielow, C., Ranninger, C., Bellwon, P., Aschauer, L., Limonciel, A., Chassaigne, H., Kristl, T., Aiche, S., Huber, C. G., Guillou, C., Hewitt, P., Leonard, M. O., Dekant, W., Bois, F., & Jennings, P. (2015). Mechanism of cisplatin proximal tubule toxicity revealed by integrating transcriptomics, proteomics, metabolomics and biokinetics. *Toxicology in Vitro: An International Journal Published in Association with BIBRA, 30*(1 Pt A), 117–127. https://doi.org/10.1016/j.tiv.2014.10.006

Yamada, T., Kawamura, T., Tsujii, S., Miura, M., Ohata, H., Katsutani, N., Matsumoto, M., & Hirose, A. (2022). Formation and evaluation of mechanism-based chemical categories for regulatory read-across assessment of repeated-dose toxicity: A case of hemolytic anemia. *Regulatory Toxicology and Pharmacology: RTP, 136*, 105275. https://doi.org/10.1016/j.yrtph.2022.105275

Yamaoka, K. A., Miyasaka, N., Inuo, G., Saito, I., Kolb, J. P., Fujita, K., & Kashiwazaki, S. (1994). 1,1′-Ethylidenebis(tryptophan) (Peak E) induces functional activation of human eosinophils and interleukin 5 production from T lymphocytes: Association of eosinophilia-myalgia syndrome with a L-tryptophan contaminant. *Journal of Clinical Immunology, 14*(1), 50–60. https://doi.org/10.1007/BF01541175

Yoon, B. I., Li, G. X., Kitada, K., Kawasaki, Y., Igarashi, K., Kodama, Y., Inoue, T., Kobayashi, K., Kanno, J., Kim, D. Y., Inoue, T., & Hirabayashi, Y. (2003). Mechanisms of benzene-induced hematotoxicity and leukemogenicity: cDNA microarray analyses using mouse bone marrow tissue. *Environmental Health Perspectives, 111*(11), 1411–1420. https://doi.org/10.1289/ehp.6164

Zhang, T., Warden, A. R., Li, Y., & Ding, X. (2020). Progress and applications of mass cytometry in sketching immune landscapes. *Clinical and Translational Medicine, 10*(6), e206. https://doi.org/10.1002/ctm2.206

Role of Flowcytometry in the Safety/Toxicity Analysis of Cosmetics and Personal Care Products

Imaan Badar Khan, Neha Verma, Shiv Poojan, Ashish Dwivedi, Subhash Dwivedi, Neera Tiwari Singh, and Bhartendu Nath Mishra

Abstract

Flow cytometry stands out as a pivotal analytical method in biomedical and clinical research, particularly in evaluating safety and toxicity. With its capacity for multiplexing, high throughput screening, and intuitive interface, flow cytometry has gained traction in the cosmetics and personal care products (PCPs) industries. This chapter delves into the imperative of scrutinizing ingredients and formulations of PCPs to ensure consumer safety, regulatory compliance, and innovation in product development. Additionally, the chapter explores how flow cytometry facilitates immunophenotypic analysis by detecting surface markers.

Keywords

Flow cytometry · Personal care products · Toxicity · Cosmetics and immunophenotyping

I. B. Khan · N. Verma · B. N. Mishra (✉)
Institute of Engineering and Technology, Department of Biotechnology, Lucknow, India
e-mail: bnmishra@ietlucknow.ac.in

S. Poojan
bioTox Solution Pvt. Ltd. NNF, Department of Biotechnology, Institute of Engineering and Technology, Lucknow, India

A. Dwivedi
Council of Scientific and Industrial Research-Indian Institute of Toxicological Research, Lucknow, India

S. Dwivedi
University of Alabama, Tuscaloosa, AL, USA

N. T. Singh
Department of Pharmacology and Toxicology, College of Osteopathic Medicine, Michigan State University, East Lansing, MI, USA

AB Pant et al. (eds.), *Flow Cytometry: Applications in Cellular and Molecular Toxicology*, https://doi.org/10.1007/978-981-97-9758-5_9

1 Introduction

As beauty and self-care become an integral part of customers' lifestyles worldwide, the cosmetics and PCPs industry keeps on thriving. With global consumption of more than 500 billion US dollars' worth of products in 2022, it is estimated to reach 600 billion US dollars in value by 2028 (https://www.statista.com/markets/415/topic/467/cosmetics-personal-care/). To meet this increasing demand of customers for extensive and effective products that also comply with the regulatory laws is a challenge to the industries. Often, the terms 'Cosmetics' and 'Personal Care Products' are used interchangeably, which is incorrect. While the **US Food and Drug Administration (FDA)** defines '**Cosmetics**' in the **Food, Drug and Cosmetics Act of 1976** as: 'articles intended to be rubbed, poured, sprinkled or sprayed on, introduced into, or otherwise applied to the human body for cleansing, beautifying, promoting attractiveness or altering the appearance' (FDA), whereas the term 'Personal Care Products' is not defined by law.

Commonly available cosmetic products meant for the purpose of—(i) maintaining personal cleanliness and hygiene, such as shampoos, soaps, deodorants, face-washes, body gels, etc., (ii) beautifying oneself such as creams, moisturizers, etc., (iii) promoting physical attractiveness such as, nail polishes, lipsticks, etc. and (iv) changing the outward appearances, such as permanent waves, hair colors fall under the category of cosmetics. Such products are not formulated to affect systemic biology, though some local, short-term effects are observed in some cases. PCPs include some products that fall under the category of cosmetics (such as perfumes, skin moisturizers, deodorants, toothpastes, shampoos) along with other products that are marketed as treatments for acne/dandruff; antiperspirants; ointments, balms and creams that claim to be skin protectants; and or other therapeutic effects, fall under the category of drugs (Fig. 1) which is defined by FDA as 'a substance that is intended for use in diagnosis, cure, mitigation, treatment, or prevention of disease.' It can also be 'a substance (other than food) intended to affect the structure or any function of the body' (FDA).

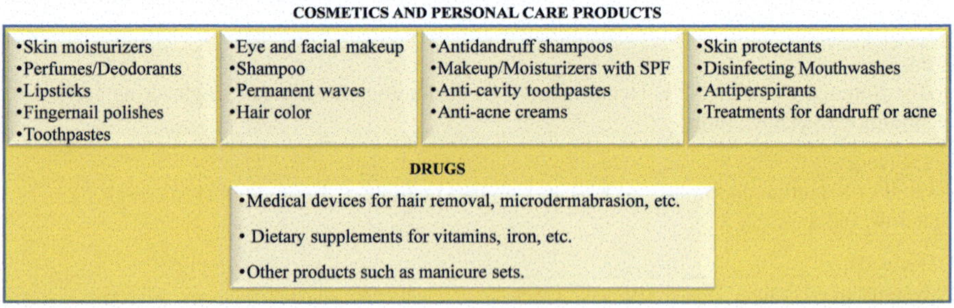

Fig. 1 Relationship between drugs, cosmetics and PCPs

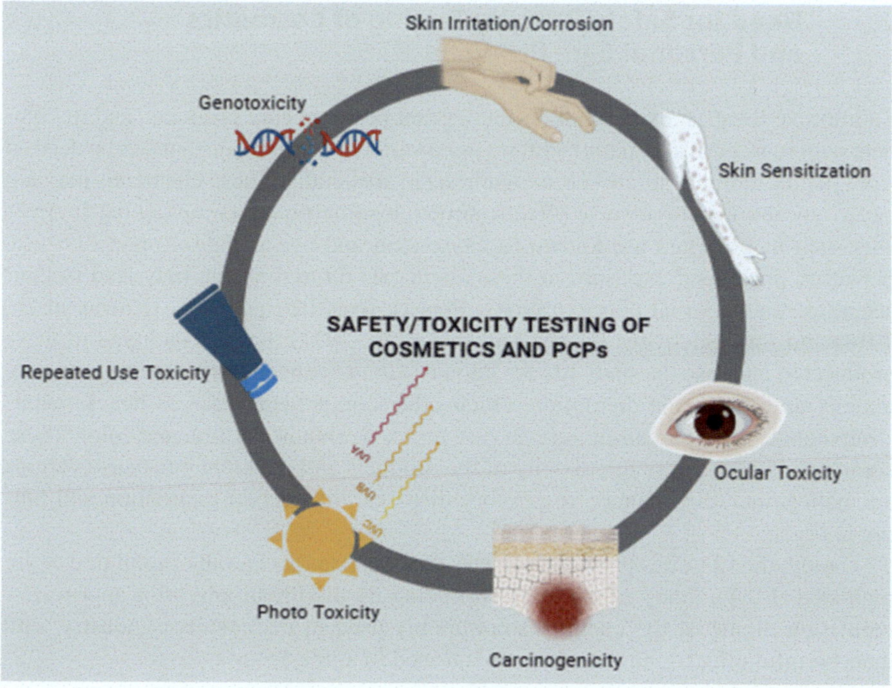

Fig. 2 Safety and toxicity testing of cosmetics and PCPs

Many of these cosmetics and personal care products contain biological components, preservatives, and fillers that can have harmful effects on human health. These chemicals can cause skin irritation or even lead to severe health issues such as cancer. Hence the assessment of the toxicity and safety of the ingredients and formulations of such products are an important aspect of their preclinical studies (Fig. 2). There are several endpoints that are considered when evaluating the toxicity/safety of ingredients and formulations of cosmetics and personal care products. The ability of products to cause ocular corrosion/irritation and skin sensitization/irritation are tested. The ability of certain compounds to absorb in or cross the dermal barrier and enter the blood stream also needs to be studied. Along with these aspects of toxicity analysis, the ability of these compounds to be potential mutagens or carcinogens should also be considered (European Commission, 2018).

2 Need for Safety/Toxicity Testing of Cosmetics and Personal Care Products

To improve the effectiveness of cosmetics and personal care products, and to compete with market rivals, manufacturers use a variety of chemicals that can be derived from plants, animals, microbes, or synthesized artificially. These chemicals may not cause any immediate adverse effects, although some products containing ingredients such as coal tars are known to cause skin and eye irritation upon first use. However, prolonged exposure to these chemicals through usage may lead to their accumulation in lethal concentrations. Preservatives like parabens (Nowak et al., 2018) and formaldehydes (López-Sánchez et al., 2021) that are included in these products to increase the shelf-life are known to cause cancer, mutation, reproductive toxicity and endocrine disruption. Often other components such as heavy metals (Naqvi et al., 2022) are used as preservatives, or to enhance aroma and color. These components also add to the toxicity of the products and are known to cause various anomalies, including cancer, respiratory diseases, intellectual retardation and failure in organ.

Though the FDA has provided a list of chemicals that are totally prohibited or are regulated (FDA), many of them are still being used without any prior approval or regulation. Some of the chemicals commonly used in the cosmetic industry with their harmful effects on humans are mentioned in Table 1.

3 Chemical Potential Health Risks

Table 1 List of some common harmful chemicals used in cosmetics and PCPs

Chemical	Potential health risks
Coal tar	Cancer
Formaldehyde and FRPs	Cancer, skin irritation, allergic reactions
Heavy metals	Cancer, neurotoxicity, organ system toxicity
Mercury compounds	May cause allergic reactions, skin irritation, nephrotic (Chan et al., 2020) or neurotoxic problems
Nanoparticles	Organ toxicity, cancer
Phthalates	Endocrine disruption, developmental and reproductive toxicity, cancer (Chi et al., 2022)
Talc	Cancer (Moline et al., 2023)

4 Regulatory Laws Compliance

Governments in many countries regulate the cosmetics and personal care products (PCPs) industries through strict laws to ensure that their products are safe for human use. Manufacturers are responsible for ensuring the safety of both the ingredients and finished cosmetic products prior to marketing. FDA has advised cosmetic firms to employ appropriate and effective testing to substantiate the safety of their products. However, the FFDCA and FDA regulations do not specify how cosmetic products and their ingredients are to be tested. On the other hand, the FDA has the authority to take enforcement action against companies or persons that break the law, as well as against items that are on the market that are not compliant. Traditional testing of cosmetic ingredients has used animal models to evaluate the safety of the ingredients on the human body. Although animal testing may be used to substantiate product safety, FDA does not specifically require such testing. The cosmetic industry's trade association, the **Personal Care Products Council (PCPC)**, has established a **Cosmetic Ingredient Review (CIR)** program to review the safety of cosmetic product ingredients, based on published and unpublished data on individual ingredients (Boyer et al., 2017). The CIR program reviews cosmetic ingredients compiled from an annual priority list of ingredients used in commercially available cosmetics. The priority list is based on the number of different products in which an ingredient is used (http://cir-safety.org/sites/default/files/2022_Final_Priorities_Active.pdf) as obtained from the **Voluntary Cosmetic Registration Program (VCRP)** (FDA).

5 Quality Control and Product Development

Quality control is mandatory to ensure the stability, safety and efficacy of the cosmetics and PCPs to validate the claims of the manufacturers by using scientific data (Chiari et al., 2012). The regular and thorough testing of these products helps the manufacturers to cater to the needs of the customers and improve the existing products and/or bring new and better products to the market. All the products are vigorously tested, inspected and evaluated, starting from their initial formulation phase to the end phase when the product can be obtained from the market. Quality control testing in the cosmetic industry involves several tests for microbes (Tan et al., 2013), allergens/irritants (Wetter et al., 2010), ingredients, sunscreen and heavy metals (Romanowski & Schueller, 2001). Also, the stability (Busch & Gassenmeier, 2000) and sensory effects (Briasco et al., 2016) of the products are evaluated. Apart from the products their packaging compatibility is also tested (European Parliament, 2009).

6 Role of Flow Cytometry

Since after the current amendment to the **European Cosmetics Directive, Cosmetic regulation No 1223/2009** which bans the use of animals as models for toxicity testing of cosmetics (2004) and cosmetic ingredients (2009) (Germano-Costa et al., 2022), there has always been a search for their alternatives.

Various in-silico and in-vitro methods have been developed and validated since then. With in-vitro being the most intrusively used, other technologies such as organ-on-chip, artificial intelligence, etc. are still in their early phases of development. The in-vitro methods of 2D and 3D cell cultures of human derived cells (Germano-Costa et al., 2022; Şenkal et al., 2022) and 3D cell cultures as human skin equivalent models (Mini et al., 2020) are commonly used for the toxicity and safety assessment of cosmetic products.

Flow cytometry is one of the most used techniques in biomedical research and clinical diagnostics that uses the physical and biochemical properties of individual cells or particles in a heterogenous population for their characterization and quantification. Equipped with features of multiplexing, high throughput analysis of entire population or subpopulation of a homogenous sample and a user-friendly interface (Fig. 3), flow cytometry can also be used for the safety and toxicity assessment of

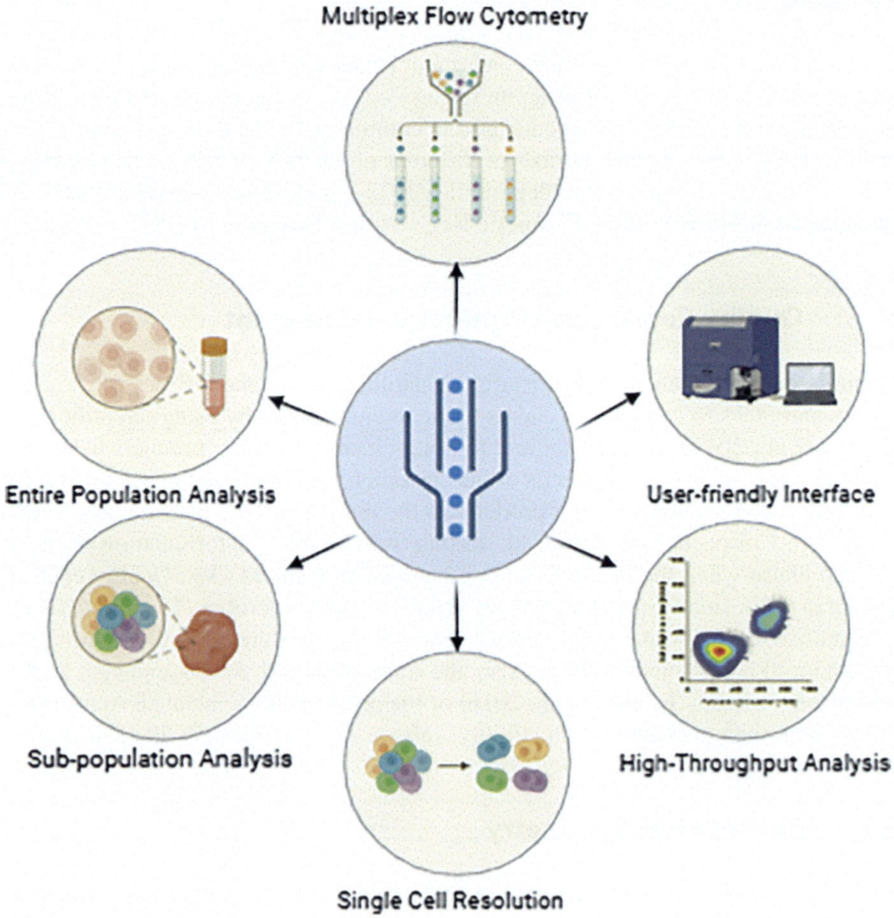

Fig. 3 Features of flow cytometry

cosmetics and PCPs in addition to its application in biomedical and clinical research. Since Flow Cytometry can measure several fluorescence parameters such as forward scatter (for size), side scatter (for granularity), intensity and polarization, it can be used to provide information related to the viability, proliferation, apoptosis, oxidative stress, inflammation, and immunogenicity of cells when tested against cosmetics and other PCPs' ingredients and formulations. Also, Flow Cytometry can be applied to evaluate the potential harmful effects of ingredients in cosmetics and PCPs such as skin sensitization (Gautam et al., 2021), irritation and phototoxicity (Mujtaba et al., 2021) by quantifying the expression of specific biomarkers—such as **CD86**, **CD54**, **MHC class II**, or **Annexin V** on the surface or inside the cells (Sakaguchi et al., 2009).

7 Uses of Flow Cytometry in Immunotoxicology of Cosmetics and PCPs

7.1 Surface Marker Detection and Immunophenotypic Analysis

In clinical immunophenotyping, flow cytometry is used to evaluate the presence or absence of multiple cell markers such as **CD4+/CD8+** on T cells (Senju et al., 1991), **CD56** for NK cells (Bryceson et al., 2009), **CD19** for B cells (D'Arena et al., 2000) and **CD45** for a subpopulation of leukocytes (Lacombe et al., 1997), this information allows for characterization of subpopulation and cell population in each sample. In the toxicity analysis of cosmetic and PCPs' ingredients and formulations, flow cytometry can be used. The alteration in the population and/or subpopulation of cells on exposure with the tested compound can be identified and sorted through detection and analysis of surface markers. These abilities of flow cytometry could easily be used to identify rare cell populations or cells with specific biomarkers (Baron et al., 2012). The intensity with which these tested compounds affect these populations of cells analyses the safety/toxicity of the products. Flow cytometry can also identify if a compound has immunoactivities or immunosuppression ability and the degree of this effect, based on the engagement of specific antibodies (**CD4+** on T cells) and the secretion of intercellular cytokines (**IL 2** and **IFN gamma** for activation and proliferation of T cell, **TNF alpha** for macrophage activation), respectively (Papenfuss, 2017).

7.2 Functional Flow Cytometry in Immunotoxicology

Apart from immunotyping analysis, flow cytometry is also used to investigate cell viability, its proliferation and signaling events among others (Dallas & Evans, 1990). The status of these events inside the cells is one of the important endpoints for the toxicity/safety analysis of cosmetic compounds and formulations. Interference with the normal functioning of the cells, disrupting or altering the signaling events, these compounds/formulations can be a 'go' or a 'no go' for industrial manufacturing. These assessments comprise the use of probes for cell activation, cell cycle progression, stimulation of proteins associated with pathways of cell damage and/or death, changes in membrane integrity or structure, membrane sulfhydryl status, and a plethora of other molecular and cellular markers on treated samples. The quick intracellular DNA fragmentation that takes place while the integrity of the cell membrane is preserved is one of the traits of an apoptotic process and can be identified and analyzed using TUNEL assay in conjugation with flow cytometry (Burchiel et al., 1997). Flow cytometry have also been used in the analysis of apoptotic cell death via cell cycle arrest in paraben treated sample (Kim et al., 2020). Such research has put forward flow cytometry as an excellent in-vitro analysis method for toxicologists in cosmetics and PCPs industries.

8 Other Applications of Flow Cytometry in Toxicity/Safety Analysis of Cosmetics and Related Products

8.1 Hematopoietic Toxicology of Cosmetics and PCPs

Flow Cytometry can be used to test hematopoietic toxicity of cosmetics' and PCPs' ingredients and formulations during clinical and in vitro studies to ensure that they meet regulatory standards and protect consumers' well-being. Toxicologists examine individual ingredients' as well as the overall formulation' potential to harmfully affect the hematopoietic system by causing bone marrow suppression, anemia, or other blood-related issues such as abnormalities in number, shape and size of blood cells (RBCs, WBCs and Platelets). This requirement of the toxicologists for a multiparametric system, capable of giving high throughput, precise analysis can be achieved by using flow cytometry (Pierzchalski et al., 2023).

8.2 Genotoxicity of Cosmetics and PCPs: Mutagenicity and Carcinogenicity

Genotoxicity is one of the major endpoints in the toxicity and safety analysis of cosmetics and other such products. Many of the ingredients of these products such as nanoparticles (Mascarenhas-Melo et al., 2023) used in sunscreens, heavy metals used in makeup (Balwierz et al., 2023), etc. are mutagenic and/or carcinogenic in nature.

These compounds adversely affect the cells by increasing their oxidative stress and damaging their DNA (Hussain et al., 2005; Xu et al., 2008). Such events define the carcinogenic fate of the cells. These abnormal cells can be identified using micronucleus assay and analyzed with a flow cytometer (Fig. 4). The result can be used as an assessment of the mutagenicity and carcinogenicity (Rodrigues et al., 2021) of the said ingredient or formulation in the clinical or in vitro toxicity analysis of the cosmetics (Buenz, 2007; Allemang et al., 2021) and PCPs.

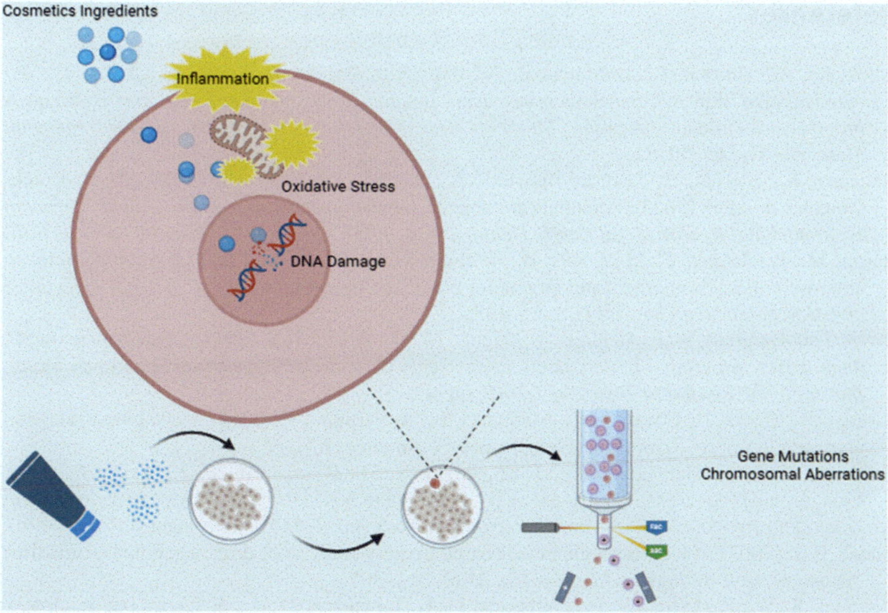

Fig. 4 Role of flow cytometry in genotoxicity testing of cosmetics/PCPs

9 Conclusion

Despite its many advantageous features for assessing the safety and toxicity of ingredients and formulations in cosmetics and PCPs, flow cytometry does have its limitations. The sensitivity and statistical significance of flow cytometric analysis results are inversely affected by the size, complexity, and heterogeneity of the sample being analyzed. While flow cytometry is adept at detecting surface markers and DNA damage, which help identify and characterize altered cells, it may not offer a complete understanding of toxicity mechanisms. Consequently, additional assays are often necessary to draw conclusive results for toxicity/safety assessments. Although flow cytometry can reveal immediate effects of ingredients, it may not capture long-term or chronic toxicity effects. In response to these limitations, researchers are endeavoring to enhance the versatility of flow cytometry for a broader range of applications in toxicity/safety studies. This involves introducing multi-laser and multiparametric versions to the market, along with other updates tailored to the specific needs of the cosmetics and PCPs industries.

References

Allemang, A., Thacker, R., DeMarco, R. A., Rodrigues, M. A., & Pfuhler, S. (2021). The 3D reconstructed skin micronucleus assay using imaging flow cytometry and deep learning: A proof-of-principle investigation. *Mutation Research/Genetic Toxicology and Environmental Mutagenesis, 865*, 50331.

Balwierz, R., Biernat, P., Jasińska-Balwierz, A., Siodłak, D., Kusakiewicz-Dawid, A., Kurek-Górecka, A., et al. (2023). Potential carcinogens in makeup cosmetics. *International Journal of Environmental Research and Public Health, 20*(6), 4780.

Baron, M., Boulanger, C. M., Staels, B., & Tailleux, A. (2012). Cell-derived microparticles in atherosclerosis: Biomarkers and targets for pharmacological modulation? *Journal of Cellular and Molecular Medicine, 16*(7), 1365–1376.

Boyer, I. J., Bergfeld, W. F., Heldreth, B., Fiume, M. M., & Gill, L. J. (2017). The cosmetic ingredient review program—Expert safety assessments of cosmetic ingredients in an open forum. *International Journal of Toxicology, 36*(5_suppl2), 5S–13S.

Briasco, B., Capra, P., Cozzi, A. C., Mannucci, B., & Perugini, P. (2016). Packaging evaluation approach to improve cosmetic product safety. *Cosmetics, 3*(3), 32.

Bryceson, Y. T., Fauriat, C., Nunes, J. M., Wood, S. M., Björkström, N. K., Long, E. O., & Ljunggren, H. G. (2009). Functional analysis of human NK cells by flow cytometry. In *Natural killer cell protocols: Cellular and molecular methods* (pp. 335–352). Springer.

Buenz, E. J. (2007). A high-throughput cell-based toxicity analysis of drug metabolites using flow cytometry. *Cell Biology and Toxicology, 23*(5), 361–365.

Burchiel, S. W., Kerkvliet, N. L., Gerberick, G. F., Lawrence, D. A., & Ladics, G. S. (1997). Assessment of immunotoxicity by multiparameter flow cytometry. *Toxicological Sciences, 38*(1), 38–54.

Busch, P., & Gassenmeier, T. (2000). Evaluation of cosmetics by sensory assessment. In *Cutaneous biometrics* (pp. 65–80). Springer US.

Chan, T. Y., Chan, A. P., & Tang, H. L. (2020). Nephrotic syndrome caused by exposures to skin-lightening cosmetic products containing inorganic mercury. *Clinical Toxicology, 58*(1), 9–15.

Chi, Z., Lin, H., Wang, X., Meng, X., Zhou, J., Xiang, L., et al. (2022). Dimethyl phthalate induces blood immunotoxicity through oxidative damage and caspase-dependent apoptosis. *Science of the Total Environment, 838*, 156047.

Chiari, B. G., Almeida, M. G. J., Corrêa, M. A., & Isaac, V. L. B. (2012). Cosmetics' quality control. In *Latest research into quality control* (pp. 337–364). InTech.

D'Arena, G., Musto, P., Cascavilla, N., Dell'Olio, M., Di Renzo, N., & Carotenuto, M. (2000). Quantitative flow cytometry for the differential diagnosis of leukemic B-cell chronic lymphoproliferative disorders. *American Journal of Hematology, 64*(4), 275–281.

Dallas, C., & Evans, D. (1990). Flow cytometry in toxicity analysis. *Nature, 345*, 557–558.

European Commission. (2018). *Scientific Committee on Consumer Safety (SCCS). The SCCS notes of guidance for the testing of cosmetic ingredients and their safety evaluation (10th revision).* SCCS/162/2018.

European Parliament. (2009). Regulation (EC) No 1223/2009 of the European Parliament and of the Council of 30 November 2009 on cosmetic products. *Official Journal of the European Union, L396*, 1–1355.

FDA. *Prohibited & restricted ingredients in cosmetics.*

FDA. *Product testing of cosmetics.*

Gautam, R., Lee, D., Maharjan, A., Acharya, M., Jo, J., Park, H., et al. (2021). Assessment of skin sensitization potential of manganese oxide and copper oxide nanoparticles through the local lymph node assay: 5-Bromo-deoxyuridine flow cytometry method. In 한국실험동물학회 학술발표대회 논문집, 183.

Germano-Costa, T., Bilesky-José, N., Guilger-Casagrande, M., Pasquoto-Stigliani, T., Rogério, C. B., Abrantes, D. C., et al. (2022). Use of 2D and co-culture cell models to assess the toxicity of zein nanoparticles loading insect repellents icaridin and geraniol. *Colloids and Surfaces B: Biointerfaces, 216*, 112564.

http://cir-safety.org/sites/default/files/2022_Final_Priorities_Active.pdf

https://www.statista.com/markets/415/topic/467/cosmetics-personal-care/

Hussain, S. M., Hess, K. L., Gearhart, J. M., Geiss, K. T., & Schlager, J. J. (2005). In vitro toxicity of nanoparticles in BRL 3A rat liver cells. *Toxicology In Vitro, 19*(7), 975–983.

Kim, M. J., Kim, C. H., An, M. J., Lee, J. H., Shin, G. S., Hwang, J. Y., et al. (2020). Propylparaben induces apoptotic cell death in human placental BeWo cells via cell cycle arrest and enhanced caspase-3 activity. *Molecular & Cellular Toxicology, 16*, 83–92.

Lacombe, F., Durrieu, F., Briais, A., Dumain, P., Belloc, F., Bascans, E., et al. (1997). Flow cytometry CD45 gating for immunophenotyping of acute myeloid leukemia. *Leukemia, 11*(11), 1878–1886.

López-Sánchez, L., Miralles, P., Salvador, A., Merino-Sanjuán, M., & Merino, V. (2021). In vitro skin penetration of bronidox, bronopol and formaldehyde from cosmetics. *Regulatory Toxicology and Pharmacology, 122*, 104888.

Mascarenhas-Melo, F., Mathur, A., Murugappan, S., Sharma, A., Tanwar, K., Dua, K., et al. (2023). Inorganic nanoparticles in dermopharmaceutical and cosmetic products: Properties, formulation development, toxicity, and regulatory issues. *European Journal of Pharmaceutics and Biopharmaceutics, 192*, 25–40.

Mini, C. A., Dorta, D. J., Maria-Engler, S. S., & Oliveira, D. P. D. (2020). Immortalized equivalent human epidermis as a platform to evaluation hair dyes toxicity: Efficiency comparison between 3D and monolayer culture. *Chemico-Biological Interactions, 330*, 109227.

Moline, J., Patel, K., & Frank, A. L. (2023). Exposure to cosmetic talc and mesothelioma. *Journal of Occupational Medicine and Toxicology, 18*(1), 1.

Mujtaba, S. F., Masih, A. P., Alqasmi, I., Alsulimani, A., Khan, F. H., & Haque, S. (2021). Oxidative-stress-induced cellular toxicity and glycoxidation of biomolecules by cosmetic products under sunlight exposure. *Antioxidants, 10*(7), 1008.

Naqvi, S. A. R., Idrees, F., Sherazi, T. A., Hassan, S. U., & Ishfaq, N. (2022). Toxicology associated with heavy metals found in cosmetics. *Journal of the Chilean Chemical Society, 67*(3), 5615–5622.

Nowak, K., Ratajczak-Wrona, W., Górska, M., & Jabłońska, E. (2018). Parabens and their effects on the endocrine system. *Molecular and Cellular Endocrinology, 474*, 238–251.

Papenfuss, T. L. (2017). Flow cytometry and immunophenotyping in drug development. In *Molecular and integrative toxicology* (pp. 343–369). Springer.

Pierzchalski, A., Zenclussen, A. C., & Herberth, G. (2023). A comprehensive battery of flow cytometric immunoassays for the in vitro testing of chemical effects in human blood cells. *Frontiers in Immunology, 14*, 1327960.

Rodrigues, M. A., Probst, C. E., Zayats, A., et al. (2021). The in vitro micronucleus assay using imaging flow cytometry and deep learning. *npj Systems Biology and Applications, 7*, 20.

Romanowski, P., & Schueller, R. (2001). Stability testing of cosmetic products. In *Handbook of cosmetic science and technology* (pp. 785–796). CRC Press.

Sakaguchi, H., Ashikaga, T., Miyazawa, M., et al. (2009). The relationship between CD86/CD54 expression and THP-1 cell viability in an in vitro skin sensitization test—Human cell line activation test (h-CLAT). *Cell Biology and Toxicology, 25*, 109–126.

Senju, M., Wu, K. C., Mahida, Y. R., & Jewell, D. P. (1991). Coexpression of CD4 and CD8 on peripheral blood T cells and lamina propria T cells in inflammatory bowel disease by two colour immunofluorescence and flow cytometric analysis. *Gut, 32*(8), 918–922.

Şenkal, S., Burukçu, D., Hayal, T. B., Kiratli, B., Şişli, H. B., Sağraç, D., et al. (2022). 3D culture of HaCaT keratinocyte cell line as an in vitro toxicity model. *Trakya University Journal of Natural Sciences, 23*(2), 211–220.

Tan, A. B., Tuysuz, M., & Otuk, G. (2013). Investigation of preservative efficacy and microbiological content of some cosmetics found on the market. *Pakistan Journal of Pharmaceutical Sciences, 26*(1), 153–157.

Wetter, D. A., Yiannias, J. A., Prakash, A. V., Davis, M. D., Farmer, S. A., & El-Azhary, R. A. (2010). Results of patch testing to personal care product allergens in a standard series and a supplemental cosmetic series: An analysis of 945 patients from the Mayo Clinic Contact Dermatitis Group, 2000-2007. *Journal of the American Academy of Dermatology, 63*(5), 789–798.

Xu, J., Lian, L. J., Wu, C., Wang, X. F., Fu, W. Y., & Xu, L. H. (2008). Lead induces oxidative stress, DNA damage and alteration of p53, Bax and Bcl-2 expressions in mice. *Food and Chemical Toxicology, 46*(5), 1488–1494.

Flow Cytometric Approaches in Clinical Diagnostics

Khaliqur Rahman and Anu Singh

Abstract

Multiparameter flowcytometry (MFC) has become an integral tool for the diagnosis and classification of diseases. The most common application of flow cytometry is immunophenotyping of normal or abnormal cells. In addition, study of DNA content, apoptosis, cell cycle analysis, reticulocyte content, red cell fragility, neutrophilic oxidative burst analysis, red cell G-6 PD deficiency etc. are some other uses. Immunophenotyping is primarily required for the diagnosis and classification of haematological neoplasm. These are also used to detect the presence of metastasis from epithelial and haematological malignancy. Additionally, they are also an indispensable tool for the disease monitoring of different haematological neoplasm. Several laboratories also utilise the power of this diagnostic tool for the diagnosis of non-neoplastic haematological disorders. This includes disorder like paroxysmal nocturnal haemoglobinuria (PNH), hereditary spherocytosis, inherited platelet disorders, chronic granulomatous disease, leukocyte adhesion defects, hemophagocytic lymphohistiocytosis, primary lymphocytic immunodeficiency disorder etc. This chapter will provide brief information on the various clinical applications of flow cytometry that are used in a day-to-day practice.

Keywords

Flowcytometry · Clinical application · Haematological neoplasm · Non-neoplastic disorders

K. Rahman (✉) · A. Singh
Department of Hematology, Sanjay Gandhi Post-Graduate Institute of Medical Sciences, Lucknow, India
e-mail: khaliq@sgpgi.ac.in

AB Pant et al. (eds.), *Flow Cytometry: Applications in Cellular and Molecular Toxicology*, https://doi.org/10.1007/978-981-97-9758-5_10

1 Introduction

Flow cytometry has been well established as a powerful technique for identification and characterisation of abnormal cell population in a wide variety of sample. It plays a pivotal role in the diagnosis, prognosis and monitoring of haematological neoplasms. Immunophenotypic characteristic of the abnormal cells are needed for a correct diagnosis and classification, which forms the basis of appropriate management of haematological disorders. The antigenic expression on the surface of each cell is unique which serves specific function for that particular lineage of cell. An overview of them is briefly tabulated below (Table 1).

FCM has also been proven as a highly sensitive and robust tool for the monitoring of residual disease in cases of various haematological neoplasms. Another area where the FCM is of pivotal importance is immunodeficiency disorders. The quantitative as well as quantitative abnormality of various leucoyte subset is assessed in order to reach a correct diagnosis. The absolute count of CD4+ helper T cells have been an integral part of anti-retroviral therapy monitoring in cases of HIV/AIDS management.

Flow cytometry is extensively utilized for a successful solid organ and hematopoietic stem cell (HSC) transplantation. Absolute enumeration of HSCs has been possible using the modified international society of hematopoiesis and graft engineering (ISHAGE) protocols. This is required for calculating the dosage for a successful engraftment of a stem cell harvest. FCM is also being used for the HLA cross match and identification of donor specific antibodies (DSA) in the recipient for optimizing graft selection and transplantation outcomes.

Pharmacogenomic studies have been one of the recent explored uses of FCM. By analysing cellular responses to pharmacological agents, optimisation of drug selection, dosage, and treatment regimens, the beneficial effect of a drug can be maximised and the toxicities can be minimised.

In conclusion, flow cytometry serves as a versatile tool in clinical practice, offering valuable insights into various aspects of disease pathogenesis, diagnosis, and treatment across various haematological and non-haematological disorders.

2 Flow Cytometric Approach to the Diagnosis and Classification of Hematological Neoplasm

After a successful morphological examination of a peripheral blood or bone marrow aspiration film, an immunophenotyping is often required for the further exploration and better understanding. MFC helps in identification and characterisation of the abnormal population, provides an overview of the maturation pattern of the normal cells, analysis of different subsets and normal/aberrant expression of antigens on the cells. Before analysing any aberrancy in the immunophenotype that can be used for a clinical diagnosis, we must be aware of the normal antigenic expression of different subsets of the leukocytes. This will help in correct identification and characterisation of these blood cells.

Table 1 Antigens normally expressed on different cells from blood/bone marrow

Cell type	Antigenic expression	Clinical utility
Neutrophils	CD10, CD11b, CD16, CD18, CD24, CD13, CD33, CD15, CD65, CD66b	CD11b/CD18 is beta integrins required for the neutrophil adhesion to extracellular matrix for their recruitment at inflammation site. Absence is seen in Leukocyte adhesion defect. Increased expression of CD64 in PMN indicates sepsis. CD16 and CD24 are GPI linked molecules, commonly used for PNH screening
Monocytes	CD14, HLADR, CD36, CD38, CD13/CD33, CD16, CD64	CD14 is GPI linked molecule and is used for PNH screening. HLADR expression and the trajectory is used for the evaluation of sepsis. CD16 is positive in a subset of monocyte, known as non-classical, the proportion being less than 10%.
NK cells	CD16, CD56, CD2, Perforin	Subsets based on the expression level of these markers. Namely, CD16bright/CD56dim cells are mostly cytotoxic cells, involved in antibody-dependent cellular cytotoxicity. CD16dim/CD56bright cells are important producers of cytokines (mainly gamma interferon), and CD16neg/CD56dim NK cells are considered antitumoral cells that are capable of direct cytotoxicity. NK cells with significantly reduced or absent perforin have been found in familial HLH 2
B-Lymphs	CD19,CD20,CD22/FMC7 CD38/CD138 and dimmer expression of CD45/CD19, loss of CD20 in Plasma cells	Their enumeration is important to rule out agammaglobulinemia. Activation status of B cells
T-Lymphs	CD3, CD4, CD8, CD7, CD2, CD5, AB, GD	Absolute CD4 count, CD4:CD* ratio, Increase in DNTs for the diagnosis of ALPS Subset analysis in PIDs
RBCs	CD235, CD59,CD55 Band 3 protein	PNH screening. EMA binding test for hereditary spherocytosis
Platelets	CD41, CD61, CD42b CD62P, CD63	Used for the diagnosis of hereditary platelet disorder, Studying activation of platelets

In cases of a hematolymphoid neoplasm the first question that a FCM analysis answers is whether the abnormal cells are mature or immature (precursors). This is the first step in the ***diagnosis*** of hematolymphoid neoplasm which is to define whether the disease is leukemia (immature precursors) or lymphoma (mature cell phenotype) extremely important as the treatment of an immature neoplasm is totally different from that of a mature neoplasm. The immunophenotypic features that defines a cell as immature is the expression of CD34, bright CD99, nuclear TdT or a dim/negative CD45 expression. The mature neoplasm on the other hand usually has a bright CD45, while being negative for CD34, CD99 and TdT.

The next answer that is needed is in terms of a lineage assignment, which helps in the ***classification*** of the neoplasm. FCM helps in assigning a lineage to the atypical cells based on certain lineage specific and lineage associated antigens expression on the cell surface or inside cell. Following table gives an overview of these antigens which helps in defining the lineage of a cell (Table 2).

Normal cells exhibit a typical maturation pattern with preserved antigenic expression corresponding to their cell lineage and stage of maturation. In contrast, abnormal cells often demonstrate either a loss or overexpression of normally expressed antigens, as well as the aberrant expression of antigens not typically associated with their lineage. The pattern of antigen expression is used to determine the lineage of blasts in cases of acute leukemia. Table 3 provides guidance in defining the lineage in challenging cases, especially when antigens from multiple lineages are expressed.

After a successful diagnosis and classification, the expression of certain antigens can predict ***prognosis***. For example, the expression of CD56 and CD7 in AMLs indicates a poor prognosis, while CD38 and ZAP70 are poor prognostic markers in CLL. Similarly, CD25 and CD123 in B-ALLs are significant. Some antigens also guide administration of ***targeted therapy*** in hematolymphoid neoplasms, such as CD38 for daratumumab therapy, CD20 for rituximab therapy, and CD52 for

Table 2 Common antigenic expression of different cell lineage

Lineage	Lineage associated markers usually expressed
Myeloid	**MPO**, CD13, CD33, CD117, CD15, CD65, CD66
Monocytic	CD14, CD64, CD11b, CD11c, lysozyme
Megakaryoctic	CD41, CD61, CD42B, CD110 (TPO), CD9
Erythroid	CD105, CD36, CD71, CD235
BPDC	CD123 (Bright), CD303, CD304, CD68, CD4, CD56
B cells	**CD19**, CD10, CD22, CD79A, CD20, CD24, CD73, PAX5
T cells	**cytCD3**, CD1A, CD2, CD3, CD4, CD5, CD7, CD8, CD43
NK cells	CD2, CD7, cytCD3, CD56, CD94 (in absence of Myeloid markers)

Antigens in bold and underline are the lineage specific markers

Table 3 Lineage assignment when there is expression of antigens from one or more lineage

Lineage	Marker
Myeloid	MPO (Cytochemistry, flow cytometry, or immunohistochemistry) ***For Monocytic differentiation-*** At least two of the following: Non Specific Esterase (NSE) cytochemistry, CD11c, CD14, CD64, Lysozyme)
B-Lineage	***Strong* CD19*** with at least 1 of the following strongly expressed- CD79a, cytoplasmic CD22, or CD10 -OR- ***Weak CD19*** with at least 2 of the following strongly expressed- CD79a, cytoplasmic CD22, or CD10
T-Lineage	Strong* cytoplasmic CD3

*Strong: as intense as in normal B- or T-cells

Table 4 Antibodies panel used for the evaluation of acute leukemic

	BV605	V500	PACB/ BV421	FITC	PE	PCP Cy5.5	PeCy7	APC	APC R700	APC H7
Acute Leukemia Orientation tube (ALOT)	CD7	CD45	cyt CD3	cyMPO	TdT	CD34	CD19	CYTCD 79A	CD13	CD38
Further tubes based on the results of A LOT tube										
B ALL/B MRD tube 1	CD20	CD45	CD123	CD86	CD73	CD34	CD19	CD10	CD304	CD38
B tube Adn	CRLF2 (TSLPR)	CD25			CD33		CD19	NG.2 (C	CD45	
Myeloid/MDS tube 1	CD15	CD45	HLADR	CD16	CD33	CD34	CD117	CD10	CD13	CD11b
AML/MDS tube 2	CD105	CD45	HLADR	CD36	CD14	CD34	CD117	CD35	CD64	CD71
T Tube	CD117	CD8	cyt CD3	CD2	CD99	CD5	CD4	CD1a	CD13	sCD3
Undifferentiated		CD45	CD123	Cyt CD42 B	Cyt CD41A	CD34	CD4	CD56	CD304	CD38

alemtuzumab therapy. The list of targeted therapies is expanding rapidly. Additionally, certain antigens serve as markers to ***monitor residual disease*** during or after treatment. When planning the panel for the evaluation of hematolymphoid neoplasms, all these factors should be considered. The panel should include antibodies that aid in diagnosis, classification, prognosis, targeted therapy, and residual disease monitoring. A sample panel used for the diagnosis of acute leukemia in our lab is shown in Table 4.

2.1 Approach for the Diagnosis of Acute Leukemia

The diagnostic sample (PB or the BM) is stained with a combination of antibodies considered to be specific or highly associated with a lineage. Usually, a combination of antibodies against B, T and myeloid lineages are used. This tube is termed as acute leukaemia orientation tube (ALOT).

Based on the result of this tube, additional tubes of corresponding lineage is applied. For example, if the ALOT tube indicate the lineage as B cell, then two additional B tubes are used which will fulfil all the five roles of FCM evaluation mentioned above. Similarly, the two myeloid tubes are added if the blasts phenotype in ALOT tube indicate a myeloid phenotype (CD13 and/Or MPO positivity) OR a T tube is added if the blasts show a T cell phenotype (cytCD3 positivity). If the lineage is uncertain, we put B1, M1, T and undifferentiated tube and assign the lineage accordingly as per the Table 3.

Immunophenotypic patterns often correlate with the associated recurrent cytogenetic abnormalities and therefore may help predict genetic mutations. The most distinctive examples include BALL associated with MLL rearrangement, which are often CD10-negative and CD15-positive, particularly when AF4::MLL is present. BALL associated with BCR::ABL1 and t(12;21)(p12;q22); TEL::AML1 also show more frequent expression of myeloid-associated antigens, especially CD13 and CD33. The BCR::ABL1 positive cases are also associated with the expression of

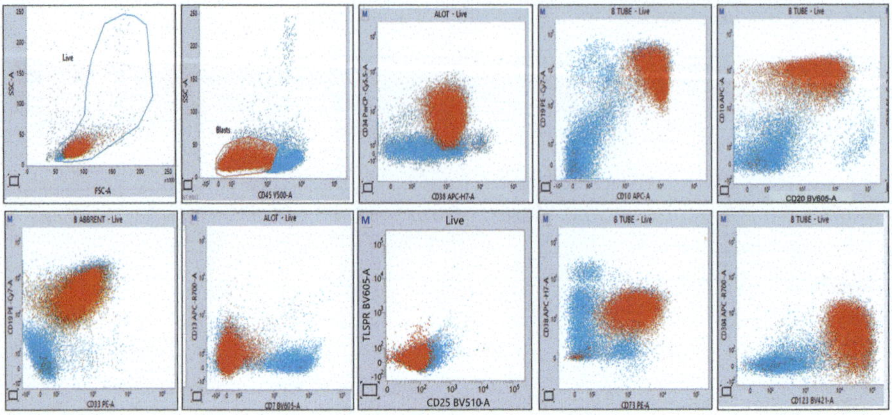

Fig. 1 Scatter plots from a typical case of B cell ALL

CD25. B-lymphoblastic leukemia with TCF3::PBX1 is reportedly associated with CD45 negativity and *positive expression of CD9* (Fig. 1).

Antigen expression patterns associated with T cell immaturity imperfectly recapitulate the early stages of T cell maturation. CD45 expression is invariably weaker than in mature lymphocytes, though not as low as is common in B-ALL/LBL. Immaturity-associated antigens, such as CD1a, CD34, CD117, TdT, and/or CD99, may be positive, demonstrating the progenitor nature of blasts. B lineage markers, such as CD10, CD79a, and rarely PAX5, may be expressed positively. Myeloid-associated antigens, especially CD13 and CD33, may also be seen (Fig. 2).

CD117 expression is rare and has been associated with FLT3-activating mutations. TALL/TLBL may further be sub classified into four major subgroups: pro T, pre T (CD1a– SCD3–), T thymic/cortical (CD1a+ SCD3– with coexpression of CD4 and CD8), T mature/medullary (SCD3+ with expression of either CD4 or CD8). Early T precursor lymphoblastic leukaemia/lymphoma (ETP-ALL) is a neoplasm composed of blasts committed to the T- cell lineage with a unique immunophenotype that includes expression of stem cell markers and/or myeloid lineage markers. Immunophenotypically it is characterised by expression of cytoplasmic CD3; absence of CD1a and CD8 expression (<5% positive blasts); absent or dim (<75% positive blasts) CD5 expression; expression (≥25% positive blasts) of one or more myeloid (CD11b, CD13, CD33, CD65, CD117) and/or stem cell (CD34, HLA-DR) markers. Many cases previously classified as pro-T or pre-T now meet the criteria for early T precursor ALL.

Most AMLs do not have a distinct immunophenotypic profile. However, some characteristic immunophenotypic patterns can provide a clue and predict particular cytogenetic/molecular abnormalities with a high degree of confidence. When present, these patterns should prompt a directed evaluation for the specific genetic aberration. AML with RUNX1::RUNX1T1 generally shows granulocytic differentiation with maturation. By flow cytometry, blasts express myeloid lineage-associated antigens, including CD13, CD15, CD33, and myeloperoxidase. CD34 is usually

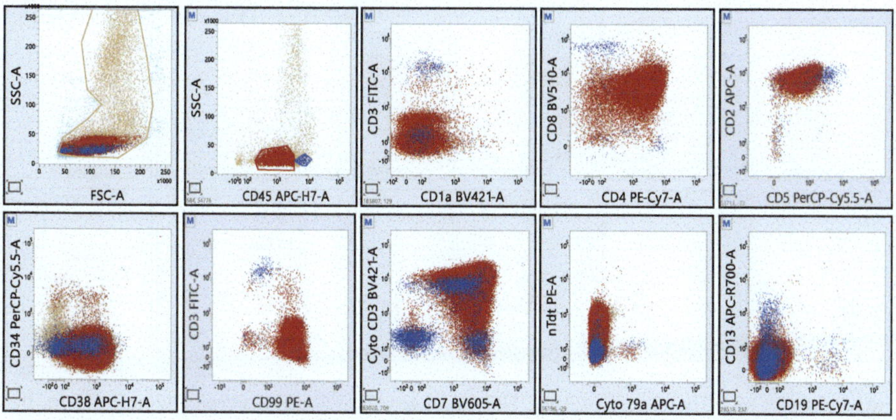

Fig. 2 Scatter plots of a typical case of T ALL

strongly positive, and TdT may or may not be positive. B lineage-associated markers, such as CD19, PAX5, and less commonly CD79a, are also expressed as common aberrancy. However, in cases of strong CD19 expression, other markers must be evaluated to rule out the possibility of MPAL, as CD19 is a 'B' lineage-defining marker. Blasts may also express CD56, which is associated with a poor prognosis (Baer et al., 1997). Dyspoiesis in the myeloid series may be apparent in the form of reduced side scatter of neutrophils and altered maturation patterns. It can be demonstrated by incorporating markers such as CD11b, CD13, CD15, CD10, CD33, and CD16 into the panel.

AML with CBFB::MYH11 are characterized by myelomonocytic differentiation, with expansion of the eosinophil population and the presence of abnormal eosinophils with immature granules. On flow cytometry (FCM), two aberrant blast populations are usually identified. A CD45-dim immature blast population expressing CD34 and myeloid markers such as CD13, CD117, and myeloperoxidase. A more mature monocytic CD45-bright population expressing monocytic antigens such as CD14, CD64, and lysozyme, in addition to myeloid markers, but lacking CD34 expression.

AML with PML::RARA has markedly increased side scatter due to its cytoplasmic complexity. These blasts typically have a triple negative profile (CD34-negative, HLA-DR-negative and CD11b Negative) and positive for myeloid markers such as CD33 and CD117 (Rahman et al., 2018a, b). However, expression of HLA and CD34, at least partly, may be seen in the microgranular variant of acute promyelocytic leukemia (APML). Aberrant expression of CD2 may also be seen in these cases, and is usually associated with FLT3 mutation (Fig. 3).

AML with RBM15::MRTFA (formerly MKL1) are characterized by megakaryocytic differentiation and may be accompanied by bone marrow fibrosis. Blasts and megakaryocytes express one or more platelet antigens: CD41, CD61, and/or CD42b. Cytoplasmic CD41 or CD61 is more specific and sensitive than surface staining. CD36 is typically positive but not specific. False interpretation of megakaryocytic

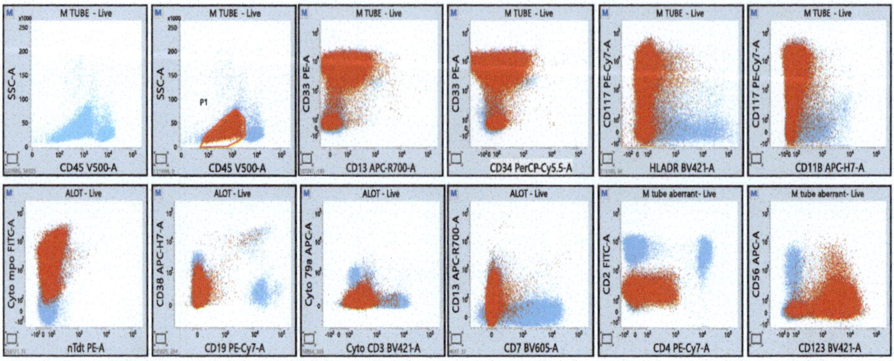

Fig. 3 Scatter plot from a case of AML NPM mutated show the blasts typically negative for the CD34 and HLADR, and expressing myeloid antigens

markers should be avoided, which could be due to platelets adhering to the blasts. Blasts may be positive for CD13 and CD33, while CD34, CD45, HLA-DR, myeloperoxidase, TdT, and lymphoid markers are negative.

AML with DEK::NUP214 shows myelomonocytic differentiation, and multilineage dysplasia is commonly seen. Blasts are usually positive for CD13, CD33, CD34, CD38, CD117, CD123, HLA-DR, and myeloperoxidase. CD7, CD9, CD15, CD64, and TdT are variably expressed, and CD34 may be negative.

AML with NPM1 mutations usually display presence of nuclear cupping in the blasts. In AML with NPM1 mutation, CD33, CD117, and CD123 expression is common, whereas CD13 is usually low. Immunophenotypically, cases may show a predominance of immature myeloid blasts (CD34+/CD117+ or CD34-/CD117+/HLA-DR+); acute promyelocytic leukemia-like features (CD34-/HLA-DR-/CD117+/myeloperoxidase+); and cases with a predominance of myelomonocytic or monocytic differentiation (CD14+/CD36+/CD64+).. AML with BCR::ABL1 fusion is a de novo AML with BCR::ABL1 detected at initial diagnosis and where no subsequent evidence of chronic myeloid leukemia can be established. >20% blasts in PB/BM is a requisite, expressing a myeloid immunophenotype in the bone marrow and/or peripheral blood. Expression of CD7, CD19, and TdT can be seen. The immunophenotype of leukemic cells with a KMT2A rearrangement varies based on cell lineage. Most often, leukemic differentiation is seen, and blasts express CD33, CD65, CD4, CD15, HLA-DR, NG2, and lysozyme. Expression of CD34 and CD117 varies depending on subtype, most often negative in monocytic cases. Expression of CD13 and CD56 may be seen in some cases. In children, KMT2A::MLLT3 and KMT2A::MLLT10 can present as acute megakaryoblastic leukemia. AML with CEBPA mutation should display at least 20% of blasts with myeloid immunophenotype in bone marrow or blood is and presence of biallelic mutations in CEBPA, or a single mutation located in the bZIP domain of the gene (smbZIPCEBPA). The leukemic blasts are usually positive for CD7, CD13, CD15, CD33, CD34, and HLA-DR.

2.2 Approach for the Diagnosis of Mature Lymphoid Neoplasm Including Multiple Myeloma

Flow cytometry immunophenotyping (FCM) allows multidimensional analysis of individual cell types based on cell size, side scatter, immunophenotypic property. It helps us to diagnose and classify the lymphomas as per the recent WHO classification, broadly as 'B' cell, 'T' cells or 'NK cells' types, based on the immunophenotypic resemblance with these cells to the different stages of lymphoid cell maturation.

The basic diagnostic approach to chronic lymphoproliferative disorders starts with an appropriate history and morphological evaluation. This is followed by immunophenotyping using a panel of carefully selected antibodies that will help in defining the lineage, stage of maturation and clonality of the these cells. First question to answer is whether the cell of interest is a B, T or NK cells and then to know whether they are clonal or not. It will help us in defining if we are dealing with a neoplastic population or a reactive proliferation of lymphoid cells. Further detailed panel is added to classify different neoplasms, based on their antigenic expression pattern. Consequently, a combination of CD45, CD3, CD20, CD19, CD10, CD4, CD8, CD16, and CD56 is used as a screening panel, commonly known as lymphoma screening tube (LST). Typical panel of antibodies used in our lab is shown in Table 5. Two antibodies, which are not co-expressed and seen in different lineage of cells are pooled together tagged with same flurophore in order to save the detection channels.

2.2.1 B Cell Lymphomas

Some of the B cell NHLs have an inherent leukemic presentation with predominantly peripheral blood or bone marrow involvement. These include chronic lymphocytic leukemia (CLL), prolymphocytic leukemia (PLL), hairy cell leukemia (HCL) and lymphoplasmacytic lymphoma (LPL). Leukemic mantle cell lymphoma and Burkitt lymphoma is also not infrequent. NHLs like follicular lymphomas and large cell lymphomas are relatively infrequent in the peripheral blood. In the latter cases, FCM can be readily applied to the lymph node fine needle aspirate (FNA) samples.

The two main roles of FCM in cases of B cell neoplasm is the confirmation of lineage and clonality. Common antibodies used CD19, CD22, CD20 and CD79. Out of all these, CD19 is the best established antibodies as the remining three show a variation in different B cell NHLs. CD20 is dim and heterogeneously expressed in CLL, CD22 is also dim expressed in many NHLs. Clonality is established by analysing the surface light chains expression in mature B cells (Orfao et al., 2004). Normally the Kappa to lambda ratio is roughly 2:1, but there is lab to lab variation. Any ratio of more than 4:1 or less than 1:2 should raise a suspicion of clonal B cell proliferation (Kaleem, 2006). The absence of surface light chain expression in lymphoid cells is also considered an indirect indicator for clonality. This phenomenon is frequently observed in CLL samples and may occasionally be seen in other lymphomas. However, a different clones of kappa lambda or preferably a polyclonal kappa lambda should be used in such situation, to prove the clonality. It should

Table 5 Antibodies panel used for the diagnosis of lymphoma

	BV605	BV510	BV421	FITC	PE	PerCP	PeCy7	APC	APCR700	APCH7
LST		CD10/CD8	CD20	LAMDA	CD200	CD5	CD19 / CD4	KAPPA/CD56	CD45	CD3
B LPD1			FMC7	CD23		CD5	CD19		CD45	CD38
BLPD 2		IgM	CD123	CD103	CD305	CD5	CD19	CD10	CD45	IgD
T LPD1	CD7	CD8	CD5	GD	TRBC1	CD5	CD4	CD2	CD45	CD3
T LPD 2		CD8	cytTCL1		CD26		CD4		CD28	CD3

however be noted that evaluation of kappa lambda among all the B cells may sometimes be erroneous. A small abnormal population with light chain restriction may be overshadowed by the background reactive population and hence the light chain ratio may be normal. This demands the addition of other antibodies which may be able to pull out and discriminate the abnormal population. These abnormal population when evaluated for light chains expression may further stand out to be light chain restricted.

The third important role of FCM is the classification of the B cell Lymphoma based on the expression pattern of different antigens that either resemble the maturation stage or specifically overexpressed/under expressed/aberrantly expressed in a particular NHL. Some f these include antibodies against CD5/CD10/CD23/CD200/CD123/CD103/CD25/CD11C/CD305/CD38/FMC7/IgM/IgD etc. Conventional approach for the classification of the B cell Lymphoma starts with the expression of CD5 and CD10, and subsequently utilizes other antigenic expression to successfully reach to a diagnosis. This is represented in Fig. 4. In recent years some of the antibodies like CD200, CD305, CD1d has evolved as crucial markers for differentiating different subtypes of B cell NHLs (Chaudhry & Karadimitris, 2014; Ghogale et al., 2023; Rahman et al., 2017a, b) (Fig. 5).

2.2.2 T-Cell Lymphomas

T-cell lymphoma are relatively uncommon lymphoid malignancies accounting for approximately 10% of all lymphoma. They are derived from post-thymic mature T-cells and are a highly diverse and heterogeneous group of entities. Unlike B cell NHLs, these disorders lack distinctive genetic markers for most disease categories, consequently, the sub-classification of T lymphomas remains a challenge. Flow cytometric evaluation of clonality has also been challenging for quite some times, which was predominantly based on the evaluation of TCR V-beta repertoire

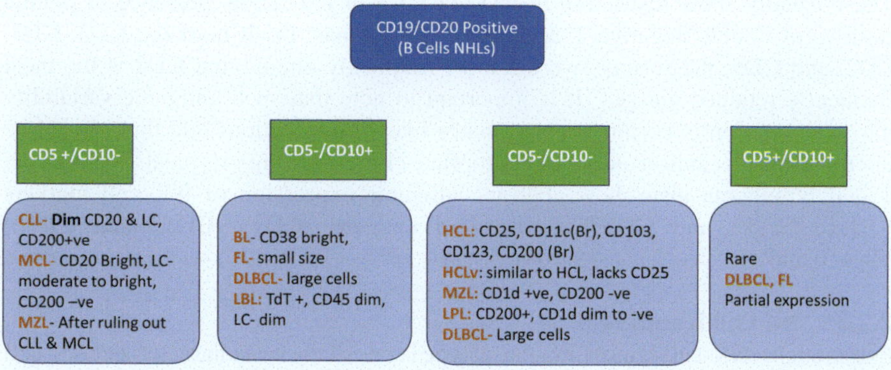

Fig. 4 Diagnostic algorithm for the B cell NHLs based on Expression of CD5 & CD10. Abbreviations: CLL Chronic Lymphocytic Leukemia, MCL Mantle cell Lymphoma, MZL Marginal zone lymphoma, BL Burkitt Lymphoma, FL Follicular Lymphoma, DLBCL Diffuse large B cell Lymphoma, LBL Lymphoblastic Lymphoma, HCL Hairy cell Leukemia, HCLv Hairy cell Leukemia variant, LPL Lymphoplasmacytic Lymphoma

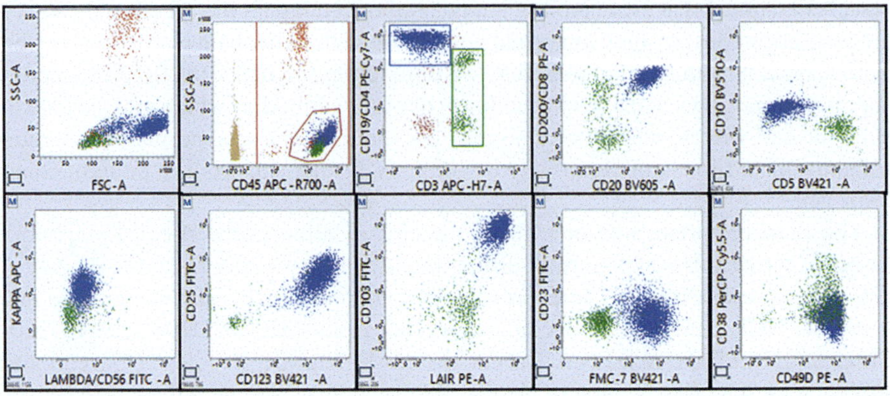

Fig. 5 Representative flow plots from a a case of Hairy cell leukemia, expressing CD19, CD20, CD123, CD25, CD103, LAIR, FMC7 and Kappa light chains

(Gorczyca et al., 2002). Thus, it was labour intensive and expensive. In the recent years, evaluation of T cell receptor beta subunit 1 (TRBC -1, clone: JOVI) has found its place similar to that of kappa lambda light chain expression of B cells. It has been found that during the T cell development the T cell receptor alpha beta (TCR-AB) unit is in turn can be composed of one of the two beta subunits, the B-1 or B-2. Researchers were successful in developing antibodies against the B subunit 1 and it was commercialized. Experience from different users have shown that an alpha beta TCR can have the expression of TRBC-1 in 30–70% cells. An expression more than 85% or less than 15%, indicates the clonal restriction of the alpha beta T cells. However, it is to be noted that this approach can only be used for the T cell neoplasm that express TCR-AB and not for those which are TCR-gamma delta.

T cell NHLs usually have dim expression or loss of one or more T cell markers. These usually show restriction of CD4 or CD8 or may show presence of double positive or double negative T cells. Amongst the pan T-cell markers, CD3, CD7, CD2, and CD5, the former two are most frequently altered and CD2 is the most frequently retained antigen. It is important to note that such antigenic variability may also be seen in certain reactive states like viral infections and thus should be interpreted with caution and in an appropriate clinical setting. Atypical T cells of the T cell lymphoma also show aberrant antigenic expression of NK cell markers (CD16, CD56, and CD57) or myeloid antigens (CD15, CD13 and CD33) (Figs. 6 and 7).

2.2.3 NK Cell Neoplasm

Extranodal T/NK-cell lymphoma–nasal type is one of the best characterized NK-cell which is CD56+, expresses cytotoxic granules TIA-1, granzyme-B, and perforin and are usually EBV+. These are negative for CD3, CD5, CD4 and CD8. Expession of other NK cell markers like CD94, CD158 etc. are also seen. In addition to CD56, these atypical NK cell proliferation also express CD57.

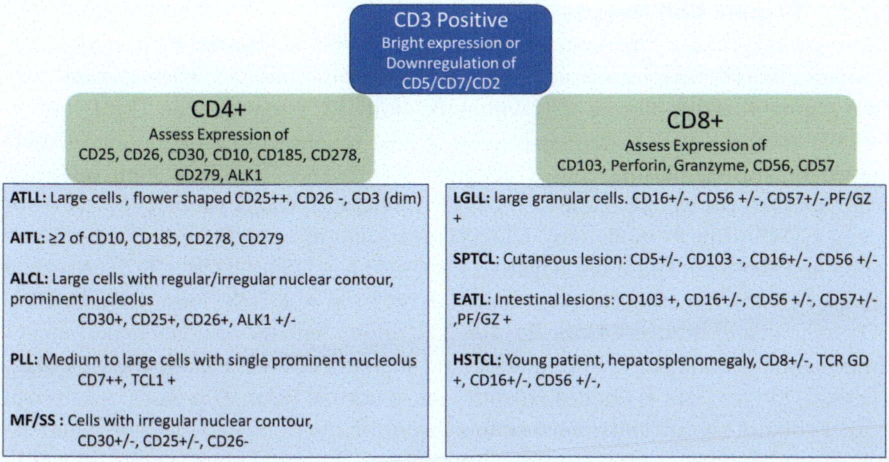

Fig. 6 Diagnostic algorithm of T cell NHL with predominantly CD4 or CD8 positive population. ATLL Adult T cell Leukemia Lymphoma, AITL Angioimmunoblastic T cell Lymphoma, ALCL Anaplastic Large cell Lymphoma, PLL Prolymphocytic Leukemia, MF/SS Mycosis fungoides/ Sezari Syndrome, LGLL Large granular cell Leukemia, SPTCL Subcutaneous panniculitis like T cell Lymphoma, EATL Enteropathy associated T cell Lymphoma, HSTCL Hepatosplenic T cell Lymphoma

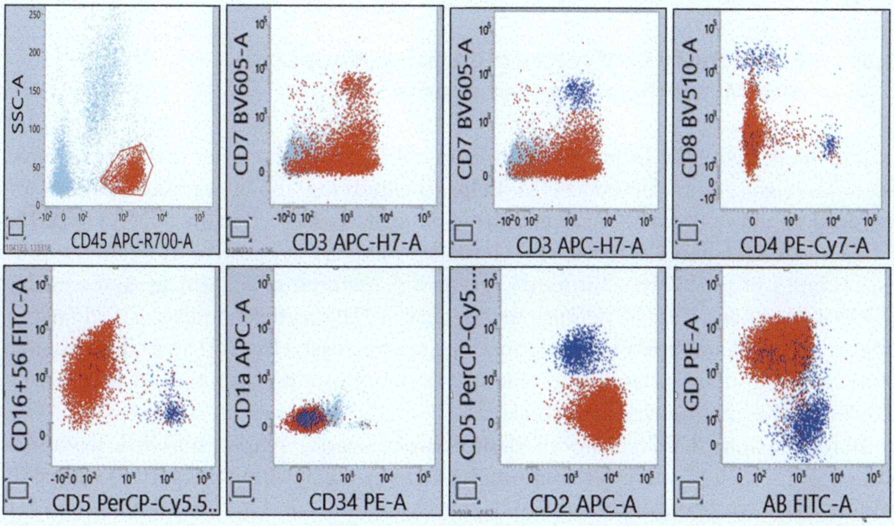

Fig. 7 A representative case of hepatosplenic T Cell lymphoma. The abnormal cells are gated in bright CD45 and CD3 positive region and are positive for CD2 and gamma delta, while being negative for CD7, CD5, CD4, CD8

2.2.4 Plasma Cell Neoplasm

Flow cytometry plays three crucial roles in plasma cell neoplasms. First, it identifies plasma cells (PCs) in informative samples. Second, it distinguishes atypical PCs from normal reactive plasma cells within the entire PC compartment. Third, it monitors the presence of atypical plasma cells in post-treatment marrow samples (MRD analysis). Normal plasma cells co-express CD38 and CD138 at high intensity, express CD19 (moderate intensity), CD27 (moderate), CD81 (homogeneous moderate), CD45 (dim to moderate), CD200 (dim) and are polytypic for cytoplasmic kappa and lambda. These are negative for CD117, CD56, CD20, CD22. Atypical plasma cells on the other hand show dim expression of CD38, loose CD27/CD81/CD19/CD45 and express CD56/CD117/CD20 along with monoclonal expression of either kappa or lambda. CD138 and other 7 markers (CD19, CD45, CD81, CD27, CD38, CD56 & CD117) is able to identify the abnormal PC in 97% cases. Additional markers do not significantly enhance this discriminatory power. However, cytoplasmic kappa/lambda confirms the clonality of these abnormal plasma cells. In nearly every case evaluated, normal plasma cells are almost always observed alongside atypical plasma cells. The proportion of these normal plasma cells has been shown to define disease progression. If normal cells constitute more than 5% of the total plasma cells, this is regarded as the MGUS phenotype, which is associated with slower progression compared to cases where normal plasma cells make up less than 5% of the total PCsThe panel used in our lab for the evaluation of plasma cells and a representative scatterplot from a case is shown below (Table 6 and Fig. 8).

2.3 FCM Utility in Myeloproliferative Neoplasm and Myelodysplastic Syndrome

In the recent years the utility of FCM in myeloid neoplasm other than AMLs have been increasingly observed. FCM helps in studying the maturation pattern of myeloid precursors. An altered maturation pattern as assessed in the CD13 vs CD16, CD11b vs CD16 and CD13 vs CD11b scatter plots are suggestive of dysmyelopoiesis (Gupta et al., 2018). Similarly, erythroid maturation pattern as assessed by CD36, CD117 and CD105 denotes the presence of dyserythrocpoiesis. The distribution of classical and non-classical monocytes, as assessed by CD16 vs CD14 scatter plots helps in differentiating the reactive vs neoplastic proliferation of monocytes (Chronic myelomonocytic leukemia).

In recent years, CD26 has been identified as a specific marker for CML leukemic stem cells. CD26-positive cells within the CD34+/CD38(dim to negative) stem cell population have been observed in all phases of CML and across all BCR::ABL1 transcript types. Notably, CD26 is not present on the LSCs in any other leukemia or in Ph+ ALL. Harnessing the importance of this fact, it can be used for the rapid diagnosis of CML (Rahman et al., 2022). Absolute quantitation of these CD26+ CML LSCs have been noted to predict the treatment response in cases of CML (Chaudhary et al., 2024). Another crucial role of FCM in an MPN is identification

Table 6 Antibodies panel used for the plasma cell evaluation

FL	BV605	BV510	BV421	FITC	PE	PERCP	PECY7	APC	APCR700	APCH7
AB	CD117	CD27	CD138	CytL	CD56	CD38	CD19	CytK	CD45	CD81

FL Flurochrome, *AB* antibodies

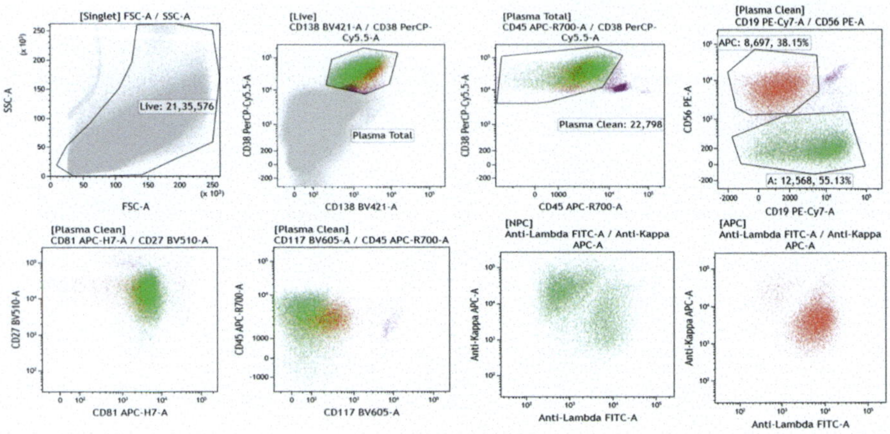

Fig. 8 A representative case of multiple myeloma MRD analysis show presence of normal (green population) as well as abnormal (red population) plasma cells

of small population of B or T lymphoid blasts which heralds a diagnosis of blast crisis as per the recent WHO classification, even if the blast count is less than 20%.

2.4 FCM for the Residual Disease Monitoring of Leukemia and Lymphoma

Minimal residual disease, or more accurately termed measurable residual disease (MRD), refers to the residual neoplastic cells that persist after treatment and are undetectable by routine diagnostic techniques such as microscopy. Detecting MRD is akin to finding a needle in a haystack; it is a challenging and a specialized task requiring advanced diagnostic methods with higher sensitivity, such as flow cytometry (FCM) or molecular diagnostic techniques. The first reports on MRD appeared in 1981, when residual T blasts were identified using immunofluorescence microscopy with a combination of anti-CD5 and anti-TdT antibodies (Brdstock et al., 1981). This groundbreaking study demonstrated that leukemic blasts could remain in the body even after morphological remission, potentially leading to relapse. Subsequent advancements included flow cytometry with three-color combinations and Southern blot analysis. With the advent of PCR (Polymerase Chain Reaction), allele-specific oligonucleotide PCR (ASO-PCR) using disease-specific Ig and TCR gene primers became a method for assessing MRD in ALL.

MRD has rapidly evolved as a fundamental and powerful prognostic indicator and has a pivotal role in treatment making decisions. It gives a lead time when the residual cells can be targeted before an overt clinical relapse occurs. It also potentially helps to target the resistant cancer stem cells and achieve a much higher and deeper response and cure.

Currently, MRD is assessed using two different methodologies. The molecular techniques which undertake an in-depth genomic analysis (tumour specific genotype) and the multicolour flow cytometry (MFC) which targets the immunophenotype of the tumor cells. Both these techniques have different applicability with some advantages and disadvantages. FCM based techniques are rapid, have wider applicability, needs lesser sophisticated equipment, and do not require the knowledge of diagnostic immunophenotype in most of the cases. However, they have lesser sensitivity as compared to the molecular techniques which are well more standardised.

FCM is essential for the diagnostic evaluation, follow-up, and treatment response assessment of hematolymphoid neoplasms. This powerful technique allows for a comprehensive assessment of expression patterns and the intensity of various surface and intracellular antigens across different cell populations. It can easily identify and characterize blasts or other atypical cells. One of the modern flow cytometer's key capabilities is its ability to rapidly acquire a substantial number of cells, making it possible to identify a few abnormal cells among millions within minutes. The advent of new, high-end software facilitates the rapid analysis of large data sets necessary for MRD analysis. Innovative software also aids in automated gating of significant populations, enabling multidimensional representation in a few plots. These technological advances, along with the improved availability of reagents, have effectively and efficiently overcome the classic limitations of MFC. The reproducibility and comparability of FCM have also improved with the help of consensus recommendations. The EuroFlow guidelines have played a significant role in terms of standardisation of sample collection, reagent selection, panel composition, technical nuances, analysis, and reporting. Most of labs follow these guidelines and use the panels which are either similar or have a reflection of EuroFlow panels in them. Common haematological neoplasms for which the MRD analysis is being routinely used includes acute myeloid leukaemia (AML), B cell Acute lymphoblastic leukaemia (B-ALL), T cell Acute lymphoblastic leukaemia (T-ALL), chronic lymphocytic leukaemia (CLL) and multiple myeloma (MM). The approach, time points, antibodies selection, number of events acquired and clinically relevant threshold for the MRD evaluation in all these diseases are little different. This is being summarised in the table below (Table 7 and Fig. 9).

3 Clinical Applications of FCM in Non Neoplastic Hematological Disorders

3.1 Platelet Assay

3.1.1 Platelet Count

MFC using antibodies targeted against CD41/CD61 has been regarded as the most sensitive and specific technique for the platelet counting. But this is not routinely used in diagnostic labs, which uses the CBC equipment based on the impedance of optical methods.

Table 7 Salient features of MRD evaluation and their comparison across different diseases

	B ALL	T ALL	AML	MM	CLL
Sample type	BM	BM/PB	BM	BM	PB/BM
No of cell required	2–5 million	2–5 million	1–3 million	5–10 million	2–5 million
Sensitivity required	0.01–0.001%	0.01–0.001%	0.1–0.001%	0.001–0.0001%	0.01–0.001%
Applicability	>99%	>99%	>95%	>99%	>95%
MRD positive threshold	0.01%	0.01%	0.1%	0.001%	0.01%
Follow up time points	EOI, EOC, Pretransplant	EOI, EOC, Pretransplant	EOI, EOC	Pretransplant, Day 100 post-transplant	Poorly standardised
Backbone antibodies of panel	CD19, CD34, CD10, CD38, CD45, CD20	CVD7, CD45, CD3, cyCD3, CD16/CD56, CD5, CD4, CD8	CD34, CD38, CD45, CD117, HLADR, CD13, CD33	CD38, CD138, CD19, CD45, CD56, CD117, cytKapp, cyt Lambda	CD19, CD5, CD20, CD23, CD79B
Additional marker	CD123, CD304, CD73, CD81, CD22, CD86	CD10, CD38, CD34, CD99	CD14, CD64, CD7, CD4, CD123, CD11B, CD133	CD200, CD229, CD319, CD81, CD28, CD27, CD20	CD200, CD43, CD81

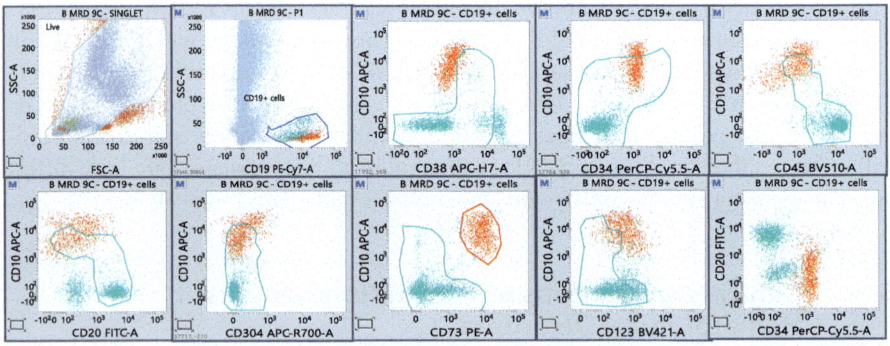

Fig. 9 Representative scatter plots from a B ALL MRD positive case. The red population represent the abnormal blasts

Inherited Platelet disorder The most common application of MFC in platelet analysis is identification of congenital deficiency of surface glycoproteins (GPIIb/IIIa; CD41/CD61 and GPIb-IX;CD42b) using respective antibodies. Absence of these surface glycoproteins are diagnostic of Glanzman's thrombasthenia and Bernard

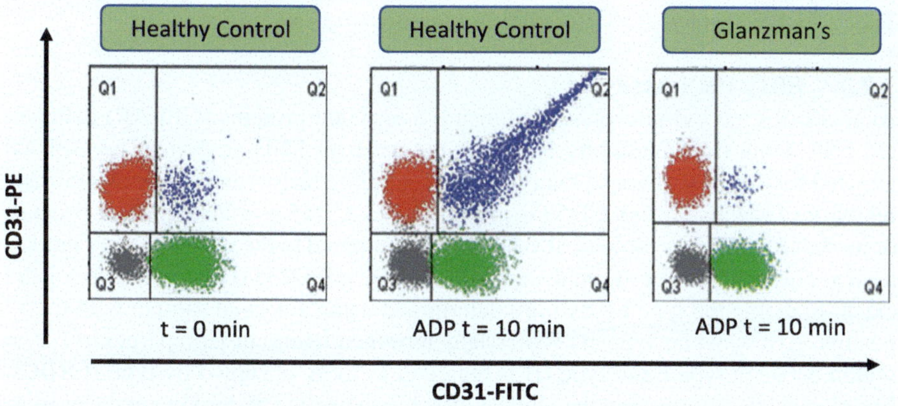

Fig. 10 Platelet aggregation studies using flow cytometry

Soulier syndrome, respectively) A very small quantity of blood is needed as compared to the platelet light transmission aggregometry based analysis.

3.1.2 Platelet Aggregation

Platelet aggregation after adding an agonist is also analysed using the flow cytometry (De Cuyper et al., 2013). This novel technique evaluates the proportion of platelet aggregates after adding an agonist. CD31 (PECAM) tagged in FITC and PE is used for this purpose. Two separate tubes are stained, each with CD31 FITC and CD31 PE, respectively. Stained tubes are fixed and then mixed together, followed by adding an agonist. The proportion of platelet aggregate is calculated by observing the double positive cells in CD31 FITC and CD31 PE bivariate plot (Fig. 10).

3.1.3 Annexin V Assay for Scott Syndrome

In Scott syndrome, the platelets are unable to expose phosphatidyl serine on their surface and thus will not bind to Annexin V. This can be targeted using flow cytometry performed on whole unlysed peripheral blood (Halliez et al., 2015). Platelets are identified using CD41. Normal platelets will be seen to co express CD41 and Annexin V after stimulation, while the platelets of Scotts syndrome will not.

3.1.4 Platelet Functional Assay

The results of utility and stability of flow cytometry based platelet aggregation and activation assay has been demonstrated by some of the good Indian literature (Dave et al., 2022). These are based on the expression of CD62P, CD63, PAC-1 etc. on the platelet activation induced by some platelet agonists. Resting platelets or inactive platelets will not show the expression, but the activated platelet will express these antigens.

3.2 RBC Assay

3.2.1 RBCs PNH Assay

Analysis of glycosyl phosphotidyl inositol associated proteins (GPI-AP) deficient red cells using flow cytometry is commonly used for PNH analysis. The GPI-AP targeted for this analysis is CD55 (Decay accelerating factor) and CD59 (Membrane inhibitor of reactive lysis). CD59 is preferred over CD55 as a PNH marker because of its stability and less biological variation as compared to the CD55. Gating marker used to purify RBCs for the analysis of PNH clone is CD235 (Glycophorin). A gate/extent to define the type I RBCs are determined using a normal sample with CD59 positive RBCs, while type III RBCs is determined using an isotype control. The region between type I and type III is regarded as those occupied by type II RBCs. The clone size, determining the clinical effect, is estimated by looking at the proportion of type III (Complete absence of CD59 expression) RBCs. The clone size is likely to be affected by a recent transfusion. Hence, a parallel assay of WBC is important (Fig. 11).

3.2.2 RBC Osmotic Fragility

When RBCs are suspended in distilled water the lysis starts, the extent of which depends on the surface to volume ratio. Spherocytes having lesser surfaces to volume ratio, get ruptured easily as compared to normal RBCs. On the other hand target cells, microcytes are capable of taking more DW inside before they get ruptured. FCM is used to measure the proportion of residual red blood cells before and after adding the DW. Number of RBC acquired during a particular time period, usually 11 s before adding DW is noted. After adding DW, some RBC get lysed and the proportion of residual RBCs are calculated. To include similar number of RBCs across different samples, a formula developed by Won and Suh is used (2009).

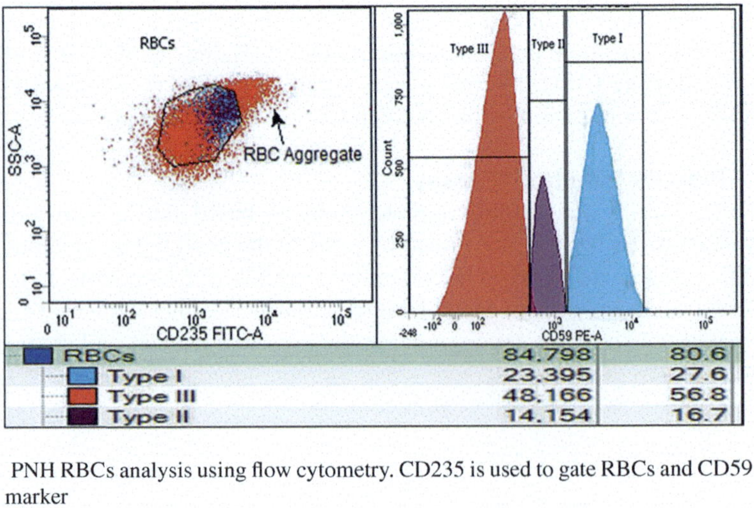

Fig. 11 PNH RBCs analysis using flow cytometry. CD235 is used to gate RBCs and CD59 is used as PNH marker

Amount of blood to be taken is derived by dividing 130 by the RBC count as obtained from a CBC analyser. This volume is added to a tube containing 1 ml of NS. After proper mixing 10 microlitre of this mixture is added to a tube containing 1.1 ml of NS. This tube is acquired for 11 s and a gate is created to include all the RBCs acquired during time. This 11 s gate is duplicated several times to include the RBCs acquired during a period of 88 s after adding 0.9 ml Distilled water. The average number of RBCs as obtained in the 7th and 8th 11 s gate is taken as the residual RBC number. This number, as a proportion of original RBC numbers as obtained in the first 11 s gate before adding the DW, and adjusting the analysis volume is the % residual RBCs. The formula to calculate the % residual RBCs is as below.

$$\left[\text{Average number of events after DW} \div \text{Number of events without DW}\right] \times 2\left[\begin{array}{l}\left(\text{final volume after adding DW;}1.1\,\text{ml NS} + 0.9\,\text{ml DW}\right) \div \\ 1.1\left(\text{Initial volume before DW}\right)\end{array}\right] \times 100$$

A laboratory control range is derived using 20–30 normal samples. With every assay 3–5 control sample is used to ascertain the test reliability. Increased osmotic fragility is determined by significantly reduced % residual RBCs after adding distilled water (Median of Controls—3SD)

3.2.3 Eosin 5 Malemide (EMA) Binding Assay

EMA is a dye usually available in the powder form from sigma aldrich. It needs to me dissolved in PBS to make a stock or working solution. The malemide moiety of EMA interacts covalently with the Lys-430 in the first extracellular loop of band 3 protein, whereas the eosin chomophore is assumed to sit in a deep pocket within the transmembrane core of band 3 (Fig. 12).

Reduction of the membrane proteins in HS causes reduced binding of EMA which can be assessed flow cytometrically as a reduced median fluorescent intensity of the FITC/FL1 channel (where the emission of EMA is detected). The EMA dye is to be prepared by dissolving it in PBS to make a concentration of 1 microgram/ml. The stock solution should be aliquoted in small volumes (100 microlitres) and kept in freezer at −20°. The utility of EMA binding for the diagnosis of hereditary spherocytosis is based on the results published by King et al. (2000) (Fig. 13).

For staining RBCs, 100 microlitre of EDTA blood is washed three times in 2 ml of PBS. The final red cell pellet is mixed well and 5 microlitre of red cell pellet is taken in an Eppendorf tube and incubated along with 25 microlitre of EMA working solution (0.5 microgram/ml) for half an hour. After half an hour the suspension is mixed once again and kept for further 30 min. The tube is centrifuged at 3500 rpm, and the excess of EMA dye is removed. Finally, the red cell pellet is washed with PBS. The pellet is dispersed and 40 microlitre of this is added to 600 microlitre of PBS for final acquisition and analysis. With every sample, 3–5 controls are also stained. MFI ratio of the test sample with that of controls is calculated. A value less than 0.8 have a very high sensitivity and specificity for hereditary spherocytosis.

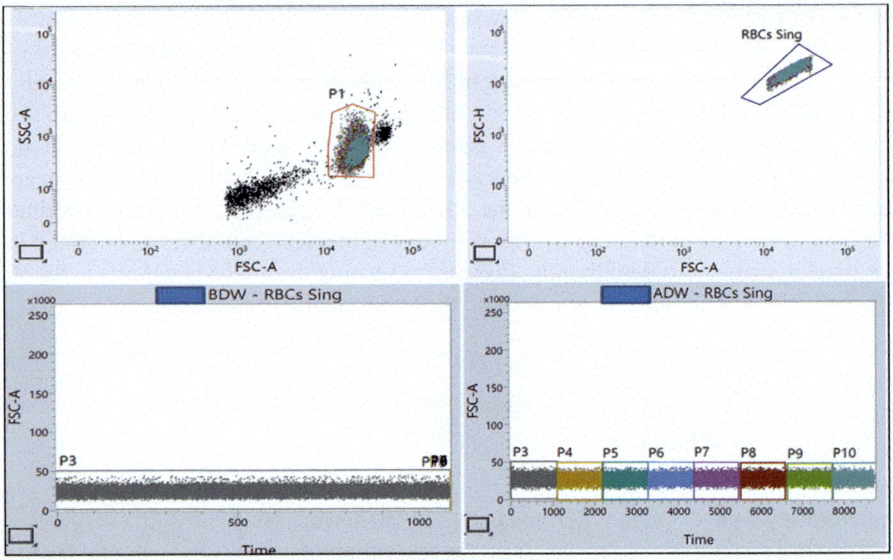

Fig. 12 FCM based OFT using time gate to quantify the proportion of residual RBCs after addition of distilled water

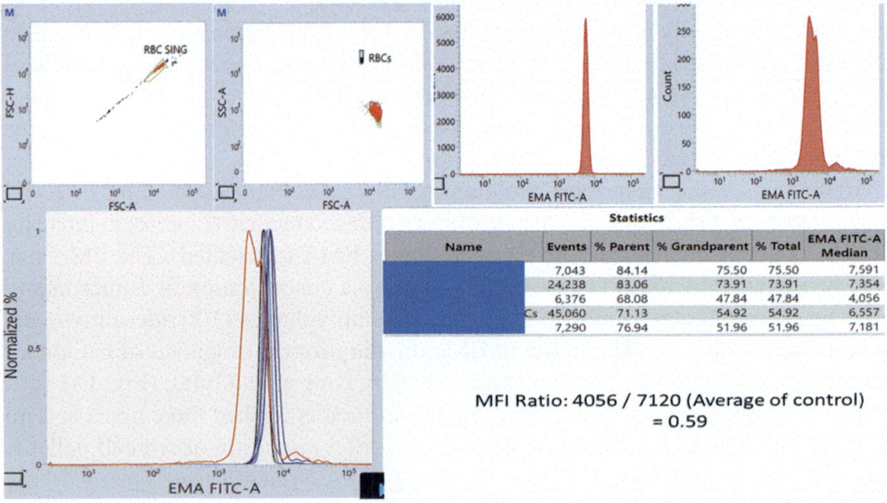

Fig. 13 EMA binding test for the diagnosis of hereditary spherocytosis. The red colored peak from a patient of HS shows a significantly reduced MFI, the ratio being less than 0.8

Hereditary pyropoikilocytosis also shows a reduced EMA binding. In addition to a reduced MFI ratio, the broadening and shouldering of peak is also suggestive of reduced EMA binding and may be suspicious of HS.

3.2.4 G-6 PD Assay

A simple, cytofluorometric method that extends the classic methemoglobin reduction test, assessing G6PD deficiency at the level of an individual erythrocyte was described by Shah et al. (Shah et al., 2012) There was strong concordance between genetic and biochemical techniques. The principle lies in the generation of ferryl hb (bright in color) or cyanmethhemoglobin (dull color) after the extended methhemoglobin reduction test in a normal or G6PD deficient individuals, respectively. In the first step, only met-Hb reacts with cyanide to generate cyanmethemoglobin (cyanmet- Hb), while oxy-Hb remains unreactive. In the second step, only oxy-Hb reacts with peroxide to generate a fluorescent product (ferryl-Hb), while cyanmet-Hb remains largely nonreactive and nonfluorescent. After these two reactions, G6PD efficient cells contain brightly fluorescent ferryl-Hb, while G6PD deficient cells contain very dimly fluorescent cyanmet-Hb, with fluorescent signal uniformly distributed throughout cells (Fig. 14).

3.3 Leukocyte Assay

3.3.1 PNH Screening

Analyses of neutrophils and monocytes are widely used for screening the presence and the clone size of GPI deficient cells (Rahman & Chandra, 2018). Lymphocytes are not used for screening purpose owing to their long lifespan as well as variable expression of the GPI-linked antigens on their surface. For analysing the presence of PNH clone, it is advisable to screen two cell lines, using two gating markers and two PNH markers. CD45 and CD15 is used as gating markers for neutrophils (as these are not GPI linked); while CD45 and CD64 is used as gating markers for monocytes. Earlier, CD66b, CD24, and CD16 were commonly used for neutrophils; while CD14 was commonly used for monocyte as PNH marker. FLAER as a marker for both neutrophils and monocyte had a great impact on leukocyte screening for PNH, especially identifying small clones which are associated with aplastic anemia (Rahman et al., 2017a, b). FLAER is a mutant form of pro-aerolysin derived from pseudomonas hydrophilia. This inactive variant is labeled with a fluorophore Alexa Fluor 488 and used for the screening of PNH. It has the binding property but lacks the channel-forming property of active aerolysin. Peripheral blood nucleated cells are able to bind to FLAER because they are able to convert pro-aerolysin to aerolysin by using proteases like furin present on their cell surface. RBCs lack these proteases and hence are unable to bind it (Brodsky et al., 2000) FLAER based assay is now regarded as 'Gold Standard' for the leukocyte screening for PNH. Recently CD157 has been identified as one of the promising reagents which is expressed on both polymorphs and monocytes (Rahman et al., 2018a, b). CD157 is also known as bone marrow stromal cell antigen 1 (BST 1). It is a single-chain GPI-AP which is helpful in pre-B cell growth. It is expressed on B cell progenitors, T progenitors, granulocytes, monocytes, and some other cells as well. It has been found as a sensitive marker with a low background rate in normal sample. CD157-based PNH

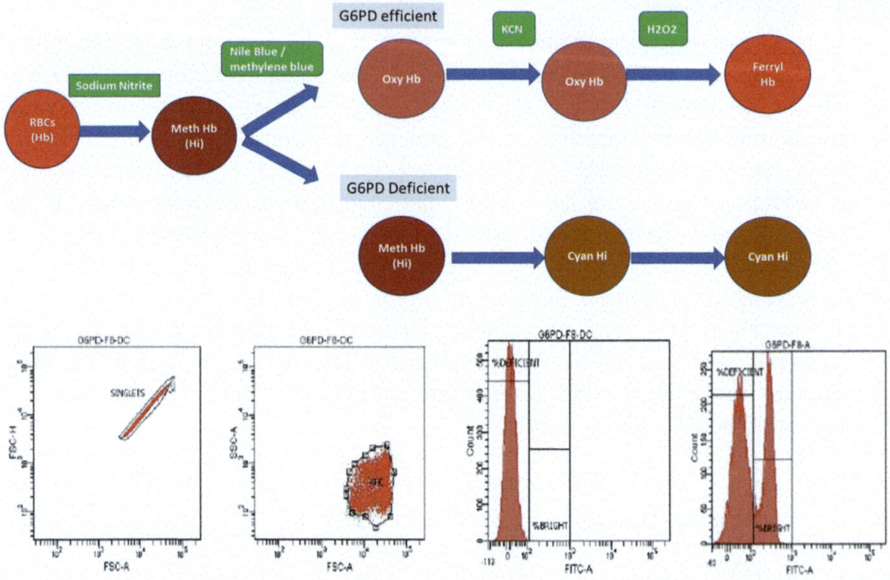

Fig. 14 FCM based G6PD assay

screening was found to be very precise with minimal inter-assay and intra-assay variations (Rahman et al., 2018a, b).

A routine sensitivity analysis is one which is sufficient enough for detecting PNH clones at 1% level, while high sensitivity analysis should be able to detect the clone sizes of 0.01%. For classic PNH patients, who present with features of hemolysis or thrombosis, a routine analysis may be sufficient enough to detect the presence of PNH clone. However, in patients with AA and MDS who usually do not have the clinical features of hemolysis, the clone size is small and in a majority of patients less than 1%. In these cases, a routine sensitivity analysis will miss the PNH clone detection in a majority of cases. Identifying these minor clones are important as ~25% of these patients may show expansion of clone size and may later present with clinical PNH. About 5000–10,000 gated events may be sufficient for routine sensitivity analysis, but for high sensitivity, analysis of 2.5 lakhs of gated population is recommended (Sutherland et al., 2014).

Routinely, a five color combination using FLAER, CD157, CD15, CD64 and CD45 is used for a high sensitivity PNH screening. A detailed overview of choosing the appropriate reagent based on the flow cytometer used is described in the literature (Illingworth et al., 2018). This single tube is sufficient for the analysis of both neutrophils and monocytes. CD45 and CD15 is used to gate neutrophils; while CD45 and CD64 is used to gate the monocyte. Gated neutrophils and monocytes are plotted on CD157 vs FLAER scatter plots. PNH clone/population is identified by the absence of CD157 and FLAER (Fig. 15).

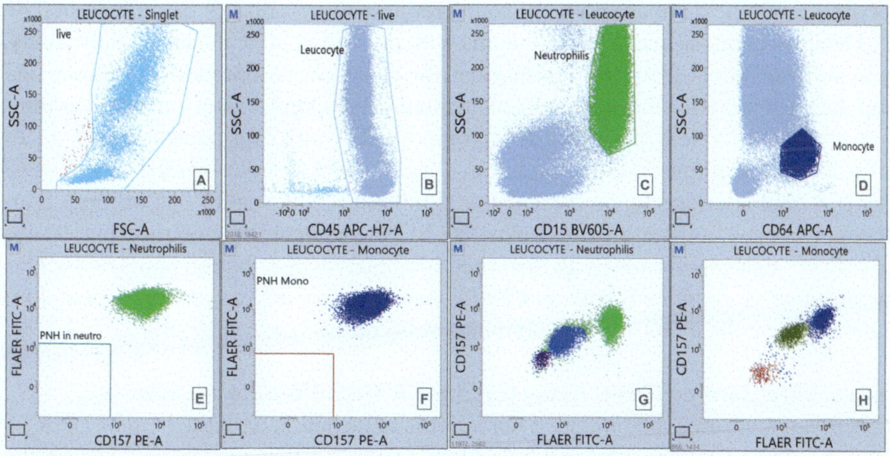

Fig. 15 FCM based PNH screening for leukocytes showing gating strategies. Plots E & F show absence of PNH clone, while Plots G&H show 'Presence' of PNH clones with all the three cell types

In order to avoid the confusion and ambiguous report, the term positive or negative is not preferred. Rather, terms like Presence/Absence or Detected/Not Detected is used. For the sake of uniformity some consensus terms are used in the reports based on the PNH clone size. If the PNH population is more than 1% it is termed as 'Presence of PNH Clone', if it is 0.1–1%, it is termed as 'minor PNH population', while PNH proportion of <0.1% is termed as 'rare GPI deficient Population'.

3.3.2 Immunodeficiency Disorders of Lymphocytes

Diagnosis of primary immunodeficiency (PID) or now better known as inborn error of immunity (IEI), is often challenging and requires identification of underlying genetic defect complemented by comprehensive evaluation of immune system. Flow cytometry, with its advances in last few decades has emerged as an indispensable tool for enumeration as well as characterization of immune cells and its subsets. Evaluation of lymphoid cell subsets and phagocytic and adhesion activities of granulocytes and monocytes, have been used classically for the evaluation of inborn errors of immunity.

Lymphocyte Subset Analysis

Euroflow has validated a single tube antibody combination to screen all the lymphoid PIDs. It offers high sensitivity to identify a lymphoid PID and to guide about the subsequent diagnostic modalities to reach to a conclusive diagnosis. It consists of a stepwise screening and characterisation of numerical alternations in lymphoid population or any other functional defects. This eight color, 12 antibodies single tube is known as primary immunodeficiency orientation tube (PIDOT) used by the euroflow (van der Burg et al., 2019). An equal importance is given to the standardised cell processing and staining steps along with the use of dedicated software incorporating the databases. The standardized EuroFlow concept has been tested on 250 healthy controls and 99 PID patients with defined genetic defects. Additionally, new software tools featuring multidimensional pattern recognition and automated population separation have been implemented. Many centers now use flow cytometry (FCM) as a screening tool for lymphoid IEI, utilizing panels similar to or inspired by EuroFlow. This method is an objective and straightforward way to identify IEI, characterized by the absence or significant reduction of one or more lymphocyte subsets, as seen in conditions like SCID and agammaglobulinemia. It also effectively detects diseases with disturbed lymphocyte distribution patterns, such as the absence of memory B cells in CD40L defects or an increase in alpha-beta double-negative cells in autoimmune lymphoproliferative syndrome (ALPS) (Table 8).

DihydroRhodamine (DHR) Assay for Chronic Granulomatous Disease

This assay measures the oxidative burst in the neutrophils with the help of DHR. A sodium heparin anticoagulated whole blood specimen is incubated at 37 °C in the presence of DHR. A stimulant Phorbol myristate acetate (PMA) is added and mixed with the whole blood specimen for additional incubation at 37 °C. The specimen is then centrifuged, and the red cell present in the cell pellet is lysed with ammonium chloride. A LIVE/DEAD viability marker and CD15 is added subsequently. Finally, cells are washed, centrifuged, and resuspended in 1% paraformaldehyde prior to analysis. Viable neutrophils are identified by the use of the viability dye and further confirmed by the presence of CD15. Approximately 20,000 viable neutrophil in the unstimulated specimen are used to set the limits for DHR expression, collected for flow cytometer. The results are derived as delta % DHR-positive neutrophils after PMA stimulation and mean fluorescence intensity for each sample. Patients with CGD have low proportion of DHR expression (O'Gorman & Corrochano, 1995).

Leukocyte Adhesion Defect Assay

This is done by observing the expression of integrins on the cell surface of neutrophils. These include CD11a, CD11b, CD11c and CD18. Most commonly the expression of CD11b and CD18 is noted using flow cytometry. Normal neutrophils express them but patient with LAD have reduced expression of these antigens.

Hemophagocytic Lymphohistiocytosis

FCM helps in analysing the expression of activated T cell profile and reduced expression of perforin in NK cells, which may be helpful in diagnosis of HLH. The proportion of highly active (HLADR and CD38 co expressing population and exhaustive (Tim 3 & PD 1 positive) subsets are increased.

Table 8 Antibodies panel used in the PIDOT

FL	BV510	BV421	FITC	PE	PerCPcy5.5	PeCy7	APC	APCH7
AB	CD45 RA	CD27	CD8/ IgD	CD16/ CD56	CD4/IgM	CD19/ TCRGD	CD3	CD45

Sepsis Evaluation

Flow cytometry plays an important role in diagnosis and prognosis of sepsis in critically ill patients. The commonly used markers are CD64 in neutrophils, HLADR expression of monocytes, CD25 expression in T cells, and TLRs on NK cells (Mahmoodpoor et al., 2018). The patients with sepsis have increased level of CD64 on neutrophils and a decreased HLADR levels on the peripheral blood monocytes, as compared to healthy controls. Among the patients with sepsis, persistent low levels of HLADR expression on monocytes have been found to be associated with poorer prognosis. Some studies have also used FCM based expression of antigens to differentiate bacterial and viral sepsis. Patients with bacterial sepsis were found to have a significantly higher CD64 levels on neutrophils than patients in viral sepsis. Conversely, patients in viral sepsis showed a significant increase in CD169 levels on monocytes compared with patients in bacterial sepsis (Bourgoin et al., 2019).

3.3.3 Flow-FISH for Telomeropathies

Flow FISH is a technique which is increasingly used for the telomere measurement (Kelesidis & Schmid, 2017). It combines the power of fluorescent in situ hybridization (FISH) with flow cytometry. White blood cells (leukocytes) are isolated from a blood sample and stained with the antibodies against the surface antigens like CD15, CD64, CD19, CD3, CD16, CD56 etc. After this, the stained sample is mixed with fluorescent peptide nucleic acid (PNA) probes that bind specifically to the telomere repeats (TTAGGG). Several controls are used to take into consideration the background staining (cellular controls). Fluoresence is measured using flow cytometer and compared to the control cells. Flow FISH telomere testing has the unique advantage over other testing protocols as it provides results on a variety of individual cell types from single blood sample. Using cell surface staining for different cells, results can be individually reported for lymphocytes, granulocytes, B-cells, naïve and memory T-cells, and NK cells. It is used for the diagnosis of telemeropathies. Very short telomere length by flow FISH identifies patients with dyskeratosis congenita (DKC).

4 Clinical Applications of FCM in Transplant Immunology

4.1 CD34 Positive Hematopoietic Stem Cell (HSC) Enumeration

Flow cytometric enumeration of CD34+ hematopoietic stem cells (HSCs) has become the preferred method for quantifying HSCs for stem cell transplantation. Reducing intra- and inter-laboratory variation in enumerating CD34+ cells is crucial, especially with the establishment of cryopreserved cord blood cell (CBC) banks to meet the growing demand for unrelated hematopoietic progenitor cell (HPC) transplants. This consistency can be reliably monitored within and between laboratories using stabilized blood preparations containing predefined numbers of CD34+ cells representative of HPC transplantation practices. The enumeration of

HSCs in G-CSF-stimulated peripheral blood, bone marrow, or the harvest product is conducted using the International Society of Hematotherapy and Graft Engineering (ISHAGE) guidelines. Accurate detection of CD34+ cells is based on a four-parameter flow cytometry method, involving forward scatter, side scatter, CD45PerCP, and CD34 PE staining. This approach discriminates HSCs (which express relatively low levels of CD45) from lymphocytes and monocytes, identifying "true" CD34+ cells as those with dim CD45 fluorescence and low side scatter (CD45dim, SSc low). The absolute count of CD34+ cells is then calculated using either the 'single platform method' or the 'double platform method'. In the single platform method, a known number of polystyrene beads is added to the sample tube during staining and acquisition. In the double platform method, the total leukocyte count (TLC) is obtained from a CBC analyzer, and the CD34+ cell count is calculated as a proportion of CD45+ leukocytes.

4.2 FCM Based HLA Cross Match

Detection of antibodies against the transplant donor present in the serum of the solid organ transplant recipient can be carried out rapidly using the FCM. A fluorescent-labeled antibody against CD3 or CD19 is used as a marker to distinguish T- from B-lymphocytes. The freshly isolated donor cells are incubated with the recipient's serum to allow for potential antibodies present in the serum of recipient to bind on the cell surface of donors. If there are donor-specific anti-HLA antibodies, the Fab fragment of the antibody will bind to the HLA antigens on the cell surface of donors cells. The detection is carried out using the Fluorescein Isothyocynate (FITC) labelled goat antihuman antibody, which in turn will bind to recipients antibodies bound to the surface of donor cells. This acts as a reporter fluorescent dye to the alloantibody. A shift in fluorescent signal before and after incubation of this antibodies is noted. A positive shift indicates the presence of anti-donor antibodies present in the recipient's serum. The flow crossmatch test is more sensitive than the conventional complement dependent cytotoxicity (CDC) crossmatch test. It is much standardized assay as compared to CDC crossmatch and is able to detect low titre antibodies. However, it cannot differentiate between for HLA and non-HLA antibodies.

5 Flow Cytometry in Non Hematological Neoplasms

The cell cycle analysis and the ploidy assay have been used in some of the very early clinical studies in reproductive biology and some human solid cancer. These approaches utilised the DNA binding dyes and the advance of FCM technique. Later on, the development of monoclonal antibodies took over and clinical utility of FCM broadened. In the early days, most antibodies were developed and categorized against blood cells, and because blood and bone marrow are naturally in liquid form, flow cytometry (FCM) became predominantly used for evaluating hematological disorders and neoplasms. This overshadowed the potential of FCM in diagnosing non-hematological disorders. However, advancements in FCM technology and the development of standardized cell processing techniques, such as enzymatic and mechanical dissociation to create single-cell suspensions, have expanded its application to solid tumors and non-hematological malignancies.

5.1 Immunophenotyping of Solid Tumors

Histopathology with immunohistochemistry (IHC) remains the cornerstone to the diagnosis of solid tumors. However, they can be used to characterise the malignant cells in the peripheral blood as the circulating tumor cells (CTC) or to diagnose the metastasis in the serous cavity effusions. Among the commonly used and well validated antigenic targets for the epithelial cancers are the EpCAM (Epithelial cell adhesion molecule), Ber EP4, CD326 and MOC31 (Pillai & Dorfman, 2016). FCM can be a rapid modality to confirm the presence of abnormal cells both in their native as well as at a metastatic site. The examination offers objective evidence and instills greater confidence compared to cytology alone. It is recommended to evaluate a large volume of serous fluid effusion. A cell pellet is prepared after gentle centrifugation and stained with a panel of antibodies. To rule out the possibility of non-specific binding and stray events, a clustered population of at least 10–20 events is required to be considered definitive.

FCM can also be used for the evaluation of pediatric round cell tumors like neuroblastoma (CD56hi/GD2+/CD81hi), primitive neuroectodermal tumors (CD271hi/CD99+), Wilms' tumor (CD56+, multiple populations), rhabdomyosarcoma (MYOD1+/myogenin+), germ cell tumors (CD56+/CD45–/NG2+/CD10+), and hemangiopericytomas (CD45–/CD34+) etc. (Ferreira-Facio et al., 2013). Though a final diagnosis needs a histopathological and genetic evaluation, a flow cytometric evaluation may be of great importance in drawing out a differential diagnosis and differentiating them from a leukemia/lymphoblastic lymphoma, which totally has a different therapeutic approach.

5.2 DNA Ploidy and Cell Cycle Analysis

As mentioned above, the DNA ploidy and the cell cycle analysis have been one of the oldest uses of flow cytometry. This have been possible with the advent of better cell processing and disaggregating methodologies. The DNA content and ploidy analysis have been used to prognosticate the tumors (Danova et al., 2018; Lin et al., 2016).

5.3 Analysis of Circulating Tumor Cells

Circulating tumor cells (CTCs) have been widely studied as a non-invasive tool for the evaluation of non-hematological malignancies. It can be used for an early diagnosis, genetic evaluation, disease monitoring, predicting the disease relapse as well as drug research (Awasthi et al., 2017). Recent years have seen development of tools that can easily enrich these cells in the peripheral blood of these patients, which can be subsequently subjected to FCM based characterisation. Their prognostic value has been studied in cancers like lung, hepatopancreatobilliary, breast, prostatic, gastrointestinal cancers.

5.4 Analysis of Circulating Endothelial Cells (CEC)

Evaluation of CEC and its progenitors have been recently indicated as potential biomarkers for angiogenesis in several types of cancer. These evaluations help us to see the disease status, evaluate the tumor progression and also act as a target for anti-angiogenic treatment. They can also be monitored for therapeutic effect (Sutherland et al., 2014).

5.5 Tumor Infiltrating Lymphocytes/Immune Cells

It has been well-established now that tumor-infiltrating lymphocytes or immune cells play a very important role in cancer development and progression. Depending on the type of interaction and the defence mechanism, these immune cells cause tumor destruction or can support the tumor progression by the immune evasion phenomenon. The tumor associated macrophages (TAMs) have been studied extensively in meningiomas, breast cancer etc. Their functionality depends on the type of tumor and their interacting signals. They can be immunosuppressive in some tumors while in other tumors they can be associated with other T and NK cells and can have a tumor cytotoxic effect. FCM has been used as an adjunct to IHC and immunofluorescence for studying these immune cells in tumors. It has the advantage of identification and characterization of heterogeneous cell populations coexisting in tumor samples (Parney et al., 2009).

5.6 Circulating Dendritic Cells

Dendritic cells (DCs) are crucial antigen-presenting cells which play an important role in antitumor immune response by controlling the initiation of the T cell-dependent immune response. They have ability to sense small changes in the tissue microenvironment and activate adaptive immune responses, whilst maintaining immunological tolerance. Three types of dendritic cells have been identified in the

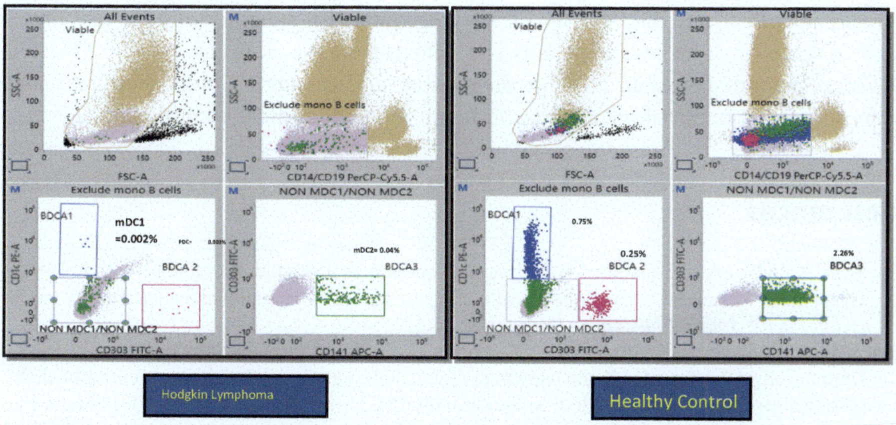

Fig. 16 FCM based enumeration of peripheral blood dendritic cell subsets (PBDC) shows a reduction in the DC subsets as compared to that of healthy donor

peripheral blood. The CD1c+ mDC1, CD141+ mDC2 and CD303 positive pDC. Several FCM studies have focused on the determination of the number and the functional alterations of DCs and their subsets in different type of human cancers like breast, lung, melanoma, Hodgkins lymphoma etc. Their numbers in peripheral blood have been found to correlate with the bulk and aggressiveness of tumors. They are recruited from peripheral blood at these sites and hence their number are reduced in the peripheral blood. On successful treatment and reduction in the tumor bulk, their number in peripheral blood return back to normal healthy levels (Galati et al., 2019). pDCs act differently depending on the microenvironment they encounter and can therefore have an inhibitory immune activity or antitumor activity. The study of pDC biology and understanding the nature of pDCs associated with several neoplasms could also have a new therapeutic possibility (Pinto et al., 2012) (Fig. 16).

Summary

In summary, FCM is a powerful tool which is being extensively used in clinical settings to evaluate various hematological and non-hematological conditions. It is an indispensable diagnostic modality for any diagnostic lab dealing with hematological neoplasm where the power of this technique is harnessed in diagnosing, classification, prognostication and monitoring of hematological neoplasms. This is particularly useful in differentiating between various types of leukemias and lymphomas, enabling tailored treatment strategies. The advancements in FCM technology and cell processing techniques have expanded its utility to solid tumors and non-hematological malignancies. Single-cell suspensions from solid tissues enables the detailed analysis of tumor cells and the tumor microenvironment. FCM also has a pivotal role in immunology, being employed to assess the qualitative and quantitative defects. It is used in transplant immunology to determine donor compatibility and to monitor for graft rejection. Furthermore, FCM plays a role in diagnosing and managing infectious diseases by identifying pathogen-specific immune responses and in assessing the overall health and function of the immune system. Overall, FCM is an indispensable tool for diagnosing and managing both neoplastic and non-neoplastic hematological and non-hematological clinical conditions.

References

Awasthi, N. P., Kumari, S., Neyaz, A., Gupta, S., Agarwal, A., Singhal, A., et al. (2017). EpCAM-based flow cytometric detection of circulating tumor cells in gallbladder carcinoma cases. *Asian Pacific Journal of Cancer Prevention, 18*(12), 3429.

Baer, M. R., Stewart, C. C., Lawrence, D., Arthur, D. C., Byrd, J. C., Davey, F. R., et al. (1997). Expression of the neural cell adhesion molecule CD56 is associated with short remission duration and survival in acute myeloid leukemia with t(8;21)(q22;q22). *Blood, 90*(4), 1643–1648.

Bourgoin, P., Soliveres, T., Ahriz, D., Arnoux, I., Meisel, C., Unterwalder, N., et al. (2019). Clinical research assessment by flow cytometry of biomarkers for infectious stratification in an emergency department. *Biomarkers in Medicine, 13*(16), 1373–1386.

Brdstock, K. F., Janossy, G., Tidman, N., Papageorgiou, E. S., Pretince, H. G., Willoughby, M., et al. (1981). Immunological monitoring of residual disease in treated thymic acute lymphoblastic leukemia. *Leukemia Research, 5*(4), 301–309.

Brodsky, R. A., Mukhina, G., Li, S., Nelson, K. L., Chiurrazi, P. L., Buckley, T., et al. (2000). Improved detection and characterization of paroxysmal nocturnal hemoglobinuria using florescent aerolysin. *American Journal of Clinical Pathology, 114*, 459–466.

Chaudhary, N., Rahman, K., Gupta, P., Gupta, R., Sarkar, M. K., Singh, M. K., et al. (2024). Peripheral blood quantitation of CD26 positive leukemic stem cells (LSCs) as a predictor of tyrosine kinase inhibitor response in chronic myeloid leukemia. *International Journal of Laboratory Hematology*. Online-Ahead of Print.

Chaudhry, M. S., & Karadimitris, A. (2014). Role and regulation of CD1d in normal and pathological B cells. *Journal of Immunology, 193*(10), 4761–4768.

Danova, M., Torchio, M., Comolli, G., Sbrana, A., Antonuzzo, A., & Mazzini, G. (2018). The role of automated cytometry in the new era of cancer immunotherapy. *Molecular and Clinical Oncology, 9*(4), 355–361.

Dave, R. G., Geevar, T., Chellaiya, G. K., Mammen, J. J., Vijayan, R., Samuel, A., et al. (2022). Stability and utility of flow cytometric platelet activation tests: A modality to bridge the gap between diagnostic demand and supply. *Platelets, 33*(7), 1043–1051.

De Cuyper, I. M., Meinders, M., van de Vijver, E., de Korte, D., Porcelijn, L., de Haas, M., et al. (2013). A novel flow cytometry-based platelet aggregation assay. *Blood, 121*(10), e70–e80.

Ferreira-Facio, C. S., Milito, C., Botafogo, V., Fontana, M., Thiago, L. S., Oliveira, E., et al. (2013). Contribution of multiparameter flow cytometry immunophenotyping to the diagnostic screening and classification of pediatric cancer. *PLoS One, 8*(3), e55534.

Galati, D., Zanotta, S., Corazzelli, G., Bruzzese, D., Capobianco, G., Morelli, E., Arcamone, M., et al. (2019). Circulating dendritic cells deficiencies as a new biomarker in classical Hodgkin lymphoma. *British Journal of Haematology, 184*(4), 594–604.

Ghogale, S., Singh, A., Deshpande, N., Girase, K., Sriram, H., Rajpal, S., et al. (2023). PB1949: role of lair (CD305) in flow cytometric detection of occult bone marrow involvement in non-CLL B-cell non-Hodgkin Lymphoma. *HemaSphere, 7*(S3), e7801452. https://doi.org/10.1097/01.HS9.0000974612.78014.52

Gorczyca, W., Weisberger, J., Liu, Z., Tsang, P., Hossein, M., Wu, C. D., et al. (2002). An approach to diagnosis of T-cell lymphoproliferative disorders by flow cytometry. *Cytometry, 50*(3), 177–190.

Gupta, R., Rahman, K., Singh, M. K., Surabhi, Y. G., & Nityanand, S. (2018). Utility of a single tube six colour flow cytometry panel for the diagnosis of myelodysplastic syndrome:

Experience of a tertiary care Centre in India. *International Journal of Hematology-Oncology and Stem Cell Research, 12*(1), 29–34.

Halliez, M., Fouassier, M., Robillard, N., Ternisien, C., Sigaud, M., Trossaert, M., et al. (2015). Detection of phosphatidyl serine on activated platelets' surface by flow cytometry in whole blood: A simpler test for the diagnosis of Scott syndrome. *British Journal of Haematology, 171*(2), 290–292.

Illingworth, A., Marinov, I., Sutherland, D. R., Wagner-Ballon, O., & DelVecchio, L. (2018). ICCS/ESCCA consensus guidelines to detect GPI-deficient cells in paroxysmal nocturnal hemoglobinuria (PNH) and related disorders part 3 – data analysis, reporting and case studies. *Cytometry. Part B, Clinical Cytometry, 94*(1), 49–66.

Kaleem, Z. (2006). Flow cytometric analysis of lymphomas: Current status and usefulness. *Archives of Pathology & Laboratory Medicine, 130*(12), 1850–1858.

Kelesidis, T., & Schmid, I. (2017). Assessment of telomere length, phenotype, and DNA content. *Current Protocols in Cytometry, 79*, 7–26.

King, M. J., Behrens, J., Rogers, C., Flynn, C., Greenwood, D., & Chambers, K. (2000). Rapid flow cytometric test for the diagnosis of membrane cytoskeleton-associated haemolytic anaemia. *British Journal of Haematology, 111*(3), 924–933.

Lin, Y. W., Tai, S. H., Chang, C. C., Chao, L. C., Wen, M. J., Huang, Y. H., et al. (2016). Application of flow cytometry for evaluating clinical prognosis and histopathological grade of human glioma. *Neurological Research, 38*(7), 625–633.

Mahmoodpoor, A., Paknezhad, S., Shadvar, K., Hamishehkar, H., Movassaghpour, A. A., Sanaie, S., et al. (2018). Flow cytometry of CD64, HLA-DR, CD25, and TLRs for diagnosis and prognosis of sepsis in critically ill patients admitted to the intensive care unit: A review article. *Anesthesiology and Pain Medicine, 8*(6), e83128.

O'Gorman, M. R., & Corrochano, V. (1995). Rapid whole-blood flow cytometry assay for diagnosis of chronic granulomatous disease. *Clinical and Diagnostic Laboratory Immunology, 2*(2), 227–232.

Orfao, A., Almeida, J., Sanchez, M. L., & San Miguel, J. F. (2004). Immunophenotypic diagnosis of leukemic B-cell chronic lymphoproliferative disorders other than chronic lymphocytic leukemia. In G. B. Faguet (Ed.), *Chronic lymphocytic leukemia* (Contemporary hematology). Humana Press.

Parney, I. F., Waldron, J. S., & Parsa, A. T. (2009). Flow cytometry and in vitro analysis of human glioma–associated macrophages. *Journal of Neurosurgery, 110*(3), 572–582.

Pillai, V., & Dorfman, D. M. (2016). Flow cytometry of nonhematopoietic neoplasms. *Acta Cytologica, 60*(4), 336–343.

Pinto, A., Rega, A., Crother, T. R., & Sorrentino, R. (2012). Plasmacytoid dendritic cells and their therapeutic activity in cancer. *Oncoimmunology, 1*(5), 726–734.

Rahman, K., & Chandra, D. (2018). Historical investigations and advances in flow cytometry-based tests in paroxysmal nocturnal. In R. Saxena & H. P. Pati (Eds.), *Hematopathology* (pp. 91–106). Elsevier.

Rahman, K., Gupta, R., Hussein, N., Yadav, G., Singh, M. K., Kumar, P., et al. (2017a). Fluorescent Aerolysin (FLAER) based paroxysmal nocturnal hemoglobinuria (PNH) screening- a single centre experience from India. *International Journal of Laboratory Hematology, 39*(3), 261–271.

Rahman, K., Kumar, P., Singh, M. K., Gupta, R., & Nityanand, S. (2017b). Role of CD200 in differential diagnosis of mature B cell neoplasm. *International Journal of Laboratory Hematology, 39*, 384–391.

Rahman, K., Gupta, R., Singh, M. K., Sarkar, M. K., Gupta, A., & Nityanand, S. (2018a). A triple negative profile (CD34/HLADR/CD11b negative), rapidly and specifically identifies acute promyelocytic leukemia. *International Journal of Laboratory Hematology, 40*(2), 144–115.

Rahman, K., Gupta, R., Harankhedkar, S., Gupta, T., Sarkar, M. K., & Nityanand, S. (2018b). Utility of CD157 as a common leukocytes marker for paroxysmal nocturnal hemoglobinuria screening in a single tube five color combination. *Indian Journal of Hematology and Blood Transfusion, 34*(2), 304–309.

Rahman, K., Singh, M. K., Chandra, D., Gupta, R., Sarkar, M. K., Gupta, P., et al. (2022). CD26 expression on circulating CD34+/CD38- progenitor population is a specific and reliable tool for the rapid flow cytometric diagnosis of chronic myeloid leukemia-A single-center validation study. *International Journal of Laboratory Hematology, 44*, 524–530.

Shah, S. S., Diakite, S. A. S., Traore, K., Diakite, M., Kwiatowski, D. P., Rockett, K. A., et al. (2012). A novel cytofluorometric assay for the detection and quantification of glucose-6-phosphate dehydrogenase deficiency. *Scientific Reports, 2*, 299.

Sutherland, D. R., Acton, E., Keeney, M., Davis, B. H., & Illingworth, A. (2014). Use of CD157 in FLAER based assay for high sensitivity PNH granulocyte and PNH monocyte detection. *Cytometry Part B, 86B*, 44–55.

van der Burg, M., Kalina, T., Perez-Andres, M., Vlkova, M., Lopez-Granados, E., Blanco, E., Bonroy, C., Sousa, A. E., Kienzler, A. K., Wentink, M., Mejstríková, E., Šinkorova, V., Stuchly, J., van Zelm, M. C., Orfao, A., & van Dongen, J. J. M. (2019). The EuroFlow PID orientation tube for flow cytometric diagnostic screening of primary immunodeficiencies of the lymphoid system. *Frontiers in Immunology, 10*, 246.

Won, & Suh. (2009). Flow cytometric detection of erythrocyte osmotic fragility. *Cytometry Part B, 76*(B), 135–141.

Cellular Uptake of Nanoparticles: A Flow Cytometry Insights

Simran Nasra and Ashutosh Kumar

Abstract

In the rapidly evolving field of cellular and molecular toxicology, understanding the cellular uptake of nanoparticles is crucial for assessing their potential toxicity and therapeutic applications. Nanoparticles, with their unique physicochemical properties, offer promising avenues for drug delivery, imaging, and targeted therapy. However, their interactions with biological systems can be complex and multifaceted. Flow cytometry has emerged as a powerful tool for unravelling the intricacies of cellular nanoparticle uptake. This chapter explores the principles and applications of flow cytometry in studying the cellular uptake of nanoparticles, shedding light on its significance in cellular and molecular toxicology.

Keywords

Nanoparticles · Drug delivery · Toxicology · Molecular toxicology

1 Introduction

The interplay between nanotechnology and cellular biology holds the promise of revolutionizing a multitude of fields, from medicine and biotechnology to environmental science and industrial applications (Jha et al., 2014). At the epicenter of this convergence lies the captivating and intricate process of cellular uptake, where nanoparticles, characterized by their minute size and unique properties, enter the living cells (Panariti et al., 2012). Understanding this phenomenon is a critical

S. Nasra · A. Kumar (✉)
Biological and Life Sciences, School of Arts & Sciences, Ahmedabad University, Central Campus, Ahmedabad, Gujarat, India
e-mail: ashutosh.kumar@ahduni.edu.in

AB Pant et al. (eds.), *Flow Cytometry: Applications in Cellular and Molecular Toxicology*, https://doi.org/10.1007/978-981-97-9758-5_11

necessity, for it is fundamental to unlocking the full potential of nanoparticles in a plethora of applications. Nanoparticles, typically ranging in size from one to 100 nanometers, exhibit unique physicochemical properties that set them apart from their bulk counterparts (Tay et al., 2014). Their size, surface area, and surface chemistry make them suitable for multifaceted applications. In medicine, they can serve as vehicles for precise drug delivery, targeting specific tissues or cells, while in industrial processes, they offer opportunities for catalysis, sensing, and materials development (Alle & Adnan, 2023). Moreover, the environmental impact of nanoparticles necessitates comprehensive studies, as their release into ecosystems has raised concerns about their interactions with organisms and potential bioaccumulation (Martínez et al., 2020). The imperative to elucidate the mechanisms of cellular uptake arises from the confluence of several factors. Firstly, understanding how nanoparticles are internalized by cells is integral to the development of targeted drug delivery systems (Chavanpatil et al., 2006). The ability to transport therapeutic agents directly to the site of action within the body, while sparing healthy tissues, hinges on our capacity to control and predict cellular uptake. In the realm of precision medicine, where tailored therapies aim to maximize efficacy and minimize side effects, the study of cellular uptake assumes pivotal importance (Chehelgerdi et al., 2023). In this context, targeted drug delivery systems represent a transformative approach. By encapsulating therapeutic agents within nanoparticles and equipping them with targeting moieties, such as ligands or antibodies, researchers can engineer drug delivery vehicles that specifically recognize and engage with diseased cells (Singh & Lillard Jr, 2009). These targeted nanoparticles offer the potential to revolutionize medicine, as they hold the promise of enhanced treatment efficacy and reduced side effects (Desai, 2012).

Secondly, the safety assessment of nanoparticles is a crucial concern, particularly in the field of biomedicine. The biocompatibility of nanoparticles—determined by how they interact with biological systems—requires meticulous investigation (Wolfram et al., 2015). This encompasses a scrutiny of potential toxicological effects and the understanding of how nanoparticles behave within living organisms, encompassing aspects such as biodistribution, clearance, and long-term consequences (Gupta & Xie, 2018). Unraveling these intricacies is a multifaceted challenge, underscoring the need for advanced analytical methodologies (Bryda & Costa, 2023). Biocompatibility assessments extend to understanding the potential toxicity of nanoparticles. When nanoparticles interact with biological systems, their physicochemical properties can influence their biocompatibility (Li et al., 2012). Size, shape, surface charge, and surface chemistry are factors that can impact the biocompatibility of nanoparticles. Moreover, the choice of materials for nanoparticle synthesis can influence their biological impact (Bilardo et al., 2022). Engineered nanoparticles, such as those used in drug delivery, diagnostics, and imaging, are intentionally designed to possess specific properties (Fig. 1c) (Patel et al., 2019). These properties may include controlled release of therapeutic agents, high loading capacity, and the ability to target specific cells or tissues (Kamaly et al., 2012). However, such modifications can influence how

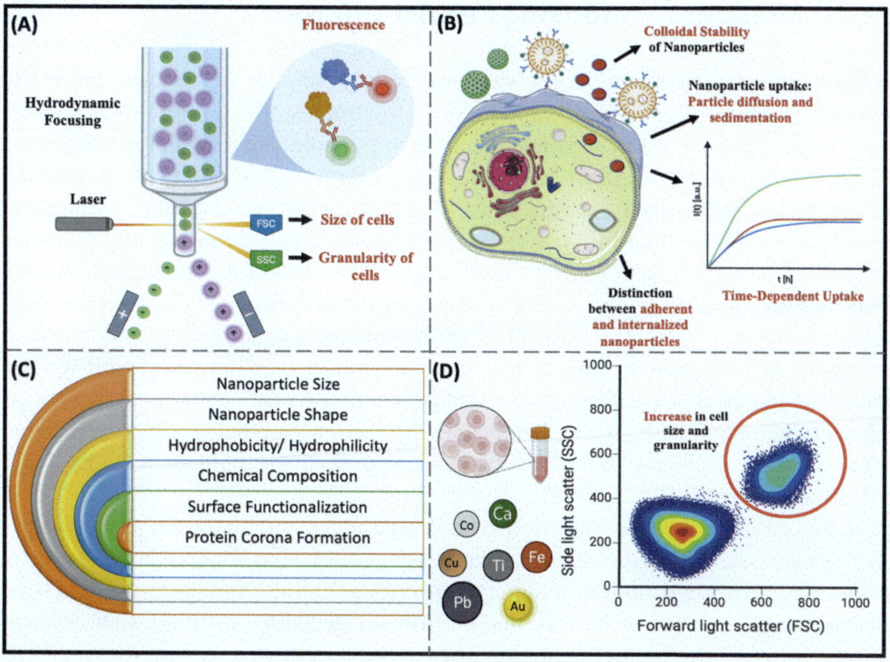

Fig. 1 (**a**) Principle of flow cytometry to analyse the nanoparticles (NPs) uptake quantifying increased size, granularity and fluorescence in cells, (**b**) Different analysis for cellular uptake of NPs, (**c**) Effect of different physical and chemical parameters of NPs on cellular uptake and (**d**) Increase in cellular size and granularity after the uptake of NPs Parts of the figure were drawn by using pictures from Servier Medical Art and Biorender. Servier Medical Art by Servier is licensed under a Creative Commons Attribution 3.0 Unported License (https://creativecommons.org/licenses/by/3.0/)

nanoparticles interact with biological systems. Understanding the intricate balance between therapeutic efficacy and potential side effects is central to the field of nanomedicine (Yetisgin et al., 2020). Lastly, in the realm of environmental science, it is imperative to comprehend how nanoparticles interact with terrestrial and aquatic organisms when they enter ecosystems (Bundschuh et al., 2019). As nanoparticles may traverse trophic levels, they have the potential to impact ecosystem health and pose a potential risk to human exposure through the food chain (Rajput et al., 2020). A thorough grasp of the cellular uptake mechanisms and biotransformation processes in environmental matrices is a fundamental prerequisite for comprehensive environmental risk assessments (Fenner et al., 2021). Environmental risk assessments require a comprehensive understanding of how nanoparticles behave in environmental matrices, their potential for bioaccumulation, and their effects on aquatic and terrestrial organisms. This understanding is essential for ensuring the sustainability of nanotechnology and preventing unintended environmental consequences (Ray et al., 2009).

2 Methods for Studying Cellular Uptake

Reliable and precise methods for studying cellular uptake are crucial, given the pivotal role that this process plays in medical treatments, the development of novel drug delivery systems, and our understanding of potential environmental risks associated with nanoparticles (Sim & Wong, 2021). Moreover, the safety assessment of nanoparticles, particularly in the field of biomedicine, necessitates meticulous investigation. Biocompatibility assessments extend to understanding the potential toxicity of nanoparticles (Kansara & Kumar, 2020). When nanoparticles interact with biological systems, their physicochemical properties can influence their biocompatibility (Egbuna et al., 2021). The size, shape, surface charge, and surface chemistry of nanoparticles are factors that can impact their biocompatibility. The choice of materials for nanoparticle synthesis can also influence their biological impact (Chaudhari et al., 2021). Understanding the intricate balance between therapeutic efficacy and potential side effects is central to the field of nanomedicine (Oktaviani, 2021).

In this context, the importance of reliable and precise methods for studying cellular uptake becomes evident. Researchers require tools that allow them to quantify, analyze, and gain a profound understanding of the cellular uptake process. These methods empower researchers to make informed decisions in the design of drug delivery systems, to assess the potential risks associated with nanoparticles, and to ensure the safety of patients receiving nanoparticle-based treatments (Johnson et al., 2021). One of the most versatile and powerful techniques for studying cellular uptake is flow cytometry. Flow cytometry is a dynamic and multidimensional analytical technique that has applications in various scientific fields. Its core function involves the analysis of particles suspended in a fluid stream (McKinnon, 2018). As these particles pass through a laser beam, they scatter and emit fluorescent signals, which are then detected and analyzed (Kumar et al., 2017). Flow cytometry offers several advantages when it comes to the study of cellular uptake. Firstly, flow cytometry permits quantitative analysis of cellular uptake events (Drescher et al., 2021). Researchers can precisely measure the percentage of cells that have internalized nanoparticles, the quantity of nanoparticles within individual cells, and the dynamics of uptake (Fig. 1b). This quantitative insight is pivotal in understanding the efficiency and kinetics of cellular uptake processes, particularly when dealing with heterogeneous cell populations (Richards et al., 2023). Secondly, flow cytometry supports high throughput analysis, enabling the rapid investigation of a substantial number of individual cells. This inherent advantage is of particular significance when confronting the inherent heterogeneity that characterizes cellular uptake events. By analyzing numerous individual cells, researchers can attain statistical significance in their findings, a feat often challenging in studies of cellular uptake (Black et al., 2011).

Furthermore, flow cytometry is a technique amenable to single-cell analysis. This feature is invaluable when considering the substantial variation that can exist between different cells within a population. Understanding the variability in cellular uptake is pivotal, as it can lead to novel insights into the cellular and molecular factors governing this phenomenon (Brehm-Stecher & Johnson, 2004). The integration of flow cytometry into the realm of cellular uptake studies has brought about a paradigm shift. Its multi-dimensional capabilities empower researchers to unravel the complexities of how nanoparticles interact with living cells (Wu et al., 2019). In this chapter, we aim to explore the principles, methodologies, and applications of flow cytometry in the context of cellular uptake, providing researchers and practitioners with the knowledge and tools to navigate the intricacies of this captivating interplay between nanotechnology and the living cell.

3 Principles of Flow Cytometry

The fundamental principle of flow cytometry revolves around hydrodynamics and optics. In practice, a sample containing cells or particles is introduced into a fluid stream, ensuring a single-file passage through the laser beam (Zhang et al., 2016). The interaction between the laser and the particles in the sample results in the scattering of light and the emission of fluorescence (as shown in Fig. 1a). To facilitate this process, flow cytometry instruments comprise several key components (McKinnon, 2018). These include a light source, typically a laser, that emits a specific wavelength of light. The laser beam is precisely focused and directed into the core of the fluid stream. Particles or cells within the sample intersect the laser beam one at a time. Scattered light, originating from the interaction between the laser and the particle, is collected by detectors at various angles, providing information on the size and granularity of the particles (Picot et al., 2012). Furthermore, if the particles are fluorescently labeled, they emit fluorescence in response to the laser excitation. This fluorescence is also captured by detectors, and its intensity is quantified. Flow cytometry instruments are equipped with a series of filters and optics to segregate the emitted fluorescence into distinct channels corresponding to specific fluorochromes (Goddard et al., 2006). By measuring the intensity of fluorescence in these channels, researchers can determine the presence and quantity of labeled molecules within or on the cells or particles. At its core, flow cytometry involves the analysis of particles suspended in a fluid stream. As particles pass through a laser beam, they scatter and emit fluorescent signals, which are then detected and analyzed (Barteneva et al., 2012). This enables the quantification of various characteristics, including size, shape, granularity, and fluorescence. For the study of cellular uptake, flow cytometry offers distinct advantages:

I. **Quantitative Analysis**: Flow cytometry allows for the quantitative analysis of cellular uptake events. Researchers can measure the percentage of cells that have internalized nanoparticles, the amount of nanoparticles per cell, and the kinetics of uptake (Khetan et al., 2019).
II. **High Throughput**: Flow cytometry can rapidly analyze a large number of individual cells, providing statistical significance to the results. This high throughput is particularly beneficial when studying the heterogeneous nature of cellular uptake (Holzner et al., 2021).

III. **Single-Cell Analysis**: Flow cytometry can provide insights at the single-cell level. This is valuable for understanding variability in uptake among different cells in a population (Thurber et al., 2013).
IV. **Fluorescent Labeling**: By labeling nanoparticles with fluorescent markers, flow cytometry can distinguish between cells that have taken up nanoparticles and those that have not. This labeling allows for precise tracking of uptake events (Zhukova et al., 2021).
V. **Dynamic Information**: Flow cytometry can capture dynamic information about uptake kinetics. Researchers can monitor how uptake changes over time, revealing valuable insights into the process (Arnoldini et al., 2013).

4 The Multifaceted Nature of Cellular Uptake: Unravelling the Complexities

Cellular uptake is far from a one-size-fits-all phenomenon, and its intricacies stem from a multitude of factors that interplay to govern this essential process. To grasp the complete landscape of cellular uptake, it is imperative to consider the intricate variables that influence the interaction between nanoparticles and their target cells (Chehelgerdi et al., 2023). These variables span a wide spectrum and encompass (as shown in Fig. 1c):

I. **Nanoparticle Size:** The size of nanoparticles emerges as a pivotal determinant in their cellular uptake dynamics. Smaller nanoparticles often display a higher propensity for cellular internalization due to their enhanced surface area and reduced diffusion barriers. Conversely, larger nanoparticles may undergo distinct mechanisms of cellular entry, necessitating careful examination to elucidate the interplay between size and uptake efficiency (Hoshyar et al., 2016).
II. **Surface Properties:** The surface properties of nanoparticles, encompassing factors such as charge and surface chemistry, wield a profound influence over their interactions with cells. Positively charged nanoparticles may exhibit distinct cellular uptake profiles compared to their negatively charged counterparts, underlining the intricate role of surface characteristics in the process (Blanco et al., 2015).
III. **Shape:** The geometry of nanoparticles, whether spherical, rod-like, or characterized by other forms, introduces another layer of complexity to cellular uptake. The nanoparticle shape is known to impact the mechanisms by which they gain access to the cellular interior, compelling researchers to explore the ramifications of shape on uptake pathways (Abbasi et al., 2023).
IV. **Surface Coatings:** Nanoparticles are frequently coated with specific molecules to bolster stability and functionality. These surface coatings, designed to serve various purposes, can wield a significant influence over cellular uptake. They may either facilitate or impede interactions with cell mem-

branes, necessitating an in-depth evaluation of the role of coatings in nanoparticle-cell engagement (Fang et al., 2018).

V. **Target Cell Type:** Distinct types of cells showcase disparities in their inclination to internalize nanoparticles. Cancer cells, for instance, may exhibit contrasting uptake patterns compared to healthy cells, hinting at the intricacies of targeted therapies (Nasra et al., 2021). Understanding these variations is pivotal for tailoring nanoparticle applications to specific cell types and biological contexts (Sanna et al., 2014).

VI. **Environmental Factors:** The milieu in which cellular uptake transpires significantly impacts the process. Factors such as pH, temperature, and the presence of biomolecules, including proteins, orchestrate intricate interplays that influence the interactions between nanoparticles and cells. These environmental conditions must be meticulously considered to decode the nuances of cellular uptake (Akhter et al., 2021).

VII. **Dose and Exposure Duration:** The concentration of nanoparticles and the duration of exposure represent fundamental variables that shape cellular uptake. Higher nanoparticle concentrations or extended exposure periods may culminate in augmented internalization by cells, underscoring the importance of dose-response relationships in nanoparticle studies (Graham et al., 2017).

VIII. **Cellular State:** The physiological state of the target cells is not to be overlooked. Actively dividing cells may exhibit distinct uptake patterns compared to quiescent cells, underscoring the influence of the cellular state on nanoparticle interactions (Marescal & Cheeseman, 2020) (Fig. 2).

5 Studying Nanoparticle Uptake

Flow cytometry excels in the quantitative analysis of nanoparticle uptake, allowing researchers to assess both the percentage of cells that have internalized nanoparticles and the amount of nanoparticles within individual cells. This is achieved through the labeling of nanoparticles with fluorescent markers, which enables their discrimination from non-internalized particles (Gottstein et al., 2013). The workflow for studying nanoparticle uptake using flow cytometry typically involves the following steps:

I. **Preparation of Samples:** The first step in studying nanoparticle uptake is the preparation of cell samples. These can be cultured cells or tissues depending on the research objectives. To initiate the study, cells or tissues are exposed to fluorescently labeled nanoparticles. These nanoparticles can be engineered to carry payloads, such as drugs (Kansara et al., 2018) or imaging agents, making them suitable for therapeutic or diagnostic purposes. Control samples should also be prepared to serve as references. These controls typically include cells or tissues without exposure to nanoparticles, allowing researchers to distinguish between cells with internalized nanoparticles and those without (Wang & Wang, 2014).

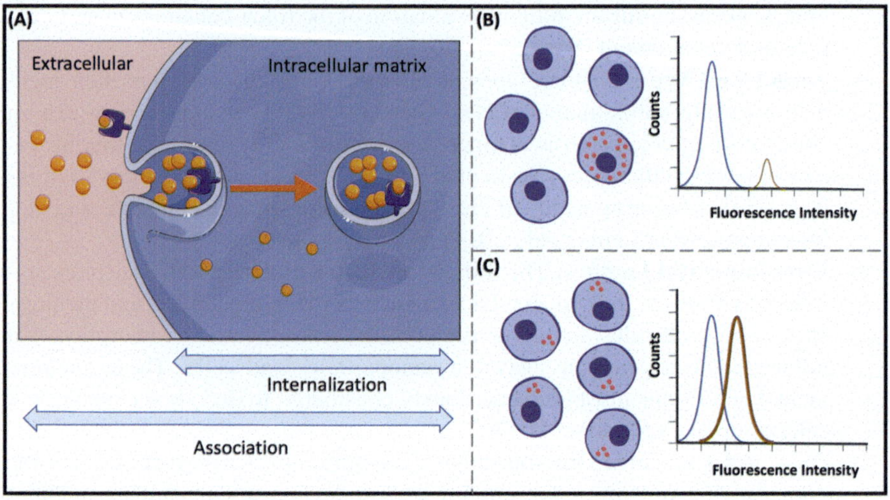

Fig. 2 The term MFI (Mean Fluorescence Index) serves as an indicator of the overall fluorescence emitted by a cell, encompassing both the material within the cell and that which remains adhered to the cell's surface. Consequently, it measures the degree of association rather than the extent of internalization. To accurately gauge internalization, it becomes necessary to employ a method capable of distinguishing between these two locations. Another relevant metric used in the context of material-cell interactions within flow cytometry is the "percent positive." This metric quantifies the percentage of cells exhibiting any increase in fluorescence compared to the background. Typically, a threshold is established at a specific fluorescence intensity value, below which 99% of cells in the untreated sample exhibit lower signals. (**a**) The collective material consists of the entire quantity found both within the cell and still affixed to the cell's surface. The term "internalization," on the other hand, exclusively pertains to material located within the cell's intracellular space. (**b**) A fluorescence histogram typically seen in flow cytometry, demonstrating strong binding in a limited number of cells and (**c**) weak binding in a substantial number of cells Parts of the figure were drawn by using pictures from Servier Medical Art and Biorender Servier Medical Art by Servier is licensed under a Creative Commons Attribution 3.0 Unported License (https://creativecommons.org/licenses/by/3.0/)

II. **Incubation and Uptake:** Following exposure to nanoparticles, cells or tissues are incubated for a predetermined period. This incubation time can vary depending on the specific research question, as it may be designed to investigate the kinetics of uptake. During this incubation period, cells interact with the nanoparticles, and if the particles are designed to enter cells, they may be internalized through processes like endocytosis or phagocytosis, depending on the cell type and nanoparticle properties (Monteiro-Riviere et al., 2013).

III. **Harvesting and Sample Preparation:** After the incubation period, the cells or tissues are harvested for analysis. This typically involves detaching adherent cells from culture dishes or collecting suspended cells or tissues. Sample preparation is a critical step. Researchers must ensure that the harvested cells or tissues are in optimal condition for flow cytometry analysis. This may involve washing the samples to remove excess nanoparticles and any remaining extracellular particles (Bi et al., 2013).

IV. **Flow Cytometry Analysis:** The prepared samples are introduced into the flow cytometer. The flow cytometer is equipped with a laser or lasers that emit specific wavelengths of light. As cells or particles pass through the focused laser beam, they scatter light and emit fluorescence. The scattering of light provides information on the size and granularity of the particles, while fluorescence signals are used to detect the presence of fluorescently labeled nanoparticles within or on the cells (McKinnon, 2018).

V. **Data Collection and Analysis:** Data collected during flow cytometry analysis includes information on the size, granularity, and fluorescence intensity of each individual cell or particle. For the specific purpose of studying nanoparticle uptake, the fluorescence signal from the labeled nanoparticles is of primary interest. The flow cytometer quantifies the fluorescence intensity, allowing researchers to assess the presence and quantity of nanoparticles within each cell. Data analysis software is then employed to interpret the collected data. Researchers can determine the percentage of cells that have internalized nanoparticles and assess the amount of nanoparticles within individual cells. This quantitative analysis provides valuable insights into the efficiency of cellular uptake, the distribution of nanoparticles among cells, and the kinetics of the uptake process (Gamucci et al., 2014).

6 Nanoparticle Uptake in Cells Analysed by Flow Cytometer (Fig. 3)

6.1 Kinetics of Uptake

The kinetics of nanoparticle uptake are essential to comprehend the pace at which cells internalize nanoparticles, shedding light on the temporal dynamics of this process. To investigate this, researchers embark on time-course experiments, meticulously measuring uptake at various time intervals throughout a designated period, such as 15 min, 30 min, 1 h, and so forth. The objective is to discern how swiftly cells embrace the nanoparticles and whether uptake reaches a point of saturation over time. Through this approach, researchers can create a kinetic curve that graphically portrays the uptake dynamics. This curve provides a visual representation of the rate at which cells internalize nanoparticles (Oh & Park, 2014). In essence, it offers a glimpse into whether uptake is swift and efficient, gradual and cumulative, or if it eventually levels off, indicating that cells have reached their capacity for internalization. The kinetic analysis is valuable for understanding the temporal aspects of cellular uptake, which is pivotal for various applications, such as drug delivery systems, where the time of payload release into cells may be critical for therapeutic efficacy. Moreover, this temporal insight can also be informative in environmental studies, where it aids in assessing the persistence and accumulation of nanoparticles within living systems over time (Klaine et al., 2008).

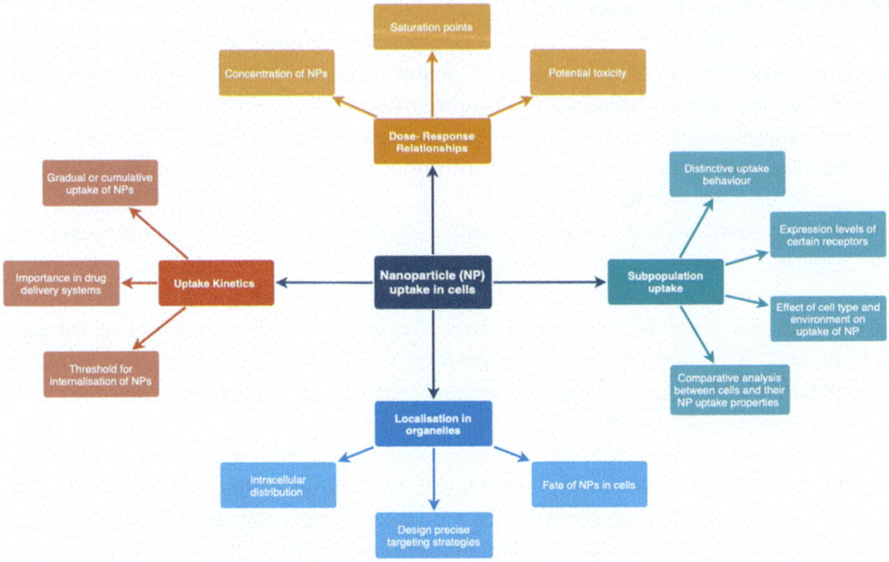

Fig. 3 Attributes associated with analysing NP uptake in the cells

6.2 Percentage of Uptake

Quantifying the percentage of cells that have internalized nanoparticles represents a fundamental aspect of studying cellular uptake. This metric offers insights into the efficiency of uptake within a cell population, allowing researchers to gauge the extent to which nanoparticles have successfully entered the cells. To achieve this, flow cytometry is employed to analyze the data, distinguishing cells with internalized nanoparticles, often indicated by fluorescence, from the total cell population (Yameen et al., 2014). This analysis yields valuable information regarding the proportion of cells that actively participate in the uptake process. A high percentage of uptake implies that a significant portion of the cell population has effectively internalized nanoparticles, indicative of an efficient uptake mechanism. Conversely, a lower percentage suggests variability in the cell population's ability to uptake nanoparticles, possibly due to variations in factors like cell health, receptor expression, or the inherent properties of individual cells (Ou et al., 2022).

6.3 Amount of Uptake

In parallel, assessing the amount of nanoparticles within individual cells delves deeper into the distribution of nanoparticle content across the cell population. This analytical parameter provides a quantitative measure of the quantity of nanoparticles that have been internalized by each cell. Flow cytometry excels in this regard by measuring the fluorescence intensity of individual cells, which correlates with the amount of internalized nanoparticles. This metric is pivotal in elucidating variations in the uptake process among different cells within the same population (Rashkow et al., 2014). Some cells may accumulate a substantial quantity of nanoparticles, while others may have minimal internalization. Such variability underscores the heterogeneity in cellular responses and is crucial for understanding the impact of nanoparticle properties, cellular health, and receptor availability on the efficiency of uptake. By combining these metrics, researchers gain comprehensive insights into the efficiency and diversity of cellular uptake within a given cell population, vital for a broad spectrum of applications spanning from drug delivery to understanding the impacts of nanoparticles in biological and environmental systems (Barua & Mitragotri, 2014).

6.4 Uptake Variability

Understanding the heterogeneity in cellular responses is a key element of nanoparticle uptake studies. Investigating the variability in uptake among different cells within the same population is instrumental in recognizing the diverse ways in which cells interact with nanoparticles (Kumar et al., 2011c). This variability may stem from factors like cell health, variations in receptor expression, or inherent disparities among individual cells. Recognizing this diversity in uptake patterns is essential for comprehending the nuanced behavior of cells when faced with nanoparticles, providing crucial insights into the complex interplay of biological and physicochemical factors in uptake dynamics (Rees et al., 2019).

6.5 Uptake in Subpopulations

The examination of subpopulations within a cell sample aims to unveil if specific subsets of cells exhibit distinctive uptake behaviors. By doing so, researchers can discern whether certain cell types or experimental conditions lead to variations in nanoparticle internalization. This exploration is particularly useful for pinpointing specific cell types that may have a heightened propensity for uptake, potentially due to differences in receptor expression or cellular characteristics (Chinen et al., 2015). Such insights help in tailoring nanoparticle-based approaches for targeted applications, where knowing which cells are more receptive to uptake is invaluable (Vallabani et al., 2014).

6.6 Effect of Nanoparticle Properties

Exploring the influence of nanoparticle properties is crucial for tailoring nanoparticle design to specific applications. Key properties such as size, surface charge, and surface chemistry can significantly impact cellular uptake. By investigating these parameters, researchers gain insights into how to engineer nanoparticles optimized for particular tasks. For instance, nanoparticles designed for drug delivery (Chaudhari et al., 2022) may need distinct properties compared to those intended for imaging or environmental remediation. (Kansara et al., 2020) Understanding the role of nanoparticle properties in uptake mechanisms is essential for crafting tailored and efficient nanoparticle-based solutions (Ridolfo et al., 2020).

6.7 Co-Localization with Cellular Structures

Assessing whether nanoparticles localize to specific cellular organelles or structures provides a deeper understanding of their intracellular distribution. By determining where within the cell nanoparticles are primarily situated, researchers can gain insights into potential targeting strategies or investigate how nanoparticles may affect specific cellular processes. This co-localization analysis is vital for applications such as drug delivery, where precise delivery to a specific cellular compartment can enhance therapeutic efficacy (Mitchell et al., 2021).

6.8 Uptake Dynamics Under Varying Conditions

Investigating how uptake changes under different experimental conditions, such as altering nanoparticle concentrations or incubation times, is instrumental in understanding the adaptability of cellular uptake mechanisms. Researchers aim to decipher how changing the experimental parameters influences the efficiency and kinetics of uptake. This insight is invaluable for optimizing conditions in drug delivery or environmental assessments, where understanding the dynamics of uptake under diverse scenarios is paramount (Salatin & Yari Khosroushahi, 2017).

6.9 Comparative Analysis

Comparing the uptake of different types of nanoparticles or formulations is an essential step in discerning which is more effective for a given application. Such analyses provide insights into the relative performance of various nanoparticles and help in selecting the most suitable candidates for specific tasks. Whether it is choosing the optimal contrast agent for imaging or an efficient drug carrier, comparative analysis guides researchers in making informed decisions for their intended applications (De Jong & Borm, 2008).

6.10 Effect of Inhibitors

Delineating the impact of specific inhibitors on uptake processes is a strategy employed to unveil the intricate mechanisms involved in nanoparticle internalization. By using inhibitors that target specific cellular pathways (e.g., endocytosis), researchers can dissect the molecular machinery at play during uptake. This knowledge is indispensable for refining nanoparticle design for precise, targeted delivery, as it illuminates the underlying cellular processes that govern nanoparticle internalization (Kuhn et al., 2014).

6.11 Cell Viability and Apoptosis

Assessing cell viability and apoptosis in the context of nanoparticle uptake is a crucial facet of research, as it directly pertains to the potential cytotoxicity of nanoparticles (Kumar et al., 2015). Evaluating the well-being of cells exposed to nanoparticles is paramount for understanding the safety and biocompatibility of these nanomaterials. Researchers perform assays to gauge cell viability, ensuring that the presence of nanoparticles does not compromise cell health. Additionally, the assessment of apoptosis, a form of programmed cell death, is vital in revealing any adverse effects on cell populations. Understanding the interplay between nanoparticle uptake and cell viability and apoptosis helps in comprehending the potential cytotoxic risks associated with nanoparticle applications, whether in drug delivery, diagnostics, or environmental studies (Mundekkad & Cho, 2022).

6.12 Dose-Response Relationships

The investigation of dose-response relationships focuses on how varying the concentration of nanoparticles influences uptake behavior. By systematically altering nanoparticle concentrations in experiments, researchers can discern how uptake changes in response to varying dosages. This analysis provides valuable insights into the saturation points, optimal dosages, and potential toxicity thresholds concerning nanoparticle uptake (Senapati et al., 2017). It helps establish a comprehensive understanding of the dose-dependent relationship between nanoparticle exposure and cellular internalization, aiding in the design of safe and effective nanoparticle-based interventions in diverse fields (Walkey et al., 2012).

6.13 Long-Term Effects

Exploring the long-term fate of internalized nanoparticles is essential to assess their persistence, release, or degradation within cells over extended periods. Researchers aim to decipher how nanoparticles evolve within the intracellular environment, addressing questions of retention, potential release, or alterations in nanoparticle properties. This long-term perspective is pivotal for understanding the enduring effects of nanoparticles, whether in the context of drug delivery, sustained therapeutic release, or environmental interactions (Kumar et al., 2011a), where long-term impact assessments are paramount (Tenchov et al., 2021).

6.14 Influence of Environmental Factors

Analyzing the influence of environmental factors, such as variations in pH or temperature, on uptake processes is vital for comprehending the adaptability of nanoparticle uptake mechanisms. Researchers systematically alter environmental conditions to assess their impact on the efficiency and kinetics of nanoparticle internalization. This exploration provides insights into the resilience of nanoparticle uptake mechanisms under changing environmental conditions, which is essential for applications in diverse settings, ranging from in vitro cell studies to in vivo environmental assessments (Gustafson et al., 2015).

6.15 Comparative Studies Across Cell Lines

Comparative studies across different cell lines offer valuable insights into how cell type influences nanoparticle uptake behavior. By contrasting uptake patterns in diverse cell types, researchers can discern variations in receptor expression, membrane properties, and intracellular processes. This knowledge is instrumental in understanding the specificity of nanoparticle-cell interactions, enabling the selection of the most suitable cell types for specific applications. Comparative studies across cell lines provide a nuanced understanding of how cellular characteristics influence nanoparticle uptake, which is essential for tailored and effective applications in areas such as drug delivery, diagnostics, and environmental science (Kattel et al., 2023).

7 Deciphering Flow Cytometry Data

7.1 Gate Setting

Setting appropriate gates in flow cytometry data analysis is a critical initial step. Gates are essentially regions within the data plot that define specific populations of cells based on their characteristics. In the context of cellular uptake studies, one commonly sets a gate to isolate cells that have internalized nanoparticles. This gate is typically determined using a control group of cells that haven't been exposed to nanoparticles. By selecting this gate, researchers can focus on the data relevant to uptake, effectively excluding non-internalized cells from analysis.

7.2 Fluorescence Intensity

The fluorescence intensity recorded by flow cytometry represents the level of fluorescence emitted by individual cells, which is directly proportional to the quantity of internalized nanoparticles. High fluorescence intensity indicates that a cell has internalized a substantial amount of nanoparticles, while low fluorescence suggests fewer or no internalized nanoparticles. This parameter is vital for understanding the extent of uptake and variations in uptake among different cells within the population.

7.3 Data Statistics

Flow cytometry data can be further analyzed statistically to extract meaningful information. Researchers often calculate the mean fluorescence intensity, which provides an average measure of uptake within the cell population. The standard deviation and coefficient of variation indicate the degree of variability among individual cells in terms of nanoparticle internalization. These statistics offer a quantitative understanding of uptake behavior, shedding light on the distribution and heterogeneity of uptake within the cell population.

7.4 Comparative Analysis

Comparative analysis involves contrasting flow cytometry data between different experimental conditions, such as varying nanoparticle properties, concentrations, or cell types. By comparing data sets, researchers can identify differences in uptake patterns and attribute them to specific factors under investigation. For instance, if one nanoparticle type exhibits higher fluorescence intensity in cells compared to another, it suggests that the former is more efficiently internalized. Comparative analysis aids in selecting the most suitable nanoparticles for specific applications based on their uptake behavior (Nasra et al., 2023).

7.5 Time-Course Analysis

In studies focusing on the kinetics of uptake, flow cytometry data is collected at multiple time points after exposing cells to nanoparticles. Analyzing this data provides insights into the temporal dynamics of uptake. Rapid increases in fluorescence intensity over time indicate swift internalization, whereas a plateau in fluorescence suggests that cells have reached their uptake saturation point. Time-course analysis is essential for understanding how quickly cells internalize nanoparticles, which is crucial for applications where the timing of payload delivery is critical, such as in drug delivery systems (Cossarizza et al., 2021).

8 Examples of Data and Their Implications

8.1 High Fluorescence Intensity

Flow cytometry data often reveals cells with high fluorescence intensity. This implies that these cells have internalized a substantial quantity of nanoparticles. The implications of high fluorescence intensity in flow cytometry data are multi-fold:

I. **Efficient Uptake Mechanism:** High fluorescence intensity suggests that the uptake process has been efficient. The cells have successfully internalized a significant number of nanoparticles. This is particularly advantageous in applications like drug delivery systems, where a high uptake rate ensures the effective delivery of therapeutic payloads. It's also valuable in cellular imaging studies, as increased internalization leads to enhanced cellular labeling for visualization.
II. **Therapeutic Significance:** In the context of drug delivery, high uptake is of paramount importance. It indicates that a large number of cells can potentially receive and respond to therapeutic agents carried by the nanoparticles. This can lead to improved treatment outcomes, especially in cases where targeted drug delivery to specific cells or tissues is essential (Sola et al., 2021).

8.2 Variability in Uptake

Flow cytometry data often exhibits variations in fluorescence intensity among individual cells within the same cell population. The implications of this variability are significant:

I. **Heterogeneity in Cellular Responses:** Variability in uptake indicates that different cells within the population are responding differently to the nanoparticles. This heterogeneity can result from various factors, including differences in cell health, variations in receptor expression, or inherent disparities in the properties of individual cells.
II. **Research Direction:** Understanding this heterogeneity is crucial as it guides researchers in investigating the underlying causes and mechanisms. It may lead to further inquiries into how nanoparticle properties, cell conditions, or signaling pathways influence the uptake process. Researchers can then tailor their experiments or interventions to account for this diversity in cellular responses (Fedr et al., 2023).

8.3 Time-Dependent Uptake

In time-course analyses of flow cytometry data, the fluorescence intensity of cells is measured at various time points after nanoparticle exposure. The implications of time-dependent uptake are as follows:

I. **Kinetics of Uptake:** Time-dependent data provides insights into the kinetics of uptake, revealing how quickly cells internalize nanoparticles. Rapid increases in fluorescence intensity over time indicate a fast and efficient internalization process.
II. **Optimizing Timing:** The knowledge of uptake kinetics is valuable in applications where the timing of nanoparticle delivery is critical. For instance, in drug delivery, understanding the rate of uptake helps in determining when the therapeutic payload is released inside the cell. This can be essential for achieving the desired therapeutic effect.
III. **Evaluating Saturation:** Time-dependent data may also indicate if cellular uptake reaches a plateau, suggesting that cells have reached their capacity for nanoparticle internalization. This is a key observation in understanding the saturation point of uptake mechanisms (Carnovale et al., 2019).

8.4 Comparative Analysis

Comparative analysis involves contrasting flow cytometry data between different nanoparticle types, formulations, or experimental conditions. The implications of this analysis are noteworthy:

I. **Selecting Optimal Nanoparticles:** Comparative analysis helps researchers select the most efficient nanoparticles for specific applications. If one type of nanoparticle exhibits significantly higher fluorescence intensity compared to others, it indicates that it is more suitable for targeted drug delivery, imaging, or any other intended purpose.
II. **Tailoring Applications:** By comparing different nanoparticle options, researchers can make informed decisions about which nanoparticles best suit their specific applications. This is crucial for customizing and optimizing nanoparticle-based interventions (George et al., 2021).

8.5 Dose-Response Relationships

Varying the concentration of nanoparticles in dose-response studies yields data that demonstrates how nanoparticle concentration influences cellular uptake. The implications of this analysis are as follows:

I. **Optimal Dosages:** Dose-response data helps establish the optimal concentration of nanoparticles for a given application. It ensures that the right concentration is used to achieve the desired cellular uptake without causing undue toxicity (Kumar et al., 2011b).
II. **Safety Considerations:** Assessing the dose-response relationship also provides insights into potential toxicity thresholds (Sharma et al., 2012). Researchers can

identify the point at which increasing nanoparticle concentration leads to adverse effects, guiding safe practices in nanoparticle applications (Bell et al., 2014).

Interpreting flow cytometry data in cellular uptake studies is essential for uncovering the mechanisms governing nanoparticle-cell interactions and for making informed decisions about nanoparticle-based applications in diverse fields, including medicine, biology, and environmental science.

9 Challenges and Future Directions in Studying Cellular Uptake Using Flow Cytometry

Studying cellular uptake through flow cytometry is a powerful approach, but it's not without its challenges. Addressing these limitations is essential for advancing this field and ensuring the reliability and applicability of the data generated. One key challenge is the potential for false positives or negatives in uptake measurements (Kansara et al., 2015). Flow cytometry relies on fluorescent labels, and sometimes, non-specific binding can lead to overestimation of uptake or mask the actual internalization. Careful selection of controls and validation methods is necessary to mitigate these issues. Another challenge lies in the need for precise quantification of uptake. Flow cytometry provides relative measures of fluorescence intensity, but translating this into exact numbers of internalized nanoparticles can be tricky. Developing robust calibration standards and reference materials is crucial for accurate quantification. Moreover, the complex interplay of factors influencing cellular uptake, such as nanoparticle properties, cell types, and environmental conditions, requires more advanced data analysis and modeling. Future research in this area should focus on integrating multi-dimensional data to establish comprehensive models of uptake mechanisms.

Future directions in cellular uptake studies using flow cytometry hold promising possibilities. Advancements in nanoparticle engineering can lead to the development of more specific and efficient carriers. This includes designing nanoparticles with tunable surface properties and tailored functionalities that enhance cellular interactions. The integration of advanced microscopy techniques with flow cytometry can provide high-resolution spatial information, offering a deeper understanding of intracellular nanoparticle localization. Furthermore, the incorporation of single-cell analysis and microfluidic systems can aid in dissecting the heterogeneity in cellular responses, shedding light on the behavior of individual cells within a population. The exploration of 3D cell cultures and organ-on-a-chip models can bridge the gap between traditional in vitro and in vivo studies, offering more physiologically relevant insights.

Overall, the future of cellular uptake studies using flow cytometry holds great potential for addressing these challenges and opening new horizons in understanding nanoparticle-cell interactions.

10 Case Studies of Successful Cellular Uptake Studies Using Flow Cytometry

To illustrate the practical applications of cellular uptake studies through flow cytometry, several case studies provide concrete examples of success in this field.

I. **Drug Delivery Nanoparticles:** In a case study, researchers employed flow cytometry to assess the cellular uptake of drug-loaded nanoparticles in cancer cells. By quantifying fluorescence intensity, they demonstrated the efficient internalization of nanoparticles, leading to targeted drug delivery and improved therapeutic outcomes (Yao et al., 2020).
II. **Cell Labeling for Tracking:** In another case, flow cytometry was utilized to investigate the uptake of labeled nanoparticles in stem cells. The study revealed high uptake efficiency, enabling the efficient tracking of these cells in vivo. This technology has potential applications in regenerative medicine and cell-based therapies (Accomasso et al., 2016).
III. **Environmental Nanoparticle Uptake:** Researchers also employed flow cytometry to analyze the uptake of nanoparticles by aquatic organisms. This case study demonstrated the effects of nanoparticle properties, environmental conditions, and exposure duration on cellular uptake (Kansara et al., 2023). Such insights are vital for assessing the environmental impact of nanoparticles (Jochums et al., 2017).

These case studies exemplify the versatility of flow cytometry in cellular uptake research. They showcase its applicability in diverse fields, ranging from medicine to environmental science (Kansara et al., 2021), and emphasize the critical role it plays in advancing our understanding of nanoparticle-cell interactions. These studies pave the way for future research endeavors, highlighting the potential for innovation and discovery in this dynamic field.

11 Conclusion: Unravelling Nanoparticle Uptake Through Flow Cytometry

In conclusion, the chapter has provided an extensive exploration of the intricate process of cellular uptake, with a primary focus on its examination through the lens of flow cytometry. We have delved into the significance of this process in various fields, from medicine to environmental science, emphasizing its pivotal role in the delivery of therapeutics, tracking of cells, and assessment of environmental impacts. As we summarize the key takeaways from this chapter, it becomes evident that understanding cellular uptake is crucial for the development of advanced nanoparticle-based applications. Flow cytometry stands as a formidable tool in this endeavor. It enables researchers to dissect the mechanisms behind cellular uptake, offering precise insights into uptake efficiency, kinetics, and heterogeneity among different cells within a population. The meticulous analysis of flow cytometry data, along with statistical measures and comparative analyses, aids in quantifying and characterizing nanoparticle internalization.

One of the central takeaways is the multifaceted nature of cellular uptake, where various factors such as nanoparticle properties, cell types, and environmental conditions interplay to influence the process. We've seen that cellular uptake is not a one-size-fits-all phenomenon; it varies widely based on these variables. This underscores the complexity of nanoparticle-cell interactions and the need for advanced analytical techniques to unravel the underlying mechanisms.

Furthermore, the chapter has elucidated that the field of cellular uptake through flow cytometry is not without challenges. False positives and negatives, precise quantification, and the integration of multi-dimensional data all present hurdles to overcome. Yet, these challenges are met with promising future directions, such as advanced microscopy techniques, single-cell analysis, and 3D cell culture models. These innovations hold the potential to refine our understanding of uptake mechanisms and provide more physiologically relevant insights.

In the context of case studies, we've seen the practical applications of cellular uptake studies using flow cytometry. These real-world examples have highlighted the versatility and applicability of flow cytometry, underscoring its pivotal role in advancing diverse fields of science and medicine. The significance of flow cytometry in understanding cellular uptake cannot be overstated. It serves as a bridge between theoretical knowledge and practical applications, offering a clear pathway to harnessing the potential of nanoparticles for targeted drug delivery, cell tracking, and environmental assessments (Nasra et al., 2022). The ability to customize and optimize nanoparticle-based interventions is intrinsically linked to our ability to decipher the complex world of cellular uptake. In this regard, flow cytometry is more than a scientific tool; it is an instrument of progress that empowers researchers to unlock the mysteries of cellular uptake, paving the way for innovative and targeted solutions to some of the most pressing challenges in science and medicine. The insights gained from studying cellular uptake through flow cytometry are, and will continue to be, indispensable in the design and application of advanced nanoparticle-based systems.

References

Abbasi, R., Shineh, G., Mobaraki, M., Doughty, S., & Tayebi, L. (2023). Structural parameters of nanoparticles affecting their toxicity for biomedical applications: A review. *Journal of Nanoparticle Research, 25*(3), 43.

Accomasso, L., Gallina, C., Turinetto, V., & Giachino, C. (2016). Stem cell tracking with nanoparticles for regenerative medicine purposes: An overview. *Stem Cells International*. https://doi.org/10.1155/2016/7920358

Akhter, M. H., Khalilullah, H., Gupta, M., Alfaleh, M. A., Alhakamy, N. A., Riadi, Y., & Md, S. (2021). Impact of protein corona on the biological identity of nanomedicine: Understanding the fate of nanomaterials in the biological milieu. *Biomedicine, 9*(10), 1496.

Alle, M., & Adnan, M. (2023). Role of nanotechnology in cancer therapies: Recent advances, current issues, and approaches. In *Advances in smart nanomaterials and their applications* (pp. 73–95). Elsevier.

Arnoldini, M., Heck, T., Blanco-Fernández, A., & Hammes, F. (2013). Monitoring of dynamic microbiological processes using real-time flow cytometry. *PLoS One, 8*(11), e80117.

Barteneva, N. S., Fasler-Kan, E., & Vorobjev, I. A. (2012). Imaging flow cytometry: Coping with heterogeneity in biological systems. *Journal of Histochemistry and Cytochemistry, 60*(10), 723–733.

Barua, S., & Mitragotri, S. (2014). Challenges associated with penetration of nanoparticles across cell and tissue barriers: A review of current status and future prospects. *Nano Today, 9*(2), 223–243.

Bell, I. R., Ives, J. A., & Wayne, B. J. (2014). Nonlinear effects of nanoparticles: Biological variability from hormetic doses, small particle sizes, and dynamic adaptive interactions. *Dose-Response, 12*(2), 202–232.

Bi, H., Krausz, K. W., Manna, S. K., Li, F., Johnson, C. H., & Gonzalez, F. J. (2013). Optimization of harvesting, extraction, and analytical protocols for UPLC-ESI-MS-based metabolomic analysis of adherent mammalian cancer cells. *Analytical and Bioanalytical Chemistry, 405*, 5279–5289.

Bilardo, R., Traldi, F., Vdovchenko, A., & Resmini, M. (2022). Influence of surface chemistry and morphology of nanoparticles on protein corona formation. *Wiley Interdisciplinary Reviews: Nanomedicine and Nanobiotechnology, 14*(4), e1788.

Black, C. B., Duensing, T. D., Trinkle, L. S., & Dunlay, R. T. (2011). Cell-based screening using high-throughput flow cytometry. *Assay and Drug Development Technologies, 9*(1), 13–20.

Blanco, E., Shen, H., & Ferrari, M. (2015). Principles of nanoparticle design for overcoming biological barriers to drug delivery. *Nature Biotechnology, 33*(9), 941–951.

Brehm-Stecher, B. F., & Johnson, E. A. (2004). Single-cell microbiology: Tools, technologies, and applications. *Microbiology and Molecular Biology Reviews, 68*(3), 538–559.

Bryda, G., & Costa, A. P. (2023). Qualitative research in digital era: Innovations, methodologies and collaborations. *Social Sciences, 12*(10), 570.

Bundschuh, M., Englert, D., Rosenfeldt, R. R., Bundschuh, R., Feckler, A., Lüderwald, S., Seitz, F., Zubrod, J. P., & Schulz, R. (2019). Nanoparticles transported from aquatic to terrestrial ecosystems via emerging aquatic insects compromise subsidy quality. *Scientific Reports, 9*(1), 15676.

Carnovale, C., Bryant, G., Shukla, R., & Bansal, V. (2019). Identifying trends in gold nanoparticle toxicity and uptake: Size, shape, capping ligand, and biological corona. *ACS Omega, 4*(1), 242–256.

Chaudhari, R., Patel, P., Meghani, N., Nasra, S., & Kumar, A. (2021). Fabrication of methotrexate-loaded gold nanoconjugates and its enhanced anticancer activity in breast cancer. *3 Biotech, 11*, 1–13.

Chaudhari, R., Nasra, S., Meghani, N., & Kumar, A. (2022). MiR-206 conjugated gold nanoparticle based targeted therapy in breast cancer cells. *Scientific Reports, 12*(1), 4713.

Chavanpatil, M. D., Khdair, A., & Panyam, J. (2006). Nanoparticles for cellular drug delivery: Mechanisms and factors influencing delivery. *Journal of Nanoscience and Nanotechnology, 6*(9–10), 2651–2663.

Chehelgerdi, M., Chehelgerdi, M., Allela, O. Q. B., Pecho, R. D. C., Jayasankar, N., Rao, D. P., Thamaraikani, T., Vasanthan, M., Viktor, P., & Lakshmaiya, N. (2023). Progressing nanotechnology to improve targeted cancer treatment: Overcoming hurdles in its clinical implementation. *Molecular Cancer, 22*(1), 169.

Chinen, A. B., Guan, C. M., Ferrer, J. R., Barnaby, S. N., Merkel, T. J., & Mirkin, C. A. (2015). Nanoparticle probes for the detection of cancer biomarkers, cells, and tissues by fluorescence. *Chemical Reviews, 115*(19), 10530–10574.

Cossarizza, A., Chang, H. D., Radbruch, A., Abrignani, S., Addo, R., Akdis, M., Andrä, I., Andreata, F., Annunziato, F., & Arranz, E. (2021). Guidelines for the use of flow cytometry and cell sorting in immunological studies. *European Journal of Immunology, 51*(12), 2708–3145.

De Jong, W. H., & Borm, P. J. (2008). Drug delivery and nanoparticles: Applications and hazards. *International Journal of Nanomedicine, 3*(2), 133–149.

Desai, N. (2012). Challenges in development of nanoparticle-based therapeutics. *The AAPS Journal, 14*(2), 282–295.

Drescher, H., Weiskirchen, S., & Weiskirchen, R. (2021). Flow cytometry: A blessing and a curse. *Biomedicine, 9*(11), 1613.

Egbuna, C., Parmar, V. K., Jeevanandam, J., Ezzat, S. M., Patrick-Iwuanyanwu, K. C., Adetunji, C. O., Khan, J., Onyeike, E. N., Uche, C. Z., & Akram, M. (2021). Toxicity of nanoparticles in biomedical application: Nanotoxicology. *Journal of Toxicology, 2021*, 1–21.

Fang, R. H., Kroll, A. V., Gao, W., & Zhang, L. (2018). Cell membrane coating nanotechnology. *Advanced Materials, 30*(23), 1706759.

Fedr, R., Kahounová, Z., Remšík, J., Reiterová, M., Kalina, T., & Souček, K. (2023). Variability of fluorescence intensity distribution measured by flow cytometry is influenced by cell size and cell cycle progression. *Scientific Reports, 13*(1), 4889.

Fenner, K., Elsner, M., Lueders, T., McLachlan, M. S., Wackett, L. P., Zimmermann, M., & Drewes, J. R. E. (2021). Methodological advances to study contaminant biotransformation: New prospects for understanding and reducing environmental persistence? *ACS ES&T Water, 1*(7), 1541–1554.

Gamucci, O., Bertero, A., Malvindi, M. A., Sabella, S., Pompa, P. P., Mazzolai, B., & Bardi, G. (2014). Detection of fluorescent nanoparticle interactions with primary immune cell subpopulations by flow cytometry. *Journal of Visualized Experiments, 85*, e51345.

George, S. K., Lauková, L., Weiss, R., Semak, V., Fendl, B., Weiss, V. U., Steinberger, S., Allmaier, G., Tripisciano, C., & Weber, V. (2021). Comparative analysis of platelet-derived extracellular vesicles using flow cytometry and nanoparticle tracking analysis. *International Journal of Molecular Sciences, 22*(8), 3839.

Goddard, G., Martin, J. C., Naivar, M., Goodwin, P. M., Graves, S. W., Habbersett, R., Nolan, J. P., & Jett, J. H. (2006). Single particle high resolution spectral analysis flow cytometry. *Cytometry Part A: The Journal of the International Society for Analytical Cytology, 69*(8), 842–851.

Gottstein, C., Wu, G., Wong, B. J., & Zasadzinski, J. A. (2013). Precise quantification of nanoparticle internalization. *ACS Nano, 7*(6), 4933–4945.

Graham, U. M., Jacobs, G., Yokel, R. A., Davis, B. H., Dozier, A. K., Birch, M. E., Tseng, M. T., Oberdörster, G., Elder, A., & DeLouise, L. (2017). From dose to response: In vivo nanoparticle processing and potential toxicity. In *Modelling the toxicity of nanoparticles* (pp. 71–100). Springer.

Gupta, R., & Xie, H. (2018). Nanoparticles in daily life: Applications, toxicity and regulations. *Journal of Environmental Pathology, Toxicology and Oncology, 37*(3), 209–230.

Gustafson, H. H., Holt-Casper, D., Grainger, D. W., & Ghandehari, H. (2015). Nanoparticle uptake: The phagocyte problem. *Nano Today, 10*(4), 487–510.

Holzner, G., Mateescu, B., van Leeuwen, D., Cereghetti, G., Dechant, R., Stavrakis, S., & deMello, A. (2021). High-throughput multiparametric imaging flow cytometry: Toward diffraction-limited sub-cellular detection and monitoring of sub-cellular processes. *Cell Reports, 34*(10), 108824.

Hoshyar, N., Gray, S., Han, H., & Bao, G. (2016). The effect of nanoparticle size on in vivo pharmacokinetics and cellular interaction. *Nanomedicine, 11*(6), 673–692.

Jha, R. K., Jha, P. K., Chaudhury, K., Rana, S. V., & Guha, S. K. (2014). An emerging interface between life science and nanotechnology: Present status and prospects of reproductive healthcare aided by nano-biotechnology. *Nano Reviews, 5*(1), 22762.

Jochums, A., Friehs, E., Sambale, F., Lavrentieva, A., Bahnemann, D., & Scheper, T. (2017). Revelation of different nanoparticle-uptake behavior in two standard cell lines NIH/3T3 and A549 by flow cytometry and time-lapse imaging. *Toxics, 5*(3), 15.

Johnson, K. B., Wei, W. Q., Weeraratne, D., Frisse, M. E., Misulis, K., Rhee, K., Zhao, J., & Snowdon, J. L. (2021). Precision medicine, AI, and the future of personalized health care. *Clinical and Translational Science, 14*(1), 86–93.

Kamaly, N., Xiao, Z., Valencia, P. M., Radovic-Moreno, A. F., & Farokhzad, O. C. (2012). Targeted polymeric therapeutic nanoparticles: Design, development and clinical translation. *Chemical Society Reviews, 41*(7), 2971–3010.

Kansara, K., & Kumar, A. (2020). In vitro methods to assess the cellular toxicity of nanoparticles. In *Nanotoxicity* (pp. 21–40). Elsevier.

Kansara, K., Patel, P., Shah, D., Shukla, R. K., Singh, S., Kumar, A., & Dhawan, A. (2015). TiO2 nanoparticles induce DNA double strand breaks and cell cycle arrest in human alveolar cells. *Environmental and Molecular Mutagenesis, 56*(2), 204–217.

Kansara, K., Patel, P., Shukla, R. K., Pandya, A., Shanker, R., Kumar, A., & Dhawan, A. (2018). Synthesis of biocompatible iron oxide nanoparticles as a drug delivery vehicle. *International Journal of Nanomedicine, 13*(sup1), 79–82.

Kansara, K., Kumar, A., & Karakoti, A. S. (2020). Combination of humic acid and clay reduce the ecotoxic effect of TiO2 NPs: A combined physico-chemical and genetic study using zebrafish embryo. *Science of the Total Environment, 698*, 134133.

Kansara, K., Sathish, C., Vinu, A., Kumar, A., & Karakoti, A. S. (2021). Assessment of the impact of abiotic factors on the stability of engineered nanomaterials in fish embryo media. *Emergent Materials, 4*(5), 1339–1350.

Kansara, K., Singh, R., Yadav, P., Mansuri, A., Kumar, A., & Bhatia, D. (2023). Lipid modification of DNA tetrahedrons enhances cellular uptake, migration, and in vivo uptake. *ACS Applied Nano Materials, 6*(14), 13443–13452.

Kattel, P., Sulthana, S., Trousil, J. I., Shrestha, D., Pearson, D., & Aryal, S. (2023). Effect of nanoparticle weight on the cellular uptake and drug delivery potential of PLGA nanoparticles. *ACS Omega, 8*(30), 27146–27155.

Khetan, J., Shahinuzzaman, M., Barua, S., & Barua, D. (2019). Quantitative analysis of the correlation between cell size and cellular uptake of particles. *Biophysical Journal, 116*(2), 347–359.

Klaine, S. J., Alvarez, P. J., Batley, G. E., Fernandes, T. F., Handy, R. D., Lyon, D. Y., Mahendra, S., McLaughlin, M. J., & Lead, J. R. (2008). Nanomaterials in the environment: Behavior, fate, bioavailability, and effects. *Environmental Toxicology and Chemistry: An International Journal, 27*(9), 1825–1851.

Kuhn, D. A., Vanhecke, D., Michen, B., Blank, F., Gehr, P., Petri-Fink, A., & Rothen-Rutishauser, B. (2014). Different endocytotic uptake mechanisms for nanoparticles in epithelial cells and macrophages. *Beilstein Journal of Nanotechnology, 5*(1), 1625–1636.

Kumar, A., Dhawan, A., & Shanker, R. (2011a). The need for novel approaches in ecotoxicity of engineered nanomaterials. *Journal of Biomedical Nanotechnology, 7*(1), 79–80.

Kumar, A., Pandey, A. K., Singh, S. S., Shanker, R., & Dhawan, A. (2011b). Engineered ZnO and TiO2 nanoparticles induce oxidative stress and DNA damage leading to reduced viability of Escherichia coli. *Free Radical Biology and Medicine, 51*(10), 1872–1881.

Kumar, A., Pandey, A. K., Singh, S. S., Shanker, R., & Dhawan, A. (2011c). A flow cytometric method to assess nanoparticle uptake in bacteria. *Cytometry Part A, 79*(9), 707–712.

Kumar, A., Najafzadeh, M., Jacob, B. K., Dhawan, A., & Anderson, D. (2015). Zinc oxide nanoparticles affect the expression of p53, Ras p21 and JNKs: An ex vivo/in vitro exposure study in respiratory disease patients. *Mutagenesis, 30*(2), 237–245.

Kumar, A., Singh, S., Shanker, R., & Dhawan, A. (2017). Nanotoxicology: Challenges for biologists. In *Issues in toxicology*. https://doi.org/10.1039/9781782623922-00001

Li, X., Wang, L., Fan, Y., Feng, Q., & Cui, F.-Z. (2012). Biocompatibility and toxicity of nanoparticles and nanotubes. *Journal of Nanomaterials, 2012*, 6.

Marescal, O., & Cheeseman, I. M. (2020). Cellular mechanisms and regulation of quiescence. *Developmental Cell, 55*(3), 259–271.

Martínez, G., Merinero, M., Pérez-Aranda, M., Pérez-Soriano, E. M., Ortiz, T., Villamor, E., Begines, B., & Alcudia, A. (2020). Environmental impact of nanoparticles' application as an emerging technology: A review. *Materials, 14*(1), 166.

McKinnon, K. M. (2018). Flow cytometry: An overview. *Current Protocols in Immunology, 120*(1), 5.1.1–5.1.11.

Mitchell, M. J., Billingsley, M. M., Haley, R. M., Wechsler, M. E., Peppas, N. A., & Langer, R. (2021). Engineering precision nanoparticles for drug delivery. *Nature Reviews Drug Discovery, 20*(2), 101–124.

Monteiro-Riviere, N. A., Samberg, M. E., Oldenburg, S. J., & Riviere, J. E. (2013). Protein binding modulates the cellular uptake of silver nanoparticles into human cells: Implications for in vitro to in vivo extrapolations? *Toxicology Letters, 220*(3), 286–293.

Mundekkad, D., & Cho, W. C. (2022). Nanoparticles in clinical translation for cancer therapy. *International Journal of Molecular Sciences, 23*(3), 1685.

Nasra, S., Chaudhari, R., & Kumar, A. (2021). Role of nanomedicine for cancer immunotherapy. In *Nanomedicine for cancer diagnosis and therapy* (pp. 115–132). Springer.

Nasra, S., Bhatia, D., & Kumar, A. (2022). Recent advances in nanoparticle-based drug delivery systems for rheumatoid arthritis treatment. *Nanoscale Advances, 4*(17), 3479–3494.

Nasra, S., Shah, T., Bhatt, M., Chaudhari, R., Bhatia, D., & Kumar, A. (2023). Reprogramming M1-to-M2 phenotype to alleviate inflammation: Using liposomal curcumin as a tool to redefine macrophage functionality. *ACS Applied Bio Materials, 6*(7), 2886–2897.

Oh, N., & Park, J.-H. (2014). Endocytosis and exocytosis of nanoparticles in mammalian cells. *International Journal of Nanomedicine, 9*(sup1), 51–63.

Oktaviani, O. (2021). Nanoparticles: Properties, applications and toxicities. *Jurnal Latihan, 1*(2), 11–20.

Ou, Y.-H., Liang, J., Chng, W. H., Muthuramalingam, R. P. K., Ng, Z. X., Lee, C. K., Neupane, Y. R., Yau, J. N. N., Zhang, S., & Lou, C. K. L. (2022). Investigations on cellular uptake mechanisms and immunogenicity profile of novel bio-hybrid nanovesicles. *Pharmaceutics, 14*(8), 1738.

Panariti, A., Miserocchi, G., & Rivolta, I. (2012). The effect of nanoparticle uptake on cellular behavior: Disrupting or enabling functions? *Nanotechnology, Science and Applications, 5*, 87–100.

Patel, P., Meghani, N., Kansara, K., & Kumar, A. (2019). Nanotherapeutics for the treatment of cancer and arthritis. *Current Drug Metabolism, 20*(6), 430–445.

Picot, J., Guerin, C. L., Le Van Kim, C., & Boulanger, C. M. (2012). Flow cytometry: Retrospective, fundamentals and recent instrumentation. *Cytotechnology, 64*, 109–130.

Rajput, V., Minkina, T., Mazarji, M., Shende, S., Sushkova, S., Mandzhieva, S., Burachevskaya, M., Chaplygin, V., Singh, A., & Jatav, H. (2020). Accumulation of nanoparticles in the soil-plant systems and their effects on human health. *Annals of Agricultural Sciences, 65*(2), 137–143.

Rashkow, J. T., Patel, S. C., Tappero, R., & Sitharaman, B. (2014). Quantification of single-cell nanoparticle concentrations and the distribution of these concentrations in cell population. *Journal of the Royal Society Interface, 11*(94), 20131152.

Ray, P. C., Yu, H., & Fu, P. P. (2009). Toxicity and environmental risks of nanomaterials: Challenges and future needs. *Journal of Environmental Science and Health Part C, 27*(1), 1–35.

Rees, P., Wills, J. W., Brown, M. R., Barnes, C. M., & Summers, H. D. (2019). The origin of heterogeneous nanoparticle uptake by cells. *Nature Communications, 10*(1), 2341.

Richards, C. J., Burgers, T. C., Vlijm, R., Roos, W. H., & Åberg, C. (2023). Rapid internalization of nanoparticles by human cells at the single particle level. *ACS Nano, 17*(17), 16517–16529.

Ridolfo, R., Tavakoli, S., Junnuthula, V., Williams, D. S., Urtti, A., & van Hest, J. C. (2020). Exploring the impact of morphology on the properties of biodegradable nanoparticles and their diffusion in complex biological medium. *Biomacromolecules, 22*(1), 126–133.

Salatin, S., & Yari Khosroushahi, A. (2017). Overviews on the cellular uptake mechanism of polysaccharide colloidal nanoparticles. *Journal of Cellular and Molecular Medicine, 21*(9), 1668–1686.

Sanna, V., Pala, N., & Sechi, M. (2014). Targeted therapy using nanotechnology: Focus on cancer. *International Journal of Nanomedicine, 9*, 467–483.

Senapati, V. A., Kansara, K., Shanker, R., Dhawan, A., & Kumar, A. (2017). Monitoring characteristics and genotoxic effects of engineered nanoparticle–protein corona. *Mutagenesis, 32*(5), 479–490.

Sharma, V., Kumar, A., & Dhawan, A. (2012). Nanomaterials: Exposure, effects and toxicity assessment. *Proceedings of the National Academy of Sciences, India Section B: Biological Sciences, 82*, 3–11.

Sim, S., & Wong, N. K. (2021). Nanotechnology and its use in imaging and drug delivery. *Biomedical Reports, 14*(5), 1–9.

Singh, R., & Lillard, J. W., Jr. (2009). Nanoparticle-based targeted drug delivery. *Experimental and Molecular Pathology, 86*(3), 215–223.

Sola, F., Canonico, B., Montanari, M., Volpe, A., Barattini, C., Pellegrino, C., Cesarini, E., Guescini, M., Battistelli, M., & Ortolani, C. (2021). Uptake and intracellular trafficking studies of multiple dye-doped core-shell silica nanoparticles in lymphoid and myeloid cells. *Nanotechnology, Science and Applications, 14*, 29–48.

Tay, C. Y., Setyawati, M. I., Xie, J., Parak, W. J., & Leong, D. T. (2014). Back to basics: Exploiting the innate physico-chemical characteristics of nanomaterials for biomedical applications. *Advanced Functional Materials, 24*(38), 5936–5955.

Tenchov, R., Bird, R., Curtze, A. E., & Zhou, Q. (2021). Lipid nanoparticles— from liposomes to mRNA vaccine delivery, a landscape of research diversity and advancement. *ACS Nano, 15*(11), 16982–17015.

Thurber, G. M., Yang, K. S., Reiner, T., Kohler, R. H., Sorger, P., Mitchison, T., & Weissleder, R. (2013). Single-cell and subcellular pharmacokinetic imaging allows insight into drug action in vivo. *Nature Communications, 4*(1), 1504.

Vallabani, N. S., Shukla, R. K., Konka, D., Kumar, A., Singh, S., & Dhawan, A. (2014). TiO2 nanoparticles induced micronucleus formation in human liver (HepG2) cells: Comparison of conventional and flow cytometry based methods. *Molecular Cytogenetics, 7*(Suppl 1), P79.

Walkey, C. D., Olsen, J. B., Guo, H., Emili, A., & Chan, W. C. (2012). Nanoparticle size and surface chemistry determine serum protein adsorption and macrophage uptake. *Journal of the American Chemical Society, 134*(4), 2139–2147.

Wang, E. C., & Wang, A. Z. (2014). Nanoparticles and their applications in cell and molecular biology. *Integrative Biology, 6*(1), 9–26.

Wolfram, J., Zhu, M., Yang, Y., Shen, J., Gentile, E., Paolino, D., Fresta, M., Nie, G., Chen, C., & Shen, H. (2015). Safety of nanoparticles in medicine. *Current Drug Targets, 16*(14), 1671–1681.

Wu, Y., Ali, M. R., Dansby, K., & El-Sayed, M. A. (2019). Improving the flow cytometry-based detection of the cellular uptake of gold nanoparticles. *Analytical Chemistry, 91*(22), 14261–14267.

Yameen, B., Choi, W. I., Vilos, C., Swami, A., Shi, J., & Farokhzad, O. C. (2014). Insight into nanoparticle cellular uptake and intracellular targeting. *Journal of Controlled Release, 190*, 485–499.

Yao, Y., Zhou, Y., Liu, L., Xu, Y., Chen, Q., Wang, Y., Wu, S., Deng, Y., Zhang, J., & Shao, A. (2020). Nanoparticle-based drug delivery in cancer therapy and its role in overcoming drug resistance. *Frontiers in Molecular Biosciences, 7*, 193.

Yetisgin, A. A., Cetinel, S., Zuvin, M., Kosar, A., & Kutlu, O. (2020). Therapeutic nanoparticles and their targeted delivery applications. *Molecules, 25*(9), 2193.

Zhang, Y., Watts, B. R., Guo, T., Zhang, Z., Xu, C., & Fang, Q. (2016). Optofluidic device based microflow cytometers for particle/cell detection: A review. *Micromachines, 7*(4), 70.

Zhukova, V., Osipova, N., Semyonkin, A., Malinovskaya, J., Melnikov, P., Valikhov, M., Porozov, Y., Solovev, Y., Kuliaev, P., & Zhang, E. (2021). Fluorescently labeled plga nanoparticles for visualization in vitro and in vivo: The importance of dye properties. *Pharmaceutics, 13*(8), 1145.

Flowcytometric Capturing of Nanoparticles Internalization in the Cell System

Ragavi Ravindran, Srividhya Nandagopal, Muthu Kumaran Panneerselvam, and Kamla Kant Shukla

Abstract

In this era of technological advancement, the role of nanoparticles (NP) is vast, mainly due to their enriched performance and pliable characteristics. In this chapter, we will be learning about the various types of cellular uptake of nanoparticles and their interactions with cell and extracellular matrix. These interactions are crucial for internalizing NPs into cells and extracellular matrix. This chapter reviews the different types and mechanisms of NP uptake and their properties and discusses factors influencing nanoparticles and cellular interactions. We also mentioned various quantification techniques, highlighting the role of flow cytometry in analyzing nanoparticle uptake and localization. The insights provided aim to enhance NP-based applications in various fields like drug delivery, imaging, and diagnostics.

Keywords

Nanoparticles · Flowcytometry · Endocytosis · Fluorescent labelled nanoparticles

1 Introduction

Nanoparticles are defined as a matter of particles with sizes from 1 to 100 nm in diameter. The invention of such nanoparticles smaller than 100 nm is the greatest achievement of nanotechnology in the twenty-first century. It has significance in

R. Ravindran · M. K. Panneerselvam
Department of Biochemistry, Jawaharlal Institute of Postgraduate Medical Education and Research, Puducherry, India

S. Nandagopal · K. K. Shukla (✉)
Department of Biochemistry, All India Institute of Medical Sciences, Jodhpur, Rajasthan, India

AB Pant et al. (eds.), *Flow Cytometry: Applications in Cellular and Molecular Toxicology*, https://doi.org/10.1007/978-981-97-9758-5_12

various fields. They are small polar molecules with increased surface area. The functional groups like metal ions influence the toxic nature of leading to increased oxidative stress (Zahra et al., 2020).

In this chapter, we are dealing with the various cellular uptake of nanoparticles (NP) processes that involve the interaction between nanoparticles with cell membranes and extracellular matrix. These interactions are required to achieve the internalization of NP inside the cell. This process is of great importance in various fields, including medicine, where nanoparticles are used for drug delivery, imaging, and diagnostics. The cellular uptake mechanisms can vary depending on the physicochemical properties of the nanoparticles, like cell chemical composition, and morphological changes, such as cell size, shape, surface charges, hydrophobicity, elasticity, and the type of cells involved. Those interactions can be quantified by using the highly sensitive flow cytometry technique, which has a high detection limit and hence provides insights into the quantification and localization of NP in intact organelles (Augustine et al., 2020).

2 Types of Nanoparticles

- Carbon based Nanoparticles has high electrical conductivity, strength, electron affinity and adaptability (Elliott et al., 2013).
- Metal-based Nanoparticles have distinct electrical properties due to the localized surface plasmon resonance (LSPR) feature. It includes copper, gold and silver nanoparticles (Khan et al., 2019).
- Ceramic nanoparticles are well known for their heat resistance properties and are hence used in coating, catalysts, and batteries (Sigmund et al., 2006).
- Lipid-based Nanoparticles have a solid lipid core and a matrix made up of lipophilic molecules 10–1000 nm in diameter.
- Semiconductive Nanoparticles are similar to metals and non-metals. They are used in making transistors in bioimaging and cancer therapy (Biju et al., 2008).

3 Properties of Nanoparticles

- Nanoparticles are usually between 1 and 100 nm in size which facilitates easy uptake into the cells.
- It has a high surface area to volume ratio
- Surface can be modified to control interactions with the biological system
- Biocompatible Nanoparticles can be engineered to be used for various applications.
- Iron oxide Nanoparticles exhibit magnetic properties which are used for imaging and therapeutic agents.
- Metal organic Nanoparticles have high stability under various biological conditions and, hence, are used for long-term transport and storage.

4 Methods of Nanoparticle Uptake into Cells

Here are some common mechanisms of cellular uptake.

Endocytosis:

(a) **Clathrin-mediated endocytosis is a specific form of endocytosis that involves the formation of clathrin-coated** pits on the cell membrane. These pits engulf specific ligands bound to receptors on the cell surface, leading to the internalization of the ligand-receptor complex. Silica NPs are internalized through this process (Foroozandeh & Aziz, 2018).

(b) **Phagocytosis** involves the engulfment of large particles, typically greater than 0.5 micrometers (Behzadi et al., 2017).

(c) **Pinocytosis** involves the engulfment of smaller particles and fluids that are predominantly polar in nature.

(d) **Macropinocytosis** is a form of endocytosis where the cell membrane forms large vesicles, or macropinosomes, to engulf extracellular fluid and particles. Macropinocytosis is a non-specific process that can be induced by various stimuli (Behzadi et al., 2017).

(e) **Caveolae-mediated endocytosis** involves caveolae which are small invaginations in the cell membrane that play a role in the internalization of small nanoparticles of size 20–100 nm (Foroozandeh & Aziz, 2018).

(f) **Receptor–mediated endocytosis** involves the binding of nanoparticles to specific receptors on the cell surface, followed by internalization. It is a highly specific process and is often exploited in targeted drug delivery systems. Gold NP are internalized through this procedure.

The entry of NP is important in determining the drug efficiency. High therapeutic efficacy is achieved by safe entry of NPs into the cells. Similarly, cytotoxicity is determined by its mode of entry and its intracellular localization.

The formation of protein corona around NPs can influence the size and cellular uptake of NP. The size of the NPs has an impact on cell membrane receptor activation and protein expression. Sometimes, the porous mesh-like extracellular matrix controls the movement of NP across them. Small NPs can cause toxicity as they can easily translocate and enter the cells but large NPs are restricted to enter through the membrane.

Using Transmission Electron Microscopy, the intracellular location of two different-sized NPS is localized. It is shown that 2 nm gold fluorescent NP was seen both inside the nucleus and cytoplasm. Gold NPs of more than 6 nm do not enter the nucleus, as they are seen only in the cytoplasm. NPs reach the inside of the nucleus by 2 distinct mechanisms ("Improving the Flow Cytometry-based Detection of the Cellular Uptake of Gold Nanoparticles | Analytical Chemistry," n.d.).

- Passive diffusion through nuclear pore complex formation that allows NPs of size up to 6–9 nm.

– Active transport through nuclear membrane pore complex that allows NPs of size up to 50 nm.

5 Factors Affecting Nanoparticles and Cellular Interaction

The morphological features like NP size, shape, texture, rigidity, surface charge, and functional groups like OH, COO, NH2, and acetamido groups play a vital role in its cell toxic effects, viability, phagocytosis, and renal excretion.

5.1 Base Materials

The base materials of NP influence the interaction of NP with the cell and its components. The most cell-friendly and bioactive base materials are chitosan and metal NPs. Other base materials include polyvinyl alcohol, polycaprolactone, and polylactic acid, which are components of mammalian cells. NPs made up of Gold and platinum are compatible with mammalian cells whereas NP made up of Cadmium and lead are toxic to mammalian cells (Shin et al., 2020).

Zinc oxide NP has higher solubility in aqueous conditions and hence shows higher toxicity. NPs made up of copper and silver interact with cell membrane constituents like nitrogen, oxygen, and sulfur, and damage the cell membranes. The functional groups surrounding the cell have an influence on cellular interaction with NPs.

The content of gold NPs is measured by 488 nm laser illumination by side scattering mode on flow cytometry. Compared to spherical-shaped gold NPs, gold rods have SPR peaks higher in the visible range. The uptake of gold NPs of different sizes (2–100 nm) was shown to be size-dependent. The size of 25–50 nm particles has high cellular internalization (Wu et al., 2019).

Side scattering signals (SSC) in FCM using a wavelength of incident light close to the SPR peak is the most promising approach for intracellular detection of gold NPs (Ibuki & Toyooka, 2012).

Titanium NPs **have higher scattering cross-sectionally in flow cytometry and enable** more sensitive detection inside the cells up to 5–10 NPs/cell. A shorter wavelength light source is used to increase the scattering and resolution compared to standard light sources emitting longer wavelength (Zucker et al., 2010).

Flowcytometry with confocal microscopy is used to study the endocytic pathway involved in SiO_2 NP uptake.

5.2 Surface Charge of NPs

Since the extracellular matrix are negatively charged due to presence of glycosaminoglycans, NPs with high positive charge can be internalized more easier than negatively charged NPs. Increasing the amino group on the surface of NPs increases the positive charge on the surface, thereby increasing the uptake of NPs into the cells (Vranic et al., 2013).

5.3 Hydrophobity of NPs

On studying the interaction of hydrophobic and hydrophilic NPs with the cell membranes, it is shown that hydrophobic NPs interact with the lipid bilayer and hydrophilic NPs interact with an aqueous solution (Augustine et al., 2020).

5.4 Elasticity of NPs

Similar to hydrophobicity, elasticity plays an important role in the internalization of NPs into the cells. Hard NPs (e.g., gold NPs, quantum dots, magnetic NPs) are NPs with high elasticity. Hydrogels, liposomes, and biodegradable polymers are soft NPs with low elasticity. Soft NPs are internalized more efficiently than hard NPs and hence used in biomedical applications like drug delivery (Foroozandeh & Aziz, 2018).

6 Quantification of Nanoparticles by Various Analytical Techniques Include

Though there are various quantification techniques available for nanoparticles, as listed below (Shin et al., 2020; Behzadi et al., 2017), in this chapter, we will be focusing on the flow cytometry quantification of nanoparticles in detail.

1. Inductively coupled plasma-based spectroscopy methods.
2. Spectroscopic methods using UV-Vis
3. Trans electron microscopy
4. Optical microscopy
5. Dark field microscopy
6. Confocal microscopy
7. Whole cell tomography
8. Scanning electron microscopy
9. Ultraviolet- visible spectroscopy
10. Nanoparticles tracking analysis
11. X-ray diffraction
12. X-ray photoelectron spectroscopy
13. Hyperspectral imaging
14. Flow cytometry

6.1 Flowcytometry Uptake into Cells

Flow cytometry is a powerful technique used to analyze and quantify the characteristics of individual cells or particles as they flow through a fluidic system. It is commonly employed in various fields, including immunology, cell biology, and medicine. In the context of nanoparticles, flow cytometry can be used to study the uptake of nanoparticles by cells (Zahavy et al., 2012; Shin et al., 2020; Pirone et al., 2021).

6.2 Fluorescent Labelled Nanoparticles

Flow cytometry is the commonly used quantification method for fluorescently labeled and non-fluorescent labeled NP uptake into the cell. It is more powerful since it requires simple sample preparation techniques. It is used for the detection of nanoparticles as small as 50–100 nm inside the cell organelles. Flowcytometry has a higher detection limit and a smaller dynamic range compared to ICP. It is used for analyzing a large number of live cells. It provides information not only on the intracellular content of NP but also on its distribution within a cell population and on its cell surface.

6.3 Incubation with Nanoparticles

Nanoparticles enter cells by active endocytosis, forming endosomes, and accumulating in the lysosomes. The cell lines are incubated with same-sized NPs for 30 min. Then nanoparticles are removed, and cell organelles are extracted at various times from 0 h to 5 h. This time variation helps to find out the colocalization of NPs within endosomes or lysosomes. When time increases from 0 h to 5 h, NPs colocalization starts decreasing in the endosomes and subsequent increase in the lysosomes. From this result, the NP's trafficking and its accumulation inside the lysosomes are determined (Garcia Romeu et al., 2021).

6.4 Flowcytometry Analysis

The uptake of NP has to be measured in two ways:

1. Percentage of cells containing NPs
2. Calibrated median fluorescence intensity of the cells.

A study done on human lung carcinoma cell lines in NPs uptake analysis showed morphological changes like losing cell-to-cell contact, rolling up of cells together rather than spreading out, and inhibiting cell proliferation. In NP uptake analysis, culture medium with serum and without serum were used in the analysis. Proteins in the serum attach to NP and form an outer layer, changing the protein size and biological effects, which has a key role in NP uptake and recognition. Flow cytometry is used to detect and generate time and space-resolved intracellular distribution of fluorescently labeled NPs.

Side scattering of nanoparticles depends not only on the cell size but also its material, complexity and cell energy. Lysosomes have higher side scattering than endosomes because endosomes are smaller in size compared to lysosomes. Hence, energy-depleted cells can be detected very rarely or less than 5% compared to normal cells (Garcia Romeu et al., 2021; Geethalakshmi et al., 2022).

6.5 Interpretation of Flowcytometry Analysis

Flowcytometry provides information on how cells have localized nanoparticles and intensity of fluorescence in each cell after NP uptake. The localization of NP can be done only with intact organelles. Leaking or damaged organelles cause poor localization in FCM. Almost 95% of fluorescent objects detected are nanoparticles enclosed in intact organelles. The presence of particles inside the cell changes the refractive index of forward scattering intensity. Side scattering intensity is affected by the internal structure of the cell. NPs located in the cell increase the scattering of light in a 90-degree direction in a dose-dependent manner. This information can be used to assess the efficiency of uptake and to compare different nanoparticle formulations or experimental conditions (Garcia Romeu et al., 2021).

7 Conclusion

Nanoparticles and their interactions with cells have a great influence on healthcare. Such interactions will have both advantages and disadvantages in the output. Hence future studies are encouraged in this field to eliminate the harmful output of their interaction with cell organelles. Similarly, the quantification of nanoparticle content and cell sorting using flowcytometry can encourage us to do further studies on proteomics and gene sequencing.

Graphical Summary

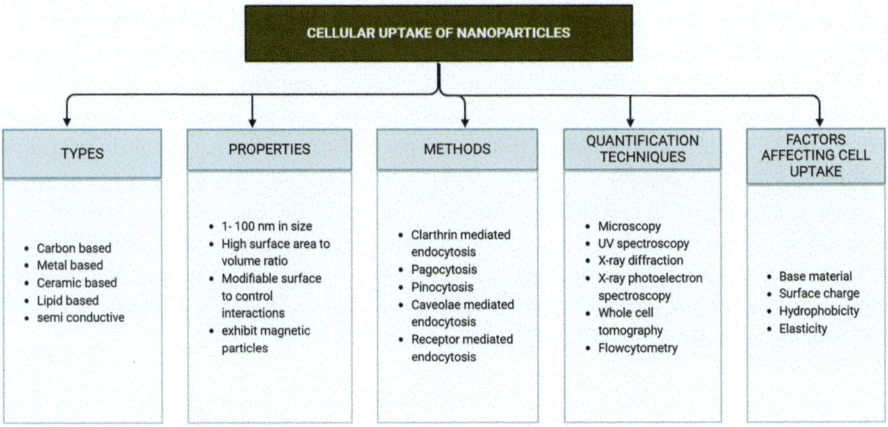

References

Augustine, R., Hasan, A., Primavera, R., Wilson, R. J., Thakor, A. S., & Kevadiya, B. D. (2020). Cellular uptake and retention of nanoparticles: Insights on particle properties and interaction with cellular components. *Materials Today Communications, 25*, 101692. https://doi.org/10.1016/j.mtcomm.2020.101692

Behzadi, S., Serpooshan, V., Tao, W., Hamaly, M. A., Alkawareek, M. Y., Dreaden, E. C., Brown, D., Alkilany, A. M., Farokhzad, O. C., & Mahmoudi, M. (2017). Cellular uptake of nanoparticles: Journey inside the cell. *Chemical Society Reviews, 46*, 4218–4244. https://doi.org/10.1039/c6cs00636a

Biju, V., Itoh, T., Anas, A., Sujith, A., & Ishikawa, M. (2008). Semiconductor quantum dots and metal nanoparticles: Syntheses, optical properties, and biological applications. *Analytical and Bioanalytical Chemistry, 391*, 2469–2495. https://doi.org/10.1007/s00216-008-2185-7

Elliott, J. A., Shibuta, Y., Amara, H., Bichara, C., & Neyts, E. C. (2013). Atomistic modelling of CVD synthesis of carbon nanotubes and graphene. *Nanoscale, 5*, 6662–6676. https://doi.org/10.1039/C3NR01925J

Foroozandeh, P., & Aziz, A. A. (2018). Insight into cellular uptake and intracellular trafficking of nanoparticles. *Nanoscale Research Letters, 13*, 339. https://doi.org/10.1186/s11671-018-2728-6

Garcia Romeu, H., Deville, S., & Salvati, A. (2021). Time- and space-resolved flow-cytometry of cell organelles to quantify nanoparticle uptake and intracellular trafficking by cells. *Small, 17*, e2100887. https://doi.org/10.1002/smll.202100887

Geethalakshmi, R., Nivaz, S., Lekshmi, G., Surendhiran, D., Hussain, C. M., & Sirajunnisa, A. R. (2022). Chapter 6 – Utilization of flow cytometry in nanomaterial/bionanomaterial detection. In C. M. Hussain (Ed.), *Handbook of microbial nanotechnology* (pp. 133–144). Academic. https://doi.org/10.1016/B978-0-12-823426-6.00016-4

Ibuki, Y., & Toyooka, T. (2012). Nanoparticle uptake measured by flow cytometry. *Methods in Molecular Biology (Clifton, N.J.), 926*, 157–166. https://doi.org/10.1007/978-1-62703-002-1_11

Improving the flow cytometry-based detection of the cellular uptake of gold nanoparticles | Analytical chemistry [WWW Document]. (n.d.). URL. https://doi.org/10.1021/acs.analchem.9b02248. (Accessed 12.7.23).

Khan, I., Saeed, K., & Khan, I. (2019). Nanoparticles: Properties, applications and toxicities. *Arabian Journal of Chemistry, 12*, 908–931. https://doi.org/10.1016/j.arabjc.2017.05.011

Pirone, D., Mugnano, M., Memmolo, P., Merola, F., Lama, G. C., Castaldo, R., Miccio, L., Bianco, V., Grilli, S., & Ferraro, P. (2021). Three-dimensional quantitative intracellular visualization of graphene oxide nanoparticles by tomographic flow cytometry. *Nano Letters, 21*, 5958–5966. https://doi.org/10.1021/acs.nanolett.1c00868

Shin, H., Kwak, M., Lee, T. G., & Lee, J. Y. (2020). Quantifying the level of nanoparticle uptake in mammalian cells using flow cytometry. *Nanoscale, 12*, 15743–15751. https://doi.org/10.1039/D0NR01627F

Sigmund, W., Yuh, J., Park, H., Maneeratana, V., Pyrgiotakis, G., Daga, A., Taylor, J., & Nino, J. C. (2006). Processing and structure relationships in electrospinning of ceramic fiber systems. *Journal of the American Ceramic Society, 89*, 395–407. https://doi.org/10.1111/j.1551-2916.2005.00807.x

Vranic, S., Boggetto, N., Contremoulins, V., Mornet, S., Reinhardt, N., Marano, F., Baeza-Squiban, A., & Boland, S. (2013). Deciphering the mechanisms of cellular uptake of engineered nanoparticles by accurate evaluation of internalization using imaging flow cytometry. *Particle and Fibre Toxicology, 10*, 2. https://doi.org/10.1186/1743-8977-10-2

Wu, Y., Ali, M., Dansby, K., & El-Sayed, M. (2019). Improving the flow cytometry-based detection of the cellular uptake of gold nanoparticles. *Analytical Chemistry, 2019*, 14261–14267. https://doi.org/10.1021/acs.analchem.9b02248

Zahavy, E., Ber, R., Gur, D., Abramovich, H., Freeman, E., Maoz, S., & Yitzhaki, S. (2012). Application of nanoparticles for the detection and sorting of pathogenic bacteria by flow-cytometry. *Advances in Experimental Medicine and Biology, 733*, 23–36. https://doi.org/10.1007/978-94-007-2555-3_3

Zahra, Z., Habib, Z., Chung, S., & Badshah, M. A. (2020). Exposure route of TiO2 NPs from industrial applications to wastewater treatment and their impacts on the agro-environment. *Nanomaterials, 10*, 1469. https://doi.org/10.3390/nano10081469

Zucker, R. M., Massaro, E. J., Sanders, K. M., Degn, L. L., & Boyes, W. K. (2010). Detection of TiO2 nanoparticles in cells by flow cytometry. *Cytometry. Part A, 77A*, 677–685. https://doi.org/10.1002/cyto.a.20927

Application of Flow Cytometry in the Development of New Alternative Model Systems in Biomedical Research

Anuj Kumar Pandey, Manish Raj Kulshrestha, Mukul Singh, Anumesh K. Pathak, and Vandana Tiwari

Abstract

Flow cytometry, a powerful cellular analysis tool, plays a primary role in developing alternative model systems for biomedical research. This chapter explores how flow cytometry aids in characterizing these systems by identifying and evaluating cellular heterogeneity, subpopulations, and surface markers. We delve into applications such as cell sorting for targeted studies, viability and cytotoxicity testing, and cell cycle analysis to understand cell proliferation. Furthermore, the chapter addresses the application of flow cytometry in well-established alternative models, viz., *C. elegans*, Zebrafish, and *Drosophila melanogaster*. It describes how flow cytometry helps identify phenotypes, functional indicators, and cellular responses to toxicants. Additionally, the chapter investigates flow cytometry's role in detecting apoptosis by examining light scattering, mitochondrial potential, and caspase activation. The detection of oxidative stress and reactive oxygen species (ROS) is also explored. The chapter looks ahead to how new types of flow cytometry, such as mass cytometry (CyTOF), image flow cytometry (IFC), and droplet microfluidics, can be combined to improve analysis. Finally, we discuss the emerging role of artificial intelligence (AI) and machine learning (ML) in automating gating strategies, revealing hidden patterns within flow data, and even assisting in model development.

A. K. Pandey · M. R. Kulshrestha · A. K. Pathak · V. Tiwari (✉)
Department of Biochemistry, Dr. Ram Manohar Lohia Institute of Medical Sciences, Lucknow, Uttar Pradesh, India

M. Singh
Department of Urology, King George's Medical University, Lucknow, Uttar Pradesh, India

AB Pant et al. (eds.), *Flow Cytometry: Applications in Cellular and Molecular Toxicology*, https://doi.org/10.1007/978-981-97-9758-5_13

Keywords
Flowcytometry · *C. elegans* · Zebrafish · *Drosophila melanogaster* · Droplet microfluidics · Artificial Intelligence · Machine Learning

1 Introduction

Animal testing has long played a primal role in traditional toxicology, effectively protecting human health for decades. Nevertheless, this approach gives rise to various ethical and practical considerations. Animal testing is frequently subject to criticism due to its inherent cruelty, expenses, and limited relevance to humans as a result of physiological differences. Consequently, in recent years, the scientific community has embraced the development of alternative model systems in toxicology. These systems offer a promising tool for assessing the safety of chemicals and drugs without the need for animal testing. As we know, the five R's principle (refuse, reduce, reuse, repurpose, recycle) emphasizes the need to minimize animal use in scientific research (Kumar, 2012). Alternative models align with this principle by offering ways to completely replace or significantly reduce the number of animals used in toxicity testing. Animal models, despite their contributions, often fail to accurately predict human responses to chemicals due to species-specific differences in physiology and metabolism. While alternative models, like-cell cultures and computational simulations, can provide more human-relevant data. Advancements in biotechnology, genomics, and computational modeling have enabled the development of sophisticated alternative models that mimic key aspects of human biology. These models offer a better understanding of how chemicals interact with human systems.

Flow cytometry has gained importance in toxicological research that involves alternate model systems (Robinson et al., 2023). This tool enables a comprehensive examination of cellular responses in various models, providing valuable insights into cell populations, dynamics, and functions. It has the ability to recognize and count various types of cells, monitor changes over a period of time, and display distinct groups within a population. Moreover, flow cytometry allows for the quantification of functional indicators, analysis of cell cycle status, and examination of intracellular activities, thereby uncovering the effects of substances and/or stimuli. Additionally, it aids in the discovery of cellular diversity, the identification of anomalies, and the measurement of differences between studies. The technology has several benefits, such as the ability to analyze individual cells with great precision, the ability to measure multiple parameters simultaneously, and the ability to process a large number of samples quickly (Kotsifaki et al., 2024; Kumari & Gupta, 2021). As the exploration of alternative models evolves, flow cytometry is anticipated to have a vital role in understanding complex biological reactions, assisting in toxicity evaluation, and helping to the development of safer items.

A major challenge in alternative model systems is ensuring their predictive power for human toxicity. Flow cytometry plays a vital role in bridging this gap by enabling the comparison of cellular responses between the model and human tissues. By analyzing the expression of key markers, signaling pathways, and functional outputs in both systems, we can assess the model's fidelity and refine its design for improved human relevance. For example, comparing the gene expression profiles of liver cells in a 3D liver model exposed to a toxicant to those from human liver biopsies can reveal similarities and discrepancies in their response pathways. This comparative analysis using flow cytometry data informs the model's optimization to better mimic human responses, ultimately leading to more reliable toxicity predictions.

Flow cytometry is not merely a powerful tool; it is a transformative approach that empowers researchers to delve deeper into the intricate workings of alternative models. By illuminating the cellular heterogeneity, quantifying functions, and bridging the gap to human relevance, flow cytometry paves the way for a new era of toxicology, one where safer products are developed with greater confidence and a reduced reliance on animal testing (Gough et al., 2017). Here, we have tried to compile the application of flow cytometry in the development of new alternative model systems in biomedical research.

2 Heterogeneity and Subpopulations

Flow cytometry enables researchers to analyze individual cells within a heterogeneous population, offering a detailed insight into the dynamics of calcium flux at a single-cell resolution. Researchers can uncover distinct subpopulations exhibiting unique calcium signaling patterns and explore the diversity among cells in a sample (Robinson et al., 2023). In addition, flow cytometry can evaluate multiple parameters by integrating calcium flux assessment with other markers or functional tests. Monitoring calcium flux via flow cytometry requires specialized tools, but it provides distinct benefits. Flow cytometry was used to measure cellular degranulation in its early stages with the help of the calcium ionophore A23187 (Phetsang et al., 2016). Valet also conducted measurements of intracellular calcium using flow cytometry (Valet et al., 1985). Calcium flux determination by flow cytometry often involves the use of fluorescent dyes like fluo-3 and indo-1 (Jennings et al., 1989).

Changes in intracellular calcium concentration in mammalian cells can occur very quickly in response to different stimuli, often happening within nanoseconds. Nevertheless, capturing these fast calcium responses can be quite difficult due to missing data caused by the introduction of compounds, leading to the loss of intricate information (Vines et al., 2010). In order to tackle this issue, flow cytometers have been adjusted to allow for uninterrupted measurement as substances are introduced to the sample. While Ca^{2+} flux measurement is a widely used application, it's worth noting that similar techniques can also be employed to monitor other ions such as magnesium, potassium, sodium, and hydrogen (Aliotta et al., 2020). These ions play crucial roles in various cellular processes, and their real-time monitoring can provide valuable insights into cell function and signaling pathways. By adapting flow cytometry techniques to measure a range of ions, researchers can gain a more comprehensive understanding of cellular responses to different stimuli.

3 Cell Sorting

The use of flow sorters for bacterial isolation and detection is quite intriguing. It allows for the quick detection and sorting of individual bacterial cells in a suspension, which is much faster than the traditional agar plating-based detection methods (Alvarez-Barrientos et al., 2000). High throughput sorting of bacterial cells allows for efficient collection at the desired outlet, facilitating subsequent analysis. Despite the impressive performance capabilities of cell sorters, their application in microbiology has been limited by challenges in distinguishing microorganisms from cell debris and background particles due to their small size. The challenge lies in distinguishing microorganisms from cell debris or background particles in the media, primarily because of their small size. Additionally, in certain instances, the issue is compounded by their pathogenic nature (Vives-Rego et al., 2000). Moreover, the challenge of effectively sorting distinct subpopulations of bacteria is exacerbated by their comparable size and shape, complicating the sorting process. Another significant issue is associated with the high concentration of bacteria in the sample. Segmenting the population of interest poses a challenge, resulting in inconsistent categorization when the bacteria concentration is extremely low. If the initial sample is impure or if there is a requirement to isolate different groups of organisms, it may be necessary to perform multiple sorting and enrichment procedures. These additional steps can have an impact on the recovery of cells (Ambriz-Avina et al., 2014).

Although whole blood can be used directly, it is common practice to lyse the red blood cells before introducing them to the cell sorter. However, in order to detect low concentrations of bacteria in the blood, it may be necessary to implement more efficient sample preparation steps (Narayana Iyengar et al., 2021). In a study conducted by Irene et al. in 2007, they demonstrated the ability to detect bacteria in cell culture production medium at concentrations as low as 1 CFU/mL by combining staining and scatter data. Nevertheless, significant amounts of bacterial protein, or DNA, were required for further analysis. For protein analysis, it is advisable to use a concentration of approximately 109 cells/mL. On the other hand, for nucleic acid analysis, a concentration of around 103 or 104 cells is recommended (McHugh & Tucker, 2007).

4 Cell Surface Marker Identification and Expression Profiling

Immunophenotyping, a widely used application of flow cytometry, focuses on identifying and characterizing immune cell populations by analyzing their surface or intracellular markers. Flow cytometry allows for the accurate identification and categorization of various immune cell populations, such as T cells, B cells, natural killer (NK) cells, dendritic cells, monocytes, macrophages, platelets, and granulocytes (Laine et al., 2019). Neurobiology utilizes flow cytometry to examine distinct neuronal subtypes, glial cells, and neural progenitor cells by utilizing surface markers such as CD133, CD56, and neuronal markers. In addition, flow cytometry is useful for studying the characteristics of endothelial cells that play a role in vascular development, angiogenesis, and diseases related to the vascular system. Through the analysis of surface markers, scientists are able to identify different populations of endothelial cells, explore their role in the body, and examine how they are affected during disease states (Pruszak et al., 2009). Flow cytometry allows for the analysis of different types of mesenchymal cells, such as mesenchymal stem cells and fibroblasts. This is done by using specific markers like CD90, CD73, and CD105 to identify and characterize these cell subsets. In addition, it can be used to analyze populations of epithelial cells, identify and describe various subtypes, examine the epithelial-mesenchymal transition (EMT), and explore how cells change in tissue development and disease progression (Boss et al., 2022).

5 Viability and Cytotoxicity Assessment

Flow cytometry offers highly accurate viability measurements, assisting in determining the percentage of viable cells within a sample. Furthermore, flow cytometry provides exceptional sensitivity for viability assays, allowing for the detection of even the most subtle alterations in cell viability (Davey & Guyot, 2020). By utilizing markers or dyes, this method can differentiate between live, apoptotic, and necrotic cells, providing valuable information about the health and state of the cells. In addition, it enables the analysis of multiple parameters, allowing for the simultaneous assessment of cell viability and other cellular characteristics. By meticulously analyzing samples at specific time points, researchers can observe variations in cell viability induced by diverse stimuli, treatments, or environmental factors. This allows for a deeper understanding of cell behavior and how it responds over time (Wlodkowic et al., 2011).

Various dyes are commonly used to detect cell viability. For instance, propidium iodide and 7-aminoactinomycin D are able to bind to DNA, but they can only penetrate cells with damaged membranes (Healy et al., 1998). As a result, these dyes make dead cells appear fluorescent (Cossarizza et al., 2017). On the other hand, Annexin V is a protein that has a high affinity for phosphatidylserine (Robinson et al., 2023). This protein is exposed on the outer surface of the plasma membrane in the initial phases of apoptosis. Flow cytometry involves the combination of fluorescent-labeled annexin V with viability dyes to distinguish between viable cells (annexin V-negative) and apoptotic cells (annexin V-positive) (Vermes et al., 1995). The viability studies also utilize cell-permeant fluorescent dyes that have the ability to enter live cells and only become fluorescent when they interact with intracellular enzymes. Some examples are calcein AM, carboxyfluorescein diacetate (CFDA), and fluorescein diacetate (FDA). The dyes are absorbed by living cells and transformed into fluorescent forms by enzymes inside the cells (Breeuwer et al., 1995). Through the use of these dyes and assays, viability assessments conducted with flow cytometry allow for the differentiation of live, apoptotic, and necrotic cell populations. This offers valuable insights into the overall health and integrity of the cells.

Flow cytometry has also been utilized to investigate the physiological reactions of bacteria to chemical substances such as phenol and toluene. An investigation was conducted on a culture of Cupriavidus necator to examine its reaction to exposure to phenol concentrations. The researchers measured the chromosomal DNA and PHB content using DAPI and Nile red, respectively. They then sorted the different subpopulations that responded differently and performed functional proteomics analyses. They conducted an assessment of toluene uptake using flow cytometry and a fluorescently labeled toluene analog dye (NBDT). After sorting the cells, proteomics analysis revealed the presence of toluene transporter porins in *P. putida* mt-2 when cultured on toluene but not glucose (Ambriz-Avina et al., 2014).

FCM has also been used to analyze the efficacy of treatments given to food goods intended for human consumption to eradicate and prevent pathogen propagation and spoiling bacteria, such as *Listeria monocytogenes* (Utratna et al., 2012; Utratna et al., 2014). This bacterium is highly adaptable to stress conditions during food processing, such as acidic pH, high salt concentration, and harsh temperatures. It has been demonstrated that, in order to cope with various pressures, this bacterium induces the expression of genes from the general stress sigma B regulon (Utratna et al., 2014). To further understand the role of during low temperature adaptation, Utratna et al. examined the production of a reporter gene fusion between the promoter of the dependent Imo2230 gene and GFP at normal and low temperatures (Utratna et al., 2014). The FCM analysis revealed a heterogeneous activation of the dependent GFP fusion expression beginning in the early exponential phase, with the highest expression occurring when cells entered the stationary phase of growth; however, this result was observed at both tested temperatures.

6 Cell Cycle Analysis and Proliferation Measurement

Flow cytometry offers diverse techniques for analyzing the cell cycle and measuring proliferation. Specific fluorescent dyes such as propidium iodide (PI), Hoechst, DAPI, 7-amino actinomycin D, mithramycin, and ethidium bromide were bound to nucleic acids. Thiazole orange and thioflavin are dyes that exhibit greater selectivity towards RNA. PI, or propidium iodide, is a dye that is commonly used in stoichiometric measurements. It has the ability to intercalate, or insert itself, between the bases of DNA. Hence, a stoichiometric correlation exists between the quantity of dye and the quantity of DNA. Due to this characteristic, PI may be utilized to precisely measure the quantity of DNA, and when meticulously processed using a flow cytometer, even minor alterations in nucleic acid content can be readily detected. Additionally, there are numerous methodologies for integrating flow cytometry with cell cycle analysis. An instance of this is the DNA/Ki67 test, which integrates phenotypic selection with cell cycle analysis for monitoring p53 cell cycle arrest, sperm cell fractionation, and detecting multidrug resistance.

The study conducted by Schwarting et al. (1986) utilized flow cytometric examination of Ki-67 to ascertain the growth fraction of lymphoma cell lines (Schwarting et al., 1986). This technique has since been employed to examine the cell cycle and cell proliferation in various cancer cells and hematopoietic stem cells. Pyronin Y was initially produced in 1889 and has since been employed as a practical dye for histology and cytochemical purposes (Eddaoudi et al., 2018). It is commonly used in conjunction with methyl green to stain RNA, whereas methyl green is utilized for DNA staining. Subsequently, a technique called double staining of Pyronin Y and Hoechst 33342 was created for the purpose of flow cytometric analysis Eddaoudi et al., 2018). This technique allows for the estimation of DNA and RNA content in intact cells (Shapiro, 1981). In addition to these methods, the dynamics of cell cycle progression can be evaluated by cell proliferation assays that quantify the amount of newly synthesized DNA and several cellular metabolism characteristics. In order to analyze cell proliferation using flow cytometry, the genomic DNA in cells that are replicating can be marked by exposing them to a thymidine analog called 5′-bromo-2′-deoxyuridine (BrdU) during the S phase of the cell cycle (Kim & Sederstrom, 2015). The BrdU that has been integrated is subsequently labeled with fluoresceinated anti-BrdU antibodies and a fluorescent DNA dye (such as propidium iodide or 7-Aminoactinomycin D) to distinguish the cells based on their respective phases in the cell cycle (G1, S, and G2/M phases). An inherent drawback of the BrdU incorporation method is the necessity for both membrane permeabilization and rigorous DNA denaturation procedures in order to facilitate antibody penetration into the inserted BrdU. 5-ethynyl-2′-deoxyuridine (EdU) has been created as a substitute for BrdU to address the limitations of the BrdU approach.

Currently, membrane-permeable fluorescent dyes are employed in dye dilution tests to evaluate cell growth. Carboxyfluorescein succinimidyl ester (CFSE or CFDA-SE) is a commonly used fluorescent dye that penetrates the cytoplasm and forms a covalent bond with intracellular amino acids. Due to the reaction's ability to generate prolonged retention of fluorescent dye within the original cell, it was initially employed for the purpose of monitoring immune cells. Given the assumption of cells having uniform size and undergoing symmetrical division, each daughter cell possesses half of the volume and cellular components of the parent cell, as well as the tagged CFSE dyes (Kim & Sederstrom, 2015). Therefore, CFSE labeling can be utilized to assess the quantity of cell generations following rapid cell divisions. Typically, the CFSE dye can be detected for up to 6–8 cell divisions using flow cytometry. Other fluorophores have been created for the dye dilution proliferation experiment, which, like CFSE dye, cover a wide range of excitation and emission spectra. These fluorescent dyes are more appropriate for multicolor applications that involve the usage of GFP derivatives, FITC, or other comparable fluorescent-conjugated antibodies. In addition, certain dilution dyes emit light in channels where cells exhibit lower levels of inherent autofluorescence, allowing their detection for up to 10 generations during cell proliferation.

7 Caenorhabditis elegans

New flow cytometry techniques have emerged as a promising and efficient way to measure protein aggregation in cellular model systems, offering a simpler and faster alternative to conventional methods (Claesson et al., 2022). A study was conducted to develop a technique that allows for the identification and analysis of fluorescent inclusions in C. *elegans*, using a standard flow cytometry setup. This approach involves using sonication to mechanically break down worms, followed by identifying fluorescent inclusions, and then comparing this number to the protein concentration in the lysate (Hernando-Rodríguez et al., 2018). This method is highly adaptable and flexible. It can be used with various C. elegans strains that mimic aggregation by expressing specific proteins. These proteins are often linked to neurodegenerative disorders like ALS and Huntington's disease and can be found in muscle or neural cells (Roussos et al., 2023). By utilizing a flow cytometry-based technique, it was possible to detect the dynamic changes in inclusion formation when certain modulators of protein homeostasis were applied. There was a significant rise in the number of FlucDM inclusions when exposed to sudden heat stress, followed by a decline during the recovery period. This pattern was observed in strains that expressed this protein in both neurons and body-wall muscle cells (Claesson et al., 2022). Based on analysis of imaging and flow cytometry data, it is clear that the number of FlucDM inclusions decreased in both strains after a period of recovery. However, study observed that the neural-expressing strain showed a more efficient recovery compared to the muscle-expressing strain (Basiji et al., 2007).

8 Zebrafish

Because zebrafish develop quickly, are genetically related to humans, and are transparent in their early life stages, they have become an invaluable tool in biological study (Rueb & Stachura, 2021). Researchers may develop flow cytometry-based zebrafish models:

8.1 Sample Preparation

Using mechanical disruption or enzymes, zebrafish larvae or embryos are separated into single cells.

8.2 Staining

To target cellular markers (proteins, receptors), viability, or both, cells are stained with certain fluorescent antibodies or dyes.

8.3 Analysis Using Flow Cytometry

An injection of the cell suspension is made into a flow cytometer. One by one, cells are exposed to a laser beam, and the amount of scattered light and fluorescence that is released is measured.

8.4 Data Analysis

Software examines the data and produces scatter plots and histograms that show cell populations according to size, granularity (internal complexity), and intensity of fluorescence.

The formation of new blood cells, or hematopoiesis, is one area in which this approach can be applied. Flow cytometry is a useful tool for researchers to distinguish and measure various blood cell types in zebrafish larvae and embryos. After then, this data can be utilized to investigate how environmental pollutants (xenobiotic compounds) or genetic changes affect the growth of red blood cells. When zebrafish are exposed to pathogens or medications, flow cytometry can be utilized to examine the growth, functioning, and reaction of their immune systems. For example, the quantity and level of activation of T-lymphocytes or macrophages can be explored. Testing for hematotoxicity involves determining whether a drug has the ability to alter blood cells. Rapid evaluation of the quantity, types, and status of red blood cells is possible with flow cytometry. Different tumor cell populations may be identified, tumor cell proliferation can be analyzed, and the effectiveness of possible cancer treatments can be evaluated using flow cytometry.

With flow cytometry, scientists can do high-throughput analysis on thousands of cells at once and obtain data on a variety of cellular properties, such as size, granularity, and several surface or internal markers (multi-parameter analysis). Because of its adaptability, the method can be used to investigate a variety of areas, such as cell division, death, and cell cycle analysis. Researchers can quickly, effectively, and informatively obtain deeper insights into biological processes by utilizing flow cytometry in conjunction with the zebrafish model.

9 Drosophila melanogaster

Drosophila melanogaster, a fruit fly, has been a keystone in scientific study for many years (Misra et al., 2017; Sharma et al., 2017). It is the perfect model organism due to its short life cycle, simple genetics, and well-understood developmental mechanisms (Pandey et al., 2022; Misra et al., 2017).

One exciting application of the Drosophila model is studying neurodevelopment. Researchers genetically modify flies to express fluorescent proteins in particular neuronal populations in order to investigate Drosophila neural development. After that, they separate the brain tissues from the developing flies and dye them using several neuronal subtype markers. They count and classify neurons at various developmental stages by analyzing the fluorescence spectra of individual cells using flow cytometry. This method provides insight into the development of neuronal populations and is useful in identifying disruptions caused by environmental or genetic mutations. For example, in Parkinson's disease, Rodriguez et al. (2018) employed flow cytometry to examine the fly brain after specific gene alterations or exposure to neurotoxins. Staining with markers specific to neurons or dopaminergic cells allows researchers to quantify cell mortality or population changes. This aids in understanding the mechanisms behind neurodegeneration and evaluating the potential efficacy of treatment therapies.

The well-known genetics of Drosophila enable researchers to produce fly lines carrying particular mutations, therefore allowing them to be studied and their effects on various cell populations (Sharma et al., 2017). This technique can be used for various research questions related to Drosophila biology, including cell death, proliferation, and immune cell function.

10 Phenotype and Functional Marker Detection: Cellular Responses to Toxicants

Efficiently segregating and isolating biological cells into precise and distinct subgroups is crucial for advancing our understanding of cellular heterogeneity and its implications for living organisms. This text describes the development of the functional phenotype flow cytometer (FPFC), a new device designed to categorize cells based on their immediate responses to stimuli. The FPFC employs a microfluidic platform to execute a sequence of cell processing procedures, facilitating the categorization of cells based on their immediate responses to stimuli. It introduces individual biological cells into a solution containing a biological reagent that serves as a stimulus. The cells are then incubated with the stimulus solution while being continuously flown. Finally, the cells are sorted into different subpopulations based on their phenotypic responses to the stimulus.

11 Apoptosis and Cell Death Detection

Flow cytometry allows for a thorough evaluation of apoptosis and cell death using several methods, such as analyzing light scattering, measuring mitochondrial transmembrane potential ($\Delta\psi m$), and detecting caspase activity.

11.1 Analysis of Laser Light Scattering

Alterations in light scattering properties occur in apoptotic cells, including cellular shrinkage and the compaction of the nucleus and cytoplasm. These changes offer vital information about the cellular morphology during the process of apoptosis. Cell shrinking leads to a decrease in forward light scatter (FSC), whereas improved light refraction and reflection often cause an increase in side scatter (SSC) (Healy et al., 1998). The alterations in light scattering properties can differentiate different phases of apoptosis and necrosis (Healy et al., 1998).

11.2 Loss of Mitochondrial Transmembrane Potential (δψm)

Evaluating the mitochondrial membrane potential (Δψm) is a highly accurate indicator of the initial stages of apoptosis. Lipophilic cationic probes, such as rhodamine 123 (Rh123) and JC-1, are used to find the drop in mitochondrial membrane potential (Δψm), which shows that the mitochondria aren't working right and is connected to apoptosis. Combining probes that are sensitive to changes in mitochondrial membrane potential (Δψm) with other markers of apoptosis can provide improved analytical capabilities for the study of programmed cell death.

11.3 Caspase Activation

Caspase activation is a distinguishing characteristic of apoptosis, which is facilitated by a group of proteases known as cysteine aspartyl-specific proteases. Fluorochrome-labeled inhibitors of caspases (FLICAs) allow for the identification of active caspases in living cells by forming a strong chemical bond. Although there have been doubts about FLICA's sensitivity towards caspase active sites, it nonetheless remains a viable tool for detecting apoptosis in living organisms, mainly because of its capacity to covalently label.

12 Oxidative Stress and Reactive Oxygen Species (ROS) Measurement

In the pursuit of chemical compounds having bactericidal properties, alternative approaches have explored the concept of inducing cellular stress and halting bacterial proliferation. One strategy that can be used is photodynamic inactivation (PDI), which involves the utilization of a non-toxic photosensitizer that is activated by safe visible light. Consequently, this leads to the generation of reactive oxygen species, which inflict lethal harm on specific bacteria. This method has undergone testing both in laboratory conditions (in vitro) and in living organisms (in vivo) (Laine et al., 2019). Due to the generic nature of PDI-induced damage to infections, the likelihood of bacteria developing resistance is quite low. Jiang et al. (2013) wrote about a study that looked at how ROS are made when hypocrellin B, a chemical found in the traditional Chinese herb H. bambuase, is exposed to light. This article describes the incubation of clinically isolated S. aureus cells with hypocrellin B, followed by photodynamic therapy. We used dichlorodihydrofluorescein diacetate (DCFH-DA) and flow cytometry (FCM) to measure the production of reactive oxygen species (ROS). Cells treated with PDI experienced a significant increase in ROS levels (Tra et al., 2011). The treatment decreased the number of living cells, as shown by CFU counts and cell damage, such as membrane damage and cytoplasmic leakage, which was confirmed by confocal laser scanning microscopy.

13 Newer Modalities in Stem Cell-Based Flow Cytometry

Flow cytometry is an effective tool in stem cell research, offering critical insights into stem cell characteristics and activities (Chen et al., 2016). Flow cytometry's capacity to examine several factors at the single-cell level makes it an essential tool in stem cell research. It makes precise characterization, sorting, and functional investigation of stem cells easier, improving our understanding of stem cell biology and accelerating the development of stem cell-based therapeutics. Here are some newer modalities in stem cell-based flow cytometry:

13.1 Mass Cytometry (CyTOF)

This approach offers the simultaneous assessment of up to 100 metal-tagged antibodies, allowing for thorough profiling of stem cell surface and internal proteins. This is especially relevant when investigating rare stem cell subpopulations with different functions (Behbehani, 2019).

13.2 Imaging Flow Cytometry (IFC)

IFC combines the capabilities of flow cytometry with microscopic imaging, allowing for the visualization of cellular morphology alongside the measurement of fluorescent markers. This is particularly useful for studying rare stem cell populations where morphology can be an important identifier.

13.3 Droplet Microfluidics

This technique allows researchers to encapsulate single stem cells in droplets containing analysis reagents. This enables not only traditional flow cytometry analysis but also simultaneous gene expression profiling at the single-cell level, providing deeper insights into stem cell heterogeneity (Macosko et al., 2015).

13.4 Fluorescence Lifetime Imaging Microscopy (FLIM)

FLIM measures the fluorescence lifetime of fluorophores, which can provide information about the microenvironment of the stem cell. This can be used to study stem cell signaling and differentiation.

14 Role of Artificial Intelligence (AI) and Machine Learning (ML) in Flow Cytometry

Flow cytometry is a powerful tool for cell analysis, but traditional approaches are slow, subjective, and error-prone (Chen et al., 2016). ML and AI are transforming flow cytometry by automating procedures and revealing hidden patterns in complex data (Lu et al., 2024). This enables researchers to create new alternative model systems for biomedical research with improved efficiency and accuracy. AI with ML may revolutionize the flow cytometry by:

14.1 Automated Gating

Cell populations can be automatically identified by ML algorithms using labeled flow cytometry data. This speeds up analysis, decreases human bias, and increases consistency. ML was employed to find uncommon immune cell groups in cancer patients' blood samples. Comparing this method to the manual method, it yielded faster analysis and more accuracy.

14.2 Uncovering Hidden Patterns

Finding Subtle patterns in flow cytometry data is challenging because it is frequently high-dimensional. These intricate datasets can be analyzed by AI systems to uncover undiscovered connections between cellular properties. An illustration is the use of AI by investigators to examine flow cytometry data from patients with various ailments. Particular cell population fingerprints found by AI could be applied to diagnose more quickly.

14.3 Developing New Model Systems

AI can assist in the creation of new in vitro (cell culture) or in silico (computer simulation) models that more closely resemble intricate biological systems by evaluating massive databases of cellular information. For example, to examine flow cytometry data from immune cells reacting to infections, researchers employed AI; a computer model that could forecast how various immune cell combinations would combat diseases was then created using this data.

Overall, AI and ML have great potential to change flow cytometry by: increasing efficiency and lowering analysis time, improving objectivity and reducing human error, and allowing researchers to extract more relevant insights from complex data. However, difficulties such as data quality and interpretability of AI models persist. As research advances, AI is poised to play an even larger role in the future of flow cytometry, leading to advances in disease detection, personalized medicine, and drug development (Pozzi et al., 2023: Li et al., 2019).

References

Aliotta, A., Bertaggia Calderara, D., & Alberio, L. (2020). Flow Cytometric Monitoring of Dynamic Cytosolic Calcium, Sodium, and Potassium Fluxes Following Platelet Activation. *Cytometry Part A: the journal of the International Society for Analytical Cytology., 97*(9), 933–944.

Alvarez-Barrientos, A., Arroyo, J., Canton, R., Nombela, C., & Sanchez-Perez, M. (2000). Applications of flow cytometry to clinical microbiology. *Clinical Microbiology Reviews, 13*(2), 167–195.

Ambriz-Avina, V., Contreras-Garduno, J. A., & Pedraza-Reyes, M. (2014). Applications of flow cytometry to characterize bacterial physiological responses. *BioMed Research International, 2014*, 461941.

Basiji, D. A., Ortyn, W. E., Liang, L., Venkatachalam, V., & Morrissey, P. (2007). Cellular image analysis and imaging by flow cytometry. *Clinics in Laboratory Medicine, 27*(3), 653–670. viii.

Behbehani, G. K. (2019). Immunophenotyping by mass cytometry. *Immunophenotyping: Methods and Protocols*, 31–51.

Boss, A. L., Damani, T., Wickman, T. J., Chamley, L. W., James, J. L., & Brooks, A. E. S. (2022). Full spectrum flow cytometry reveals mesenchymal heterogeneity in first trimester placentae and phenotypic convergence in culture, providing insight into the origins of placental mesenchymal stromal cells. *eLife, 11*, e76622.

Breeuwer, P., Drocourt, J. L., Bunschoten, N., Zwietering, M. H., Rombouts, F. M., & Abee, T. (1995). Characterization of uptake and hydrolysis of fluorescein diacetate and carboxyfluorescein diacetate by intracellular esterases in Saccharomyces cerevisiae, which result in the accumulation of fluorescent products. *Applied and Environmental Microbiology, 61*(4), 1614–1619.

Chen, C. L., Mahjoubfar, A., Tai, L. C., Blaby, I. K., Huang, A., Niazi, K. R., & Jalali, B. (2016). Deep learning in label-free cell classification. *Scientific Reports, 6*(1), 21471.

Claesson, K., Chew, Y. L., & Ecroyd, H. (2022). Exploiting flow cytometry for the unbiased quantification of protein inclusions in Caenorhabditis elegans. *Journal of Neurochemistry, 161*(3), 281–292.

Cossarizza, A., Chang, H. D., Radbruch, A., Akdis, M., Andra, I., Annunziato, F., et al. (2017). Guidelines for the use of flow cytometry and cell sorting in immunological studies. *European Journal of Immunology, 47*(10), 1584–1797.

Davey, H., & Guyot, S. (2020). Estimation of Microbial Viability Using Flow Cytometry. *Current Protocols in Cytometry, 93*(1), e72.

Eddaoudi, A., Canning, S. L., & Kato, I. (2018). Flow Cytometric Detection of G0 in Live Cells by Hoechst 33342 and Pyronin Y Staining. *Methods in Molecular Biology, 1686*, 49–57.

Gough, A., Stern, A. M., Maier, J., Lezon, T., Shun, T. Y., Chennubhotla, C., Schurdak, M. E., Haney, S. A., & Taylor, D. L. (2017). Biologically relevant heterogeneity: metrics and practical insights. *Slas Discovery: Advancing Life Sciences R&D, 22*(3), 213–237.

Healy, E., Dempsey, M., Lally, C., & Ryan, M. P. (1998). Apoptosis and necrosis: mechanisms of cell death induced by cyclosporine A in a renal proximal tubular cell line. *Kidney International, 54*(6), 1955–1966.

Hernando-Rodríguez, B., Erinjeri, A. P., Rodríguez-Palero, M. J., Millar, V., González-Hernández, S., Olmedo, M., Schulze, B., Baumeister, R., Muñoz, M. J., Askjaer, P., & Artal-Sanz, M. (2018). Combined flow cytometry and high-throughput image analysis for the study of essential genes in Caenorhabditis elegans. *BMC Biology, 16*(1), 36.

Jiang, Y., Leung, A. W., Wang, X., Zhang, H., & Xu, C. (2013 Dec 1). Inactivation of Staphylococcus aureus by photodynamic action of hypocrellin B. Photodiagnosis and photodynamic therapy. 10(4):600–606. https://doi.org/10.1016/j.pdpdt.2013.06.004

Jennings, L. K., Dockter, M. E., Wall, C. D., Fox, C. F., & Kennedy, D. M. (1989). Calcium mobilization in human platelets using indo-1 and flow cytometry. *Blood, 74*, 2674–2680.

Kim, K. H., & Sederstrom, J. M. (2015). Assaying cell cycle status using flow cytometry. *Current Protocols in Molecular Biology, 111*(1), 28–26.

Kotsifaki, A., Maroulaki, S., & Armakolas, A. (2024). Exploring the Immunological Profile in Breast Cancer: Recent Advances in Diagnosis and Prognosis through Circulating Tumor Cells. *International Journal of Molecular Sciences, 25*(9), 4832.

Kumar, A. H. (2012). Effectively communicating the 5R's (replace, reduce, refine, reuse, and rehabilitate) of research ethics, biomedical waste, personalized medicines and the rest. *Journal of Natural Science, Biology and Medicine, 3*(1), 1–2.

Kumari, P., & Gupta, A. (2021). Assays to assess the proliferative behavior of cancer cells. In *Protocol handbook for cancer biology* (pp. 23–41). Academic.

Laine, R. F., Sinnige, T., Ma, K. Y., Haack, A. J., Poudel, C., Gaida, P., Curry, N., Perni, M., Nollen, E. A., Dobson, C. M., & Vendruscolo, M. (2019). Fast fluorescence lifetime imaging reveals the aggregation processes of α-synuclein and polyglutamine in aging Caenorhabditis elegans. *ACS Chemical Biology, 14*(7), 1628–1636.

Li, Y., Mahjoubfar, A., Chen, C. L., et al. (2019). Deep Cytometry: Deep learning with Real-time Inference in Cell Sorting and Flow Cytometry. *Scientific Reports, 9*, 11088. https://doi.org/10.1038/s41598-019-47193-6

Lu, Z., Morita, M., Yeager, T. S., Lyu, Y., Wang, S. Y., Wang, Z., & Fan, G. (2024). Validation of Artificial Intelligence (AI)-Assisted Flow Cytometry Analysis for Immunological Disorders. *Diagnostics, 14*(4), 420.

Macosko, E. Z., Basu, A., Satija, R., Nemesh, J., Shekhar, K., Goldman, M., Tirosh, I., Bialas, A. R., Kamitaki, N., Martersteck, E. M., Trombetta, J. J., Weitz, D. A., Sanes, J. R., Shalek, A. K., Regev, A., & McCarroll, S. A. (2015). Highly Parallel Genome-wide Expression Profiling of Individual Cells Using Nanoliter Droplets. *Cell, 161*(5), 1202–1214.

McHugh, I. O., & Tucker, A. L. (2007). Flow cytometry for the rapid detection of bacteria in cell culture production medium. *Cytometry Part A: the journal of the International Society for Analytical Cytology., 71*(12), 1019–1026.

Misra, S., Pandey, A. K., Gupta, S., Kumar, A., Khanna, P., & Kristipati, R. R. (2017). Estrogen related receptor is required for the testicular development and for the normal sperm axoneme/mitochondrial derivatives in Drosophila males. *Scientific Reports, 7*(1), 1–5.

Narayana Iyengar, S., Kumar, T., Martensson, G., & Russom, A. (2021). High resolution and rapid separation of bacteria from blood using elasto-inertial microfluidics. *Electrophoresis, 42*(23), 2538–2551.

Pandey, A. K., Sharma, V., & Ravi, R. K. (2022). Drosophila ecdysone receptor activity-based ex vivo assay to assess the endocrine disruption potential of environmental chemicals. *Environmental Science and Pollution Research, 29*, 56430–56441.

Phetsang, W., Pelingon, R., Butler, M. S., Kc, S., Pitt, M. E., Kaeslin, G., Cooper, M. A., & Blaskovich, M. A. (2016). Fluorescent trimethoprim conjugate probes to assess drug accumulation in wild type and mutant Escherichia coli. *ACS Infectious Diseases, 2*(10), 688–701.

Pozzi, P., Candeo, A., Paiè, P., Bragheri, F., & Bassi, A. (2023). Artificial intelligence in imaging flow cytometry. *Frontiers in Bioinformatics, 3*, 1229052.

Pruszak, J., Ludwig, W., Blak, A., Alavian, K., & Isaacson, O. (2009). CD15, CD24, and CD29 define a surface biomarker code for neural lineage differentiation of stem cells. *Stem Cells, 27*(12), 2928–2940.

Robinson, J. P., Ostafe, R., Iyengar, S. N., Rajwa, B., & Fischer, R. (2023). Flow cytometry: the next revolution. *Cells, 12*(14), 1875.

Rodriguez, T. P., Mast, J. D., Hartl, T., Lee, T., Sand, P., & Perlstein, E. O. (2018). Defects in the neuroendocrine axis contribute to global development delay in a Drosophila model of NGLY1 deficiency. G3: Genes, Genomes. *Genetics, 8*(7), 2193–2204.

Roussos, A., Kitopoulou, K., Borbolis, F., & Palikaras, K. (2023). Caenorhabditis elegans as a Model System to Study Human Neurodegenerative Disorders. *Biomolecules, 13*(3), 478.

Rueb, K. F., & Stachura, D. L. (2021). Using flow cytometry to detect and quantitate altered blood formation in the developing zebrafish. *Journal of Visualized Experiments*, 170. https://doi.org/10.3791/61035

Schwarting, R., Gerdes, J., Niehus, J., Jaeschke, L., & Stein, H. (1986). Determination of the growth fraction in cell suspensions by flow cytometry using the monoclonal antibody Ki-67. *Journal of Immunological Methods, 90*, 65–70.

Shapiro, H. M. (1981). Flow cytometric estimation of DNA and RNA content in intact cells stained with Hoechst 33342 and pyronin Y. *Cytometry, 2*(3), 143–150.

Sharma, V., Pandey, A. K., Kumar, A., Misra, S., Gupta, H. P., Gupta, S., Singh, A., Buehner, N. A., & Ram, K. R. (2017). Functional male accessory glands and fertility in Drosophila require novel ecdysone receptor. *PLoS Genetics, 13*(5), e1006788.

Tra, T., Gong, L., Kao, L. P., Li, X. L., Grandela, C., Devenish, R. J., Wolvetang, E., & Prescott, M. (2011). Autophagy in human embryonic stem cells. *PLoS One, 6*(11), e27485.

Utratna, M., Cosgrave, E., Baustian, C., Ceredig, R., & O'Byrne, C. (2012). Development and optimization of an EGFP-based reporter for measuring the general stress response in Listeria monocytogenes. *Bioengineered Bugs, 3*(2), 93–103.

Utratna, M., Cosgrave, E., Baustian, C., Ceredig, R. H., & O'Byrne, C. P. (2014). Effects of growth phase and temperature on σB activity within a Listeria monocytogenes population: evidence for RsbV-independent activation of σB at refrigeration temperatures. *BioMed Research International, 2014*, 641647.

Valet, G. K., Raffael, A., & Russmann, L. (1985). Determination of intracellular calcium in vital cells by flow cytometry. *Die Naturwissenschaften, 72*, 600–602.

Vermes, I., Haanen, C., Steffens-Nakken, H., & Reutelingsperger, C. (1995). A novel assay for apoptosis. Flow cytometric detection of phosphatidylserine expression on early apoptotic cells using fluorescein-labeled Annexin V. *Journal of Immunological Methods, 184*(1), 39–51.

Vines, A., McBean, G. J., & Blanco-Fernandez, A. (2010). A flow-cytometric method for continuous measurement of intracellular Ca(2+) concentration. *Cytometry Part A:Tthe Journal of the International Society for Analytical Cytology, 77*(11), 1091–1097.

Vives-Rego, J., Lebaron, P., & Nebe-von, C. G. (2000). Current and future applications of flow cytometry in aquatic microbiology. *FEMS Microbiology Reviews, 24*(4), 429–448.

Wlodkowic, D., Telford, W., Skommer, J., & Darzynkiewicz, Z. (2011). Apoptosis and beyond: cytometry in studies of programmed cell death. *Methods in Cell Biology, 103*, 55–98.

The manufacturer's authorised representative in the EU is Springer Nature Customer Service Centre GmbH, Europaplatz 3, 69115 Heidelberg, Germany. If you have any concerns regarding our products, please contact ProductSafety@springernature.com

Printed and bound by CPI Group (UK) Ltd, Croydon, CR0 4YY
01/10/2026
02225958-0005